Biological and Medical Physics, Biomedical Engineering

This series is intended to be comprehensive, covering a broad range of topics important to the study of the physical, chemical and biological sciences. Its goal is to provide scientists and engineers with monographs- both, authored and contributed volumes to address the growing need for information. The fields of biological and medical physics and biomedical engineering are broad, multidisciplinary and dynamic. They lie at the crossroads of frontier research in physics, biology, chemistry, and medicine.

Books in the series emphasize established and emergent areas of science including molecular, membrane, and mathematical biophysics; photosynthetic energy harvesting and conversion; information processing; physical principles of genetics; sensory communications; automata networks, neural networks, and cellular automata. Equally important is coverage of applied aspects of biological and medical physics and biomedical engineering such as molecular electronic components and devices, biosensors, medicine, imaging, physical principles of renewable energy production, advanced prostheses, and environmental control and engineering.

Ankur Gogoi · Nirmal Mazumder
Editors

Biomedical Imaging

Advances in Artificial Intelligence and Machine Learning

Springer

Editors
Ankur Gogoi
Department of Physics
Jagannath Barooah University
Jorhat, Assam, India

Nirmal Mazumder
Department of Biophysics, Manipal School of Life Sciences
Manipal Academy of Higher Education
Manipal, India

ISSN 1618-7210 ISSN 2197-5647 (electronic)
Biological and Medical Physics, Biomedical Engineering
ISBN 978-981-97-5344-4 ISBN 978-981-97-5345-1 (eBook)
https://doi.org/10.1007/978-981-97-5345-1

This Springer imprint is published by the registered company Springer Nature Singapore Pte Ltd.
The registered company address is: 152 Beach Road, #21-01/04 Gateway East, Singapore 189721, Singapore

Preface

In recent years, there has been an explosion of interest and investment in the application of artificial intelligence (AI) and machine learning (ML) techniques in the field of biomedical imaging. This can be attributed to the exponential growth in the availability of medical imaging data, driven by advances in imaging technologies, as well as the increasing need to analyze and extract insights from such data in a timely, accurate, and efficient manner. The book, *Biomedical Imaging: Advances in Artificial Intelligence and Machine Learning*, is a collection of state-of-the-art research and developments in this exciting and rapidly evolving area. The book covers a range of topics, from the fundamentals of AI, ML, and deep learning (DL) to the latest applications of these techniques in the analysis and interpretation of various medical imaging modalities. The authors, all leading experts in their respective disciplines, provide comprehensive overviews of the key concepts and challenges, along with detailed discussions of the most successful and promising approaches, and numerous examples of how these techniques are being used to improve patient care, drive medical discovery, and enhance the understanding of disease and treatment responses.

In '*Biomedical Imaging: Advances in Artificial Intelligence and Machine Learning*', we emphasize the importance of close collaboration between computer scientists, engineers, and biomedical researchers, in order to fully realize the potential of AI and ML to address the complex and pressing challenges in the field of biomedical imaging. Chapter 1 by Kalita et al., titled "Artificial Intelligence in Diagnostic Medical Image Processing for Advanced Healthcare Applications" highlights recent advancements in biomedical imaging modalities, including Magnetic Resonance Imaging (MRI), X-ray, ultrasound, Computed Tomography (CT), Electrical Impedance Tomography (EIT), Positron Emission Tomography (PET), and microscopic imaging. The chapter explores the transformative impact of AI in medical imaging, highlighting its role in various tasks, including image reconstruction, analysis, and interpretation, and addressing its benefits, implications, and future directions. Biomedical imaging enhances healthcare quality by enabling non-invasive diagnosis and treatment and to this front, AI plays a crucial role in healthcare, particularly in biomedical imaging. However, there are a few concerns and limitations of AI applications in biomedical imaging. Chapter 2 by Aiyappa et al., titled "From Pixels to Predictions: Exploring the Role of Artificial Intelligence in Radiology" discusses the advantages and ethical implications of AI in radiology, showcasing the benefits of collaboration between AI and radiologists in diagnostic imaging. Spectroscopic and microscopic techniques offer real-time analysis and work with different sample sizes. However, the complexity of spectroscopic peaks and the massive amounts of generated data highlight the need for automated algorithms. Chapter 3 by Sharma et al., titled "Challenges in Accurately Using Artificial Intelligence Machine Learning in Biomedical Imaging" highlights the challenges of implementing AI in biomedical imaging, with emphasis on the limitations and potential pitfalls, to aid researchers in utilizing AI effectively in this context. AI, ML, and DL are progressively finding applications in radiology for educational purposes, image data interpretation, and augmenting patient outcomes and radiologists' quality of life. Despite apprehensions among some radiologists regarding AI replacing their role, recent advancements advocate for the implementation of collaborative intelligence between humans and technology, necessitating radiologists' understanding and acknowledgment. In the following Chap. 4 authored by Karnati et al., entitled "Tracing Historical Connections: The Interwoven Evolution of Artificial Intelligence, Confocal Microscopy, and Marvin Minsky (1927–2016)," the enduring legacy of Marvin Lee Minsky is brought to light, emphasizing the profound influence of his endeavors on the development of both AI and confocal microscope. Undoubtedly, Minsky's interdisciplinary prowess and dedication to scientific advancement extended beyond theory, shaping practical innovations that still influence AI and advanced microscopy today. In Chap. 5 titled "Types of Optical Microscopic Analysis for Cell Death Using Artificial Intelligence'", Das et al., elucidates the application of AI for the analysis of data on cell death from advanced optical microscopy and spectroscopy techniques such as Coherent anti-Stokes Raman scattering, Stimulated Raman scattering, Raman spectroscopy and others. Chapter 6 by Kudva et al., titled "Applications of Artificial Intelligence in the Analysis of Images of the Oral Cavity for Cancer Detection" explores the use of AI in analyzing oral cavity images for cancer detection, highlighting the advantages of remote consultations, early detection, and leveraging AI-powered image analysis. The authors discuss key AI methods, strategies, challenges, and limitations, including image quality, biases, human-computer interface, and ethical considerations such as data privacy, informed consent, and regulatory frameworks in healthcare settings. In Chap. 7 titled "Leveraging Machine Learning for Advanced Biomedical Imaging: Insights from Speckle Pattern Analysis", Kalyuzhner and Zalevsky discuss the use of back scattered laser light in biomedical imaging and sensing, creating a self-interference pattern called secondary speckle pattern. Furthermore, the machine learning techniques that are employed to extract the spatial and temporal variations of these speckle patterns are also elucidated in the chapter. The chapter concludes by discussing the challenges and future perspectives of this novel sensing methodology. Chronic diseases continue to pose a significant challenge to humanity, with the biomedical field constantly evolving to address these issues. However, early diagnosis at the cellular or molecular level remains an unresolved problem. In Chap. 8 titled "Two Photon Fluorescence Integrated Machine Learning for Data Analysis and Interpretation", Raju et al., describe the various applications of two photon fluorescence integrated ML. This chapter explores the transformative potential of this integrated approach across multiple domains, particularly in cancer biology, stem cell research, and brain science. By harnessing the high-resolution imaging capabilities of two-photon microscopy and the analytical prowess of machine learning algorithms, researchers can dissect the intricacies of tumor microenvironments, predict tumor behavior, and identify therapeutic targets with enhanced accuracy. Chapter 9 by Anjikar et al., titled "Deep Learning in Biomedical Applications of Raman Spectroscopy" focuses on the use of Raman spectroscopy, a vibrational spectroscopic tool, as a promising diagnostic tool for disease detection. The authors provide an in-depth discussion on the analysis of Raman spectroscopy data using deep learning techniques. The chapter aims to inspire further research and contribute to the clinical transition of Raman spectroscopy, leading to a revolution in the biomedical field and a reduction in the burden of diseases. Chapter 10 by Roy titled "COVID-19 and Pneumonia Detection from Chest X-Ray Images by Deep Learning Model" presents an overview of automatic COVID-19 and pneumonia detection from Chest X-Ray (CXR) images using deep learning models, specifically analyzing the use of Convolution Neural Network (CNN) models for COVID-19 detection and addressing the challenge of data imbalance. The author puts forth the argument that the Xception model and Mobile Freeze-Net with Attention-based loss function outperform other existing pre-trained models and improve generalization ability on the CXR dataset. Chapter 11 by Karnati et al., titled "Deep Learning Integrated Multiphoton Microscopy" elucidates the integration of deep learning techniques for a variety of image processing tasks, thereby aiding in the enhancement of optical microscope images. The chapter also summarizes recent studies that elucidate the design of data-driven instruments that combine microscopy and computing techniques for current and future biomedical applications. Deep Learning techniques and algorithms are further discussed in Chap. 12 by Rajput et al., titled "U-Net: A Versatile Deep Learning Architecture for Multi-Disease Detection". The authors explore the application of U-Net, a robust convolutional neural network (CNN) architecture, for semantic segmentation of medical images of brain tumors, lung nodules, and liver lesions, enabling pixel-level understanding of imaging of the diseases. Chapter 13 titled "Memristor-Based Neuromorphic Computing and Artificial Neural Networks for Computer Vison and AI Applications" by Patel et al. summarizes memristive-based neuromorphic computing to address the efficiency and data transfer rate limitations of traditional digital processors in emulating brain functions. The authors also discuss the memristive concepts relevant to deep neural networks and spiking neural networks and the main technological obstacles and potential directions for neuromorphic computing. Chapter 14 titled "Generations of CT—The Evolution" by Aiyappa et al., elucidates in detail the development generations, and technical advances of computerized axial transverse scanning, otherwise known as computed tomography (CT). The authors also discuss the impact and relevance of CT imaging on biomedical imaging and advancement in recent times.

The book will be useful for researchers, students, and professionals who are interested in the latest developments and the future direction of this field. The meticulous efforts of the authors have established a robust knowledge base while offering a glimpse into future prospects of the field. We believe that this comprehensive resource will be warmly received by the research and development community engaged in biomedical imaging and AI. We extend our heartfelt gratitude to all those who have contributed to the realization of this publication. We would like to express our sincere appreciation to the authors for their valuable insights and prompt response to the reviewers' feedback. Our heartfelt thanks also go to the reviewers for their thoughtful recommendations and comments, which have greatly enhanced the quality of the chapters. We are indebted to our colleagues for their invaluable suggestions and support throughout the preparation of this volume. Special thanks are due to Loyola D'Silva and Saranya Devi Balasubramanian of Springer Nature Singapore for their exceptional professional support. Furthermore, we acknowledge the unwavering support received from the Manipal Academy of Higher Education, Manipal, India, and Jagannath Barooah University, Jorhat, India. Finally, we express our deepest gratitude to our families for their understanding, encouragement, and unwavering support throughout the entire editing process. Most importantly, we wish our readers a pleasant and productive browsing experience.

Jorhat, Assam, India — Ankur Gogoi
Manipal, Karnataka, India — Nirmal Mazumder

Contents

Contributors

Jini Abraham Department of Radiodiagnosis, AJ Institute of Medical Sciences and Research Centre, Mangalore, India

M. J. Akshit Aiyappa Department of Radiodiagnosis, AJ Institute of Medical Sciences and Research Centre, Mangalore, India

Ajinkya Anjikar The United Graduate School of Agricultural Sciences, Tottori University, Tottori, Japan

Abhijit Boruah Department of Computer Science and Engineering, Dibrugarh University, Dibrugarh, Assam, India

Ishita Chakraborty Department of Biophysics, Manipal School of Life Sciences, Manipal Academy of Higher Education, Manipal, Karnataka, India

Divyani Chatterjee Department of Biophysics, Manipal School of Life Sciences, Manipal Academy of Higher Education, Manipal, Karnataka, India

Baishali Das Department of Biophysics, Manipal School of Life Sciences, Manipal Academy of Higher Education, Manipal, Karnataka, India

Tapan Das Department of Power Electronics and Instrumentation Engineering, Jorhat Institute of Science and Technology, Jorhat, Assam, India

Ankur Gogoi Department of Physics, Jagannath Barooah University, Jorhat, Assam, India

Bidisha Goswami National Institute of Electronics and Information Technology (NIELIT), Guwahati, Assam, India

Nabajyoti Goswami National Institute of Electronics and Information Technology (NIELIT), Guwahati, Assam, India

Indira Govindaraju Department of Biophysics, Manipal School of Life Sciences, Manipal Academy of Higher Education, Manipal, Karnataka, India

Shyam K. Jaiswal KIMS-Sunshine Hospitals, Begumpet, Hyderabad, Telangana, India

Kolleboyina Jayaramulu Hybrid Porous Materials Laboratory, Department of Chemistry, Indian Institute of Technology Jammu, Jammu & Kashmir, India

Nayan M. Kakoty Embedded Systems and Robotics Lab, School of Engineering, Tezpur University, Tezpur, Assam, India

Amlan Jyoti Kalita Embedded Systems and Robotics Lab, School of Engineering, Tezpur University, Tezpur, Assam, India

Zeev Kalyuzhner Faculty of Engineering and The Nano-Technology Center, Bar-Ilan University, Ramat-Gan, Israel

Fu-Jen Kao Institute of Biophotonics, National Yang-Ming Chiao Tung University, Taipei, Taiwan

Rupal Kapdi Institute of Technology, Nirma University, Ahmedabad, India

Manikanth Karnati Department of Biophysics, Manipal School of Life Sciences, Manipal Academy of Higher Education, Manipal, Karnataka, India

Yury V. Kistenev Laboratory LMIML, Tomsk State University, Tomsk, Russia

Adarsh Kudva Department of Oral and Maxillofacial Surgery, Manipal College of Dental Sciences, Manipal. Manipal Academy of Higher Education (MAHE). India, Manipal, Karnataka, India

Vidya Kudva NITTE (Deemed to be University), Department of Electronics and Communication Engineering, NMAM Institute of Technology, Nitte, Karnataka, India

Sanika Kulkarni Department of Biophysics, Manipal School of Life Sciences, Manipal Academy of Higher Education, Manipal, Karnataka, India

Saurov Mahanta National Institute of Electronics and Information Technology (NIELIT), Guwahati, Assam, India

Nirmal Mazumder Department of Biophysics, Manipal School of Life Sciences, Manipal Academy of Higher Education, Manipal, Karnataka, India

Sindhoora Kaniyala Melanthota Department of Biophysics, Manipal School of Life Sciences, Manipal Academy of Higher Education, Manipal, Karnataka, India

Neha Mohan Manipal College of Dental Sciences,Manipal. Manipal Academy of Higher Education, Manipal, Karnataka, India

Chandrabhas Narayana Chemistry and Physics of Materials Unit, International Centre of Materials Science and School of Advanced Materials, Jawaharlal Nehru Centre for Advanced Scientific Research, Bangalore, India;
Rajiv Gandhi Centre for Biotechnology, Thycaud, Thiruvananthapuram, India

Hemanth Noothalapati Faculty of Life and Environmental Sciences, Shimane University, Matsue, Japan;
Department of Chemical Engineering, Indian Institute of Technology Hyderabad, Hyderabad, India

Rajapandian Paneerselvam Department of Chemistry, SRM University AP, Amaravati, Andhra Pradesh, India

Mansi Patel Flextronics Lab, Department of Physics, School of Energy Technology, Pandit Deendayal Energy University, Gandhinagar, India

Prince Patel Department of Information and Communication Technology, School of Technology, Pandit Deendayal Energy University, Gandhinagar, India

Vathsala Patil Department of Oral Medicine and Radiology, Manipal College of Dental Sciences, Manipal, Manipal Academy of Higher Education, Manipal, Karnataka, India

N. Pooja Department of Biophysics, Manipal School of Life Sciences, Manipal Academy of Higher Education, Manipal, Karnataka, India

Archi Rai Manipal College of Dental Sciences,Manipal. Manipal Academy of Higher Education, Manipal, Karnataka, India

Snehal Rajput School of Computer Science Engineering & Technology, Bennett University, Greater Noida, India

Gagan Raju Department of Biophysics, Manipal School of Life Sciences, Manipal Academy of Higher Education, Manipal, India

Nidhi Prahlad Rao Faculty of Life and Environmental Sciences, Shimane University, Matsue, Japan

Mehul S. Raval School of Engineering and Applied Science, Ahmedabad University, Ahmedabad, India

Mohendra Roy School of Technology, Pandit Deendayal Energy University, Gandhinagar, India

Mohendra Roy Department of Information and Communication Technology, School of Technology, Pandit Deendayal Energy University, Gandhinagar, India

Santanu Roy Computer Science and Engineering, Pandit Deendayal Energy University (PDEU), Gandhinagar, India;
NIIT University, Rajasthan, India

Mayuri Sharma National Institute of Electronics and Information Technology (NIELIT), Guwahati, Assam, India

Preethi R. Shenoy Manipal College of Dental Sciences,Manipal. Manipal Academy of Higher Education, Manipal, Karnataka, India

B. Suresh Kumar Shetty Department of Forensic Medicine, Kasturba Medical College, Mangalore, India;
Manipal Academy of Higher Education, Manipal, Karnataka, India

Yumnam Jayanta Singh National Institute of Electronics and Information Technology (NIELIT), Guwahati, Assam, India

Ankur Solanki Flextronics Lab, Department of Physics, School of Energy Technology, Pandit Deendayal Energy University, Gandhinagar, India

Tatsuyuki Yamamoto Faculty of Life and Environmental Sciences, Shimane University, Matsue, Japan

Zeev Zalevsky Faculty of Engineering and The Nano-Technology Center, Bar-Ilan University, Ramat-Gan, Israel

Guan-Yu Zhuo Institute of Translational Medicine and New Drug Development, China Medical University, Taichung, Taiwan

Abbreviations

3D	Three-Dimensional
ACM	Association for Computing Machinery
AD	Alzheimer's Disease
AI	Artificial Intelligence
AICD	Activation-Induced Cell Death
ANN	Artificial Neural Network
AR	Augmented Reality
ARG	Antibiotic Resistant Genes
ASPECTS	Alberta Stroke Programme Early CT Score
AST	Antimicrobial Susceptibility Test
AUC	Area Under the Curve
AUC-PR	Area Under the Precision-Recall Curve
AUC-ROC	Area Under the Receiver Operating Characteristic Curve
AWS	Amazon Web Services
BDL	Bile Duct Ligated
BF	Blood Flow
BMI	Brain-Machine Interface
BPH	Benign Prostatic Hyperplasia
BraTS	Brain Tumor Segmentation
BWCCE	Balanced Weighted Categorical Cross Entropy
CAD	Computer-Aided Diagnosis
CARS	Coherent Anti-Stokes Raman scattering
CAT	Computerized Axial Tomography
CBWCCE	Custom Weighted Balanced Categorical Cross Entropy
CC	Correlation Coefficient
CHO	Chinese Hamster Ovary cell
CLAHE	Control Enhancement Adaptive Local Histogram Equalization
CLARA	CLAssification through wavelet transform of RAman spectra
CLSM	Confocal Laser Scanning Microscopy
CM	Confocal Microscopy
CMOS	Complementary Metal-Oxide-Semiconductor

CNN	Convolutional Neural Networks
CoRR	Computing Research Repository
CT	Computed Tomography
CTA	CT Angiography
CTC	Circulating Tumor Cell
CXR	Chest X-Ray
DA	Data Augmentation
DBN	Deep Belief Network Denoising Autoencoder (SDAE)
DCNN	Deep CNN
de-CNN	De-Convolutional Neural Network
DICE	Dice Coefficient
DL	Deep Learning
DLR	Deep Learning Reconstruction
DLSS	Deep-Learning-Based SERS Screening Strategy
dMRI	Diffusion-Weighted Magnetic Resonance Imaging
DNA	Deoxyribonucleic Acid
DNN	Deep Neural Networks
DOT	Diffuse Optical Tomography
DSC	Dice Similarity Coefficient
DT	Decision Tree
EBCT	Electron Beam CT
ECM	Extracellular Matrix
EEG	Electroencephalography
EIR	Equivalent Imaging Rate
EIS	Electrochemical Impedance Spectroscopy
EIT	Electrical Impedance Tomography
ELISA	Enzyme-Linked Immuno-Sorbent Assay
EM	Expectation-Maximization
EMI	Electric and Musical Industries
FC	Fully Connected
FCN	Fully Convolutional Networks
FLAIR	Fluid Attenuated Inversion Recovery
FLI	Fluorescence Lifetime Imaging
FLIM	Fluorescence Lifetime Imaging Microscopy
FLI-Net	Fluorescence Lifetime Imaging Network
fMRI	Functional Magnetic Resonance Imaging
FN	False Negative
FP	False Positive
FPRED	False Positive Reduction
FrCN	Full resolution Convolutional Network
FRET	Fluorescence Resonance Energy Transfer
GAN	Generative Adversarial Network
GAP	Global Average Pooling
GD	Gradient Descent
GLCM	Gray Level Cooccurrence Matrix

GRIN	Gradient Refractive Index
GRU	Gated Recurrent Unit
H&E	Haematoxylin and Eosin
HC	Head Circumference
HCC	Hepatocellular Carcinoma
HD	Haudorff Distance
HER2	Human Epidermal Growth Factor Receptor
HIS	Hyperspectral Imaging
HNS	Human Nasopharyngeal Swab
HOG	Histogram of Oriented Gradients
HOG	Histogram-Oriented-Gradient
HP	Histopathology
HR	High Resolution
HRS	High-Resistance State
IBC	Immuno Histo-Chemistry
ICU	Intensive Care Units
IoU	Intersection Over Union
IRF	Instrument Response function
KiTS	Kidney and Kidney Tumor Segmentatio
KNN	K-Nearest Neighbors
LDA	Linear Discriminant Analysis
LiTS	Liver Tumor Segmentation
LR	Logistic Regression
LRS	Low-Resistance State
LSM	Least Square Mean
LSTM	Long Short-Term Memory
LUNA	Lung Nodule Analysis
MCI	Moderate Cognitive Impairment
MDR	Multidrug Resistance
MeSH	Medical Subject Headings
MIC	Minimum Inhibitory Concentration
ML	Machine Learning
ML-NN	Multi-Layer Network
MLP	Multilayer Perceptron
MPM	Multiphoton Microscopy
MRCP	Magnetic Resonance Cholangiopancreatography
MRI	Magnetic Resonance Imaging
MSC	Mesenchymal Stem Cell
MSS	Microsatellite Stable
MSSA	Methicillin-Sensitive Staphylococcus Aureus
MTF	Modulation Transfer Functions
MTT	Mean Transit Time
NADH	Nicotinamide Adenine Dinucleotide + Hydrogen
NAFLD	Non-alcoholic Fatty Liver Disease
NASH	Non-alcoholic Steatohepatitis

NB	Naive Bayes
NBC	Nano-Plasmonics Biosensing Chips
NC	Normal Controls
NDET	Nodule Detection
NIR	Near-Infrared
NLO	Nonlinear Optical
NLP	Natural Language Processing
NMR	Nuclear Magnetic Resonance
NN	Neural Network
NOM	Nonlinear Optical Microscopy
NPC	Nasopharyngeal Carcinoma
NS1	Non-Structural Protein 1
NT	Nuchal Translucency
OCT	Optical Coherence Tomography
OM	Optical Microscopy
OSCC	Oral Squamous Cell Carcinoma
P&D	Potentiation and Depression
PACS	Picture Archiving and Communication System (PACS)
PALM	Photoactivated Localization Microscopy
PBMC	Peripheral Blood Mononuclear Cell
PCA	Principal Component Analysis
PCa	Prostate Cancer
PC	Aprostate Cancer
PCA.Net	Principal Component Analysis Net
PE	Percentage Error
PET	Positron Emission Tomography
PSA	Prostate-Specific Antigen Test
PSNR	Peak Signal-to-Noise Ratio
PT	Post-Treatment
PTPN	Parameter-Tuning Policy Network
PTV	Planning Target Volume
RadLex	Radiology Lexicon
RANO	Response Assessment for Neuro-oncology
RBM	Gaussian restricted Boltzmann machine
RCM	Reflectance Confocal Microscopy
RECIST	Response Evaluation Criteria in Solid Tumours
ReLU	Rectified Linear Unit
RF	Random Forests
RIS	Radiology Information Systems
RMSE	Room Mean Square Error
RMSEC	Root Mean Square Error of Calibration
RMSECV	Root Mean Square Error of Cross Validation
RMSEP	Root Mean Square Error of Prediction
RNN	Recurrent Neural Network
ROC	Receiver Operating Characteristic

ROI	Region of Interest
RS	Raman Spectroscopy
RS-CNN	Raman-Specified Convolutional Neural Network
RTPCR	Reverse Transcriptase-Polymerase Chain Reaction
RUS	Random Under Sampling
RUSRF	RUS Random Forest
SAE	Stacked Autoencoder
SARS-2	Severe Acute Respiratory Syndrome-2
SBR	Signal to Background
SCC	Squamous Cell Carcinoma
sCT	Synthetic Computed Tomography
SDAE	Stacked Denoising Autoencoder
SERS	Surface Enhanced Raman Spectroscopy
SHG	Second Harmonic Generation
SLAM	Simultaneous Label-Free Autofluorescence Multiharmonic
SMOTE	Synthetic Minor Oversampling Technique
SNARC	Stochastic Neural Analog Reinforcement Calculator
SNN	Spiking Neural Network
SPECT	Single-Photon Emission Computed Tomography
SRS	Stimulated Raman scattering
STED	Stimulated Emission Depletion Microscopy
STORM	Stochastic Optical Reconstruction Microscopy
SUV	Standardized Uptake Value
SVM	Support Vector Machine
TACE	Transarterial Chemoembolization
TB	Tuberculosis
TBI	Traumatic Brain Injury
TC	Tumor Core
TCSPC	Time Correlated Single Photon Counting
TL	Transfer Learning
TN	True Negative
TOF-MRA	Time-of-flight MR Angiography
TP	True Positive
TPE	Two Photon Excitatio
TPM	Two Photon Fluorescence Microscopy
TPM	Two-Photon Microscopy
t-SNE	T-Distributed Stochastic Neighbor Embedding
U-HRCT	Ultra-High-Resolution Computed Tomography
USG	Ultrasonography
UTI	Urinary Tract Infection
VoxRes-Net	Voxel-Wise Residual Network
VP	Viral Pneumonia
VR	Virtual Reality
VST	Variance Stabilization Transform
WCCE	Weighted Categorical Cross Entropy

WT	Whole tumor
YOLO	You-Only-Look-Once

Chapter 1
Artificial Intelligence in Diagnostic Medical Image Processing for Advanced Healthcare Applications

Amlan Jyoti Kalita, Abhijit Boruah, Tapan Das, Nirmal Mazumder, Shyam K. Jaiswal, Guan-Yu Zhuo, Ankur Gogoi, Nayan M. Kakoty, and Fu-Jen Kao

Abstract Recent advancements in biomedical imaging modalities, including magnetic resonance imaging (MRI), X-ray, ultrasound, computed tomography (CT), electrical impedance tomography (EIT), positron emission tomography (PET), and (optical) microscopic imaging, have significantly propelled healthcare by elucidating intricate details of human tissues and organs. Despite these achievements, challenges

---

Amlan Jyoti Kalita and Abhijit Boruah: These authors contributed equally to this work.

A. J. Kalita · N. M. Kakoty (✉)
Embedded Systems and Robotics Lab, School of Engineering, Tezpur University, Tezpur, Assam 784028, India
e-mail: nkakoty@tezu.ernet.in

A. Boruah
Department of Computer Science and Engineering, Dibrugarh University, Dibrugarh, Assam 786004, India

T. Das
Department of Power Electronics and Instrumentation Engineering, Jorhat Institute of Science and Technology, Jorhat, Assam 785010, India

N. Mazumder
Department of Biophysics, Manipal School of Life Sciences, Manipal Academy of Higher Education, Manipal, Karnataka 576104, India

S. K. Jaiswal
KIMS-Sunshine Hospitals, Begumpet, Hyderabad, Telangana 500016, India

G.-Y. Zhuo
Institute of Translational Medicine and New Drug Development, China Medical University, Taichung 404328, Taiwan

A. Gogoi (✉)
Department of Physics, Jagannath Barooah University, Jorhat, Assam 785001, India
e-mail: ankurgogoi@gmail.com

F.-J. Kao
Institute of Biophotonics, National Yang-Ming Chiao Tung University, Taipei 11221, Taiwan

A. Gogoi and N. Mazumder (eds.), *Biomedical Imaging*, Biological and Medical Physics, Biomedical Engineering, https://doi.org/10.1007/978-981-97-5345-1_1

persist in terms of image acquisition, processing, big medical data management, operator-dependent variabilities, and image processing intricacies. In this context, artificial intelligence (AI) has emerged as a transformative solution, alleviating operator-dependent variabilities and augmenting diagnostic, treatment planning, and patient care capabilities. In this chapter, we present a detailed exploration of AI applications across various imaging modalities, tracing their historical evolution in medical imaging. We study the workflow of AI in tasks such as image reconstruction, analysis, interpretation, segmentation, and enhancement. Additionally, we explore its role in computer-aided diagnosis (CAD), treatment planning, patient group classification, and predicting and reducing radiotherapy doses. This chapter encompasses discussions on benefits, challenges, ethical considerations, and future directions in the field of AI in biomedical imaging. This chapter underscores the pivotal role of AI in enhancing the efficiency, objectivity, reliability, and accuracy of medical imaging services.

1.1 Introduction

Unveiling the structural (de)formation and (dys)function of different tissues and organs in the human body is the keystone in diagnosing abnormalities, treating diseases, and ensuring a long and healthy life for humankind. In recent years, tremendous advancements in the fields of medical sciences and healthcare systems have revolutionized the very paradigm of disease prevention, diagnosis, and treatment [1–3]. In particular, the evolution of advanced biomedical imaging modalities, such as magnetic resonance imaging (MRI), X-ray, ultrasound, computed tomography (CT), electrical impedance tomography (EIT), positron emission tomography (PET), and (optical) microscopic imaging, has enabled the precise visualization of not only the physical anatomy but also intricate biochemical and physiological processes within the human body [4].

Undoubtedly, such advancements in medical imaging have enhanced diagnostic and research capabilities, opening new frontiers for healthcare professionals and researchers. It enables them to explore, analyze, and comprehend the intricate workings of the human body in unprecedented detail and depth at an enhanced spatiotemporal resolution. Despite their pivotal role, biomedical imaging typically generates a substantial amount of big data, necessitating proper management and analysis to extract meaningful information about diseases [5, 6]. Moreover, the precision of diagnostic outcomes is heavily dependent on the expertise of radiologists and/or analysts, introducing inherent susceptibility to both inter- and intra-operator variations. In addition, inherent challenges associated with image acquisition, such as image artifacts, intensity fluctuations, and insufficient contrast differentiation between tissues, as well as those related to image pre- and post-processing (including contrast enhancement, noise reduction, image sharpening, image compression, image segmentation, feature extraction, and object classification), significantly compromise the diagnostic potential, clinical relevance, and accuracy of analysis results [7, 8].

Addressing these challenges is essential for advancing the field and ensuring that biomedical imaging remains a reliable and powerful tool in the pursuit of improved diagnostics and personalized medical approaches. In this context, AI has emerged as a transformative force in the field of medical imaging, revolutionizing the way healthcare is delivered by devising tailored biomedical image analytical methodologies capable of ameliorating the impact of operator-dependent variabilities [9, 10]. The fusion of diverse AI technologies with medical imaging modalities has ushered in a paradigm shift in diagnosis, treatment strategizing, and patient care, revealing novel avenues for enhanced precision and efficiency.

Notably, the potential for AI to assist in the triage and interpretation of conventional radiographs (X-ray images) holds great significance, given that radiographs constitute the most frequently conducted imaging studies in most radiology departments [11]. Incorporating AI into the digitization of images and automating protocols can enhance the efficiency of acquiring MRI images and improve overall work performance and image-based disease detection [12, 13]. AI methods such as machine learning (ML) and deep learning (DL) have enhanced the early diagnosis of health issues such as brain disorders [14], Parkinson's disease [15], autism [16], and schizophrenia [17], which involves functional MRI (fMRI) image analysis. AI methods have been used for the analysis of CT scans for the clinical care of traumatic brain injury (TBI) patients [18], pulmonary nodule detection in the lungs [19] and the diagnosis and treatment of COVID-19 through the analysis of CT scans of the chest [20]. The applications of AI techniques in solving issues such as lesion segmentation, object detection, anatomical localization, and computer-aided diagnosis (CAD) of various anatomical regions using ultrasound modalities have been widely discussed [21]. AI techniques, mostly DL approaches, have recently been used in PET images for lesion detection [22], differential diagnosis and cancer staging [23] and diagnosis of spinal metastases [24]. Recently, DL algorithms have also gained considerable attention in solving the nonlinear, ill-posed diffuse optical tomography (DOT) reconstruction problem for biomedical imaging. The convoluted neural networks (CNNs) have been used to estimate the bulk optical properties more accurately and efficiently than conventional reconstruction methods in DOT [25, 26]. The best and worst attribute of DL methods is their dependency on the data and the data alone to formulate their solution. Thus, DL can be used for almost any problem or task for which sufficient training data are available for the network. The DL algorithm acts as a black box and does not have any prior knowledge of the physical process it is trying to resolve. Thus, efforts are now toward understanding the underlying principle of the problem and including methodologies and previous knowledge from analytical approaches into DL algorithms to help decrease the complexity and computational burden of the problem. In a recent work, contrary to the traditional black-box DL approach, a physics-enabled deep network is considered to invert the Lippmann–Schwinger integral equation using a mathematical model of deep convolution and obtain an accurate three-dimensional distribution of optical anomalies [27].

Thus, the incorporation of AI into medical imaging has fundamentally altered the landscape of diagnostics, treatment strategies, and patient management. By employing techniques such as ML and DL, AI has enhanced the effectiveness of image acquisition and enabled prompt disease detection across a spectrum of imaging modalities. Current endeavors focus on integrating analytical approaches and prior insights into DL algorithms, fostering opportunities for continued progress in the discipline and ultimately leading to improved patient results to improve the overall efficiency, objectivity, accuracy, and reliability of medical imaging services [28]. In this chapter, we embark on a comprehensive exploration of the intersection between AI and medical imaging, delving into the profound implications, challenges, and potential future directions of this field. In the subsequent sections of this chapter, we will discuss manifold applications of AI in various biomedical imaging modalities and scrutinize the benefits and implications of this transformative fusion. Furthermore, we will discuss the challenges and ethical considerations, examine applications, and finally, contemplate the exciting future directions and prospects of AI-assisted biomedical imaging.

1.2 Medical Imaging in Modern Healthcare

The human body is characterized by intricate multiscale and multilayered organizational systems encompassing a diverse array of structures [29, 30]. At the macroscopic level, organs exhibit millimeter-scale dimensions and are composed of various tissues with specialized functions. At the microscopic scale, cellular entities manifest dimensions in the micrometer range, contributing to the specialized functions of tissues and organs. Furthermore, within the confines of the cellular milieu, intricate structures at the nanometer scale, including organelles and molecular complexes, define the subcellular landscape. For example, at the macroscopic level, the brain is an organ with millimeter-scale dimensions, housing distinct regions such as the cerebral cortex, cerebellum, and brainstem. Each of these regions comprises various tissues with specialized functions, forming the basis for cognitive, sensory, and motor processes. At the microscopic scale, neural tissue within the brain consists of neurons, glial cells, and intricate networks of blood vessels, all with dimensions in the micrometer range. Neurons, the fundamental units of the nervous system, play a crucial role in transmitting electrical signals, while glial cells support and modulate neuronal function. The complex organization of these microscopic elements contributes to the overall functionality of the brain. Furthermore, at the nanoscale, the structure and function of synaptic connections and neurotransmitters become prominent [30, 31]. Likewise, the heart is a complex organ spanning multiple scales, encompassing various structures [32–35]. These range from millimeter-sized elements such as the muscular structure and distinct chambers, including the atria and ventricles, to micrometer-sized cardiac tissues. These tissues include cardiomyocytes (heart muscle cells), blood vessels, and connective tissues. Additionally, the cardiac nanoenvironment involves nanoscale features of the extracellular matrix (ECM), which play a vital role in cellular interactions within the heart [36, 37]. Beyond spatial scales, numerous activities and functions of diverse tissues and organs extend across various temporal scales. Figures 1.1a and b show multiple spatial and temporal scales of the

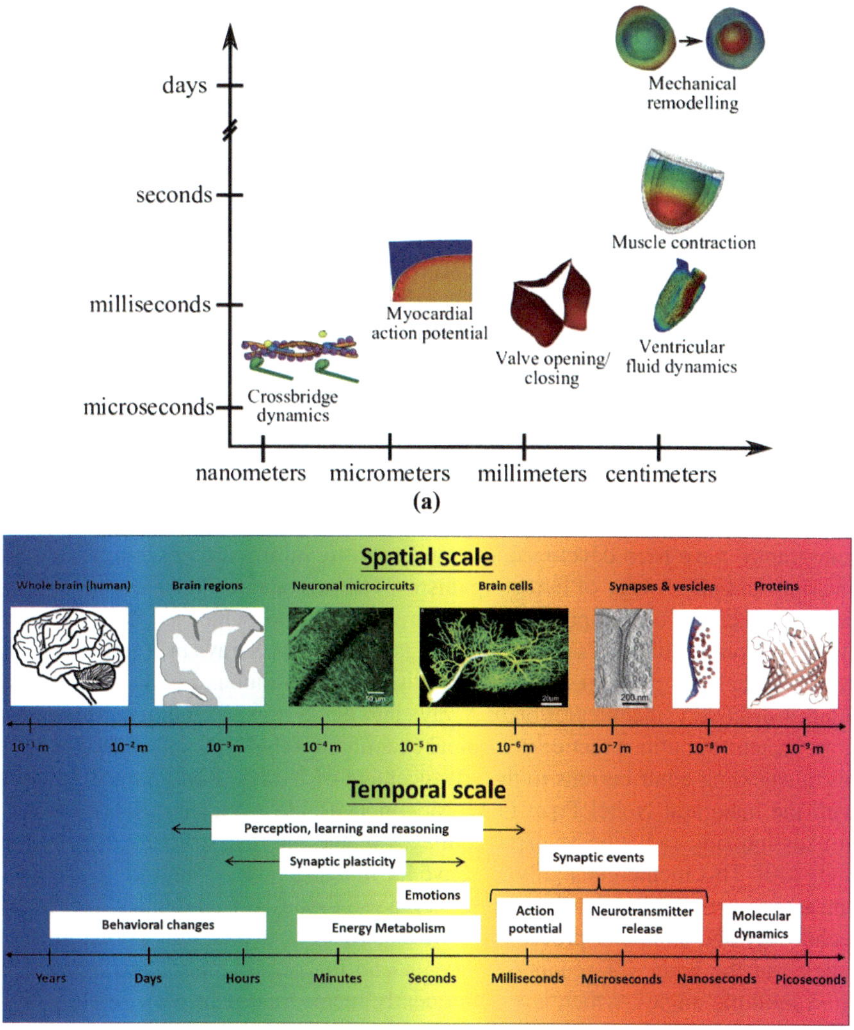

(b)

Fig. 1.1 **a** Spatial and temporal scales of various mechanisms of heart function are shown on the horizontal and vertical axes, respectively. The figure has been reproduced with permission from Quarteroni et al. [41], Copyright © 2016 Elsevier B.V.; **b** Diverse spatial and temporal scale of the structure and the function of the human brain. The figure has been reproduced with permission from Gogoi et al. [30], Copyright © 2019, Springer Nature Singapore Pte Ltd. Notably, the figure for the neuronal microcircuit Hippocampus CA1 is adapted from Asrican et al. [42], Copyright © 2013 Asrican et al., Creative Commons Attribution License (CC BY); the figure for brain cells (Purkinje cell) is adapted with permission from Denk et al. [43], Copyright © 1997 Cell Press; the figure for synapses and vesicles is adapted from Korogod et al. [44], Copyright © 2015, Korogod et al., Creative Commons Attribution License (CC BY); the figure for the protein (transmembrane protein: porin) is adapted from Hammond et al. [45], Copyright © 2013, The Author(s), Creative Commons Attribution License (CC BY 3.0) brain and heart, respectively. Notably, such hierarchical spatial and temporal arrangement of structures across multiple scales is fundamental to the complexity and functionality of biological systems. Understanding these multiscale dimensions is pivotal for elucidating physiological processes, pathological conditions, and the interplay of components within the intricate framework of the human body. Importantly, the evolution of biomedical imaging techniques over the past century has introduced a high level of vision to medical science, resulting in a paradigm shift in the healthcare system. Medical imaging has become indispensable for diagnosis, treatment planning, monitoring disease progression/regression, and treatment effectiveness over time because it allows healthcare professionals to visualize internal structures and detect issues beyond the scope of physical examinations alone [38–40].

1.2.1 Evolution of Medical Imaging Modalities

Over the years, diverse modalities, such as MRI, CT, nuclear imaging, and ultrasonography, have been developed, offering versatile information encompassing the anatomy and physiology of living organisms. These modalities have become essential components of contemporary healthcare practices worldwide [46]. Notably, since Wilhelm Conrad Röntgen's groundbreaking discovery of X-rays or Röntgen-rays in 1895, the field of medical imaging has witnessed continuous innovation. Shortly after its discovery, the integration of X-rays into diagnostic and therapeutic applications became a routine practice in medical examinations [47]. In acknowledgment of his outstanding contributions to the advancement of X-rays, Röntgen was honored with the inaugural Nobel Prize in Physics in 1901 [48]. Notably, X-ray imaging utilizes ionizing radiation to capture images of the internal structure of a human body [49]. This method is highly effective for imaging bones, identifying fractures, and assessing the chest, abdomen, and specific soft tissues. CT imaging utilizes X-ray technology and computer processing to reconstruct detailed cross-sectional images (slices) of the body. CT scanners rotate around the patient, collecting X-ray data from multiple angles, which are subsequently reconstructed into 3D images [50]. Notably, Allan M. Cormack and Godfrey N. Hounsfield were jointly awarded the Nobel Prize in Physiology or Medicine in 1979 for their significant contributions to the development of CT [51].

Based on the laws of nuclear magnetic resonance (NMR), another widely used imaging modality, MRI, employs radiofrequency signals capable of penetrating various parts of the human body to visualize internal body structures and functions, including neurological disorders and musculoskeletal abnormalities [52–55]. It excels in soft tissue contrast and is particularly valuable for imaging the brain, spinal cord, joints, and soft tissues, aiding in the diagnosis of neurological issues, musculoskeletal injuries, and malignancies. This encompasses the identification of typical abnormalities such as tumors and traumatic injuries, establishing MRI as a crucial instrument in both oncology research and the clinical care of cancer patients [56]. In the field of neuroscience, diffusion-weighted MRI (dMRI), one of the most widely utilized MRI modalities employed to observe the diffusion of water molecules in brain tissues, thereby revealing their microarchitectures, is an integral part of the Human Connectome Project [57, 58]. Similarly, fMRI captures functional brain patterns by monitoring variations in blood flow and oxygen metabolism resulting from changes in neural activity [59]. Recently, 7-Tesla MRI has demonstrated enormous potential in the diagnosis of neurodegenerative diseases, such as Parkinson's disease, Alzheimer's disease, and Amyotrophic Lateral Sclerosis [60, 61]. Notably, Paul C. Lauterbur and Peter Mansfield were jointly awarded the Nobel Prize in Physiology or Medicine in 2003 for their groundbreaking discoveries that established MRI as a practical imaging method for use in disease diagnosis and clinical medicine [62, 63].

Ultrasound imaging utilizes high-frequency sound waves (typically $\geq$ 16 kHz) to generate real-time images of internal organs and structures [64]. Despite being discovered 12 years prior to X-rays in 1883, the medical application of ultrasound commenced in the 1950s. Currently, it is employed in various medical contexts, including abdominal, cardiac, maternity, gynecological, urological, and cerebrovascular examinations [65, 66]. On the other hand, nuclear medicine imaging involves radioactive tracers that emit gamma rays, which are detected by specialized cameras. Tracers visualize metabolic and functional processes, focusing on specific organs or tissues [46, 67]. Common nuclear medicine techniques include PET [68, 69] and single-photon emission computed tomography (SPECT) [70], which are widely employed for cancer staging, heart imaging, and neurological assessments.

In addition to these non-optical imaging techniques, various advanced optical methodologies, including autofluorescence imaging, DOT, optical coherence tomography (OCT), narrow band imaging, confocal laser scanning microscopy (CLSM), multiphoton excitation microscopy, stimulated emission depletion microscopy (STED), stochastic optical reconstruction microscopy (STORM), photoactivated localization microscopy (PALM), etc., have been widely applied in biomedical research, particularly in biopsy and surgical specimens [71–75]. Such optical methods have demonstrated their ability to provide visual depictions of tissue morphology and extract crucial information about overall tissue quality and underlying molecular signatures, aiding pathologists in accurate observation and interpretation [76]. These cutting-edge techniques are instrumental across diverse domains, including ophthalmology (for high-resolution retinal imaging) [77], cardiology (enabling precise vascular visualization) [78], surveillance of gastrointestinal and respiratory conditions [79], and oncology (providing a sophisticated approach to detecting malignant lesions and intricately monitoring cancer progression and regression) [80]. DOT is used as an alternate imaging modality for breast cancer detection owing to its outstanding contrast to hemoglobin oxidation (the first clinical trial using optical light was conducted in ACRIN 6691 to predict the clinical outcome in breast cancer patients undergoing neoadjuvant chemotherapy [81]). Notably, the Nobel Prize in Chemistry for the year 2014 was collectively awarded to Eric Betzig, Stefan Hell, and William E. Moerner in recognition of their significant contributions to the advancement of super-resolved fluorescence microscopy [82]. These imaging techniques developed to date have played a pivotal role in advancing our understanding

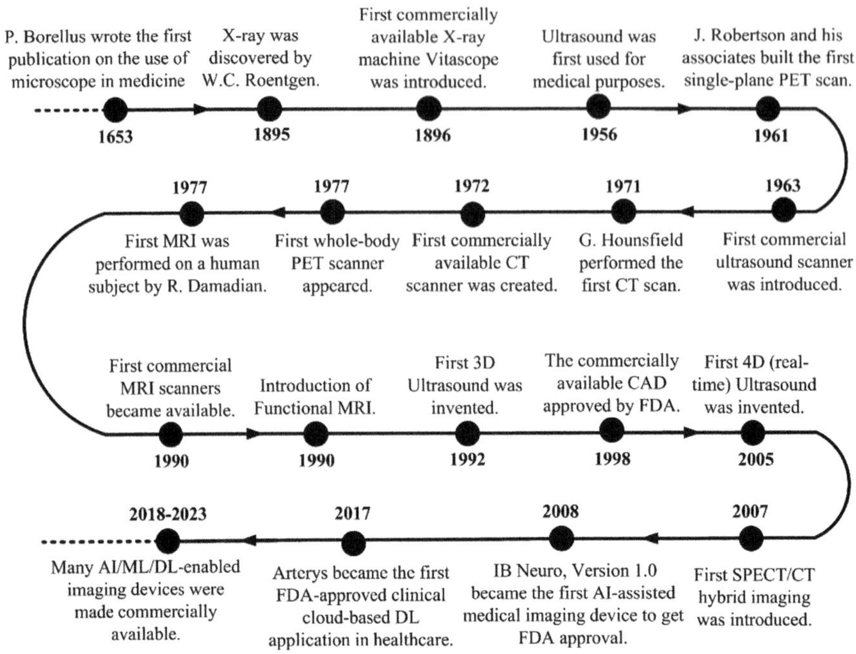

Fig. 1.2 A historical timeline of the key discoveries in medical imaging

of the human body, facilitating accurate disease diagnosis and improving prevention strategies. A representation of the evolution of medical imaging modalities over the past century is shown in Fig. 1.2.

1.2.2 Role of Multiscale Multiparametric Imaging in Healthcare

Notwithstanding the enormous capacity and importance of standalone imaging modalities, considering the multiscale, multidimensional, and multiparametric nature of diseases, it is often necessary to integrate images or results from multiple imaging modalities that are applicable to both preclinical and clinical settings [83, 84]. This integration proves imperative for achieving enhanced diagnostic accuracy, improved patient convenience, and optimized workflow efficiency [85]. In this context, recent years have witnessed the development of several sophisticated multimodal (or hybrid) imaging modalities [86–91]. These modalities strategically integrate at least two traditionally distinct yet powerful microscopy platforms, fostering synergies that contribute to a comprehensive understanding and identification of disease signatures by extracting and correlating valuable information from the same targeted sample using each platform, thereby benefiting the overall analytical capability of the multi-modal system [92, 93]. For example, PET is frequently used in conjunction with CT (PET/CT) or MRI (PET/MR) to offer supplementary information in the development of a possible problem [94, 95]. Furthermore, the introduction of multiparametric imaging techniques capable of amalgamating quantitative data represented by various parameters, e.g., standardized uptake value (SUV), regional tumor blood flow (BF), mean transit time (MTT), etc., from a single imaging modality or multimodal imaging system has facilitated intricate and multifaceted visualization of diseases, particularly in the fields of cardiology, oncology, immunology, etc. [96–101].

1.2.3 Primary Challenges in Medical Imaging: Importance of AI

Although biomedical imaging holds immense promise for advancing diagnostics and patient care, it encounters several primary challenges and limitations. For instance, the acquisition of high-quality medical images is a formidable task due to inherent factors such as noise and artifacts, which can compromise the quality, clarity, and precision of the images [102]. Additionally, conventional image reconstruction techniques are often computationally intensive, leading to delayed image acquisition and diagnosis. Moreover, medical images, such as those from MRI, CT, or PET, often have high resolution and complexity, leading to large datasets [103]. Additionally, medical image data are often stored in disparate systems and databases, making seamless integration and interoperability challenging [104]. Furthermore, interpreting medical images can be subjective, and there may be variability in interpretations with differences in expertise, experience, and subjective judgment [105]. In numerous instances, the availability of specialized medical proficiency for the interpretation of these images may be restricted, particularly in remote or underserved areas. This limitation can adversely affect diagnostic accuracy, potentially giving rise to inconsistencies in patient management and influencing decisions regarding treatment strategies.

In this context, the importance of AI in addressing these issues becomes increasingly evident as it offers innovative solutions to enhance the capabilities of medical image processing. AI algorithms, encompassing ML techniques and DL models, exhibit remarkable efficiency in noise reduction, artifact correction, and rapid processing of large and complex datasets compared to conventional systems [106, 107]. AI tools are also capable of improving the study of large and intricate medical image datasets by learning intricate patterns and relationships [38]. In addition, AI algorithms can easily be designed to facilitate interoperability, enabling a more cohesive workflow by integrating not only existing healthcare systems but also medical image data with electronic health records and other pertinent patient information. The potential for human error in data analysis can be minimized with AI, specifically CAD, as detailed in Sect. 3.3.4. Moreover, AI tools can provide support to healthcare professionals by automating routine tasks and aiding in preliminary analyses. This enhances accessibility to expert-level insights, especially in regions with a shortage of specialized personnel [108].

Although the role of AI in medical imaging holds considerable promise, persistent challenges such as ethical concerns, reliability, and generalizability need to be addressed. The subsequent sections provide a comprehensive description of AI, covering basic concepts, the overall workflow, applications, and associated limitations.

1.3 AI in Medical Imaging

AI-powered solutions are changing this landscape by automating and augmenting many aspects of image interpretation, offering rapid and reliable insights into complex imaging data. Medical image analysis usually includes delineation of the objects of interest (segmentation) or description of labels (classification). In the following two sections, the basics of AI and key techniques and algorithms in AI for medical imaging are described.

1.3.1 Basics of Artificial Intelligence

Intelligence refers to the ability to resolve challenging innovative problems when previous knowledge or gathered experience offers limited assistance [109]. It is a natural cognitive ability in humans and is often evaluated through tools such as advanced progressive matrices. AI, on the other hand, is the emulation of human intelligence processes by computers. The commencement of the AI discourse can be traced back to Vannevar Bush's seminal work, "As We May Think [110]", and Alan Turing's groundbreaking work, "Computing Machinery and Intelligence," published in 1950 [111, 112]. Soon after, John McCarthy, along with Marvin Minsky, Nathaniel Rochester, and Claude Shannon, who are acknowledged as the pioneers of AI, introduced the term "Artificial Intelligence" in a proposal submitted to the Rockefeller Foundation on August 31, 1955, titled 'A Proposal for the Dartmouth Summer Research Project on Artificial Intelligence' [113–116]. The renowned Dartmouth Workshop, organized in the summer of 1956, is often regarded as the birthplace of AI [117, 118]. McCarthy coined the term "Artificial Intelligence" [113], defining it as "the science and engineering of making intelligent machines" [119]. In essence, AI is a method aimed at enabling machines to function and exhibit behavior similar to humans [120, 121].

The operational framework of AI involves the utilization of algorithms and computational models aimed at emulating human intelligence, thereby enabling the execution of tasks that traditionally demand human predictive capabilities [122]. ML is a branch of AI that deals with prediction, specifically in two categories: classification and regression. The ML process begins with data collection, followed by data preprocessing, including refinement and formatting. Subsequently, significant features are selected from the data, which are used to train a model using various learning algorithms. Model evaluation ensures its accuracy and effectiveness, after which the model can be deployed for inference on new data. The process often involves a feedback loop for continual improvement, incorporating new data or user feedback for enhancing the accuracies of the algorithms while training on the models. To illustrate, consider the scenario where the task of an ML model is to identify a dog in an image. The process begins by building a dataset of images with dogs as one of the objects, followed by the annotation of the images. These images will be used to train the models using ML algorithms. When a new image is fed to the model, identification involves extracting relevant structural features, such as height, shape, nose, eyes, tails, etc., from contours. It then compares these features against a vast dataset of known objects, identifying patterns and similarities. Throughout this process, the ML system narrows its options, gradually refining its understanding until it can confidently recognize the image as a dog. The more data the ML system is exposed to, the better it becomes at accurately identifying them, mirroring the way our brains improve with experience and exposure to different stimuli. Thus, ML systems continuously learn and evolve, becoming more adept at tasks through exposure to additional information.

AI technologies expand beyond traditional ML and DL techniques. Most prominent of them are expert systems, logic, knowledge representation and reasoning, planning, searching and problem solving, decision making and optimization techniques. Figure 1.3 shows a Venn diagram illustrating the interrelationship between AI, ML and DL, showing common ML and DL algorithms. ML and DL are subsets of AI with a focus on enabling predictive decision-making capabilities in applications such as natural language processing, computer vision, and signal processing. Within ML, algorithms iteratively learn patterns and relationships from extensive datasets, allowing them to improve their performance progressively over time [123]. This learning process involves the adjustment of model parameters or weights based on feedback from the data to minimize prediction errors or maximize predictive accuracy. ML can be categorized into three major categories (supervised, unsupervised and reinforcement) based on the nature of the datasets available for training. Supervised learning techniques in ML include models that learn from labeled data to make predictions [124]. Supervised algorithms include linear regression [125], logistic regression [126], Naïve Bayes [127], k-nearest neighbors [128], decision trees [129], random forest [130], and support vector machine (SVM) [131]. Unsupervised learning models, on the other hand, identify patterns and structures in unlabeled data by employing techniques such as principal component analysis (PCA) [132], hierarchical clustering [133], manifold learning [134], and K-means [135]. In reinforcement learning, models learn through trial and error based on feedback from the environment [136].

In 1957, Frank Rosenblatt introduced the perceptron model as an attempt to mimic a biological neuron [137]. A perceptron consists of a single layer of artificial neurons,

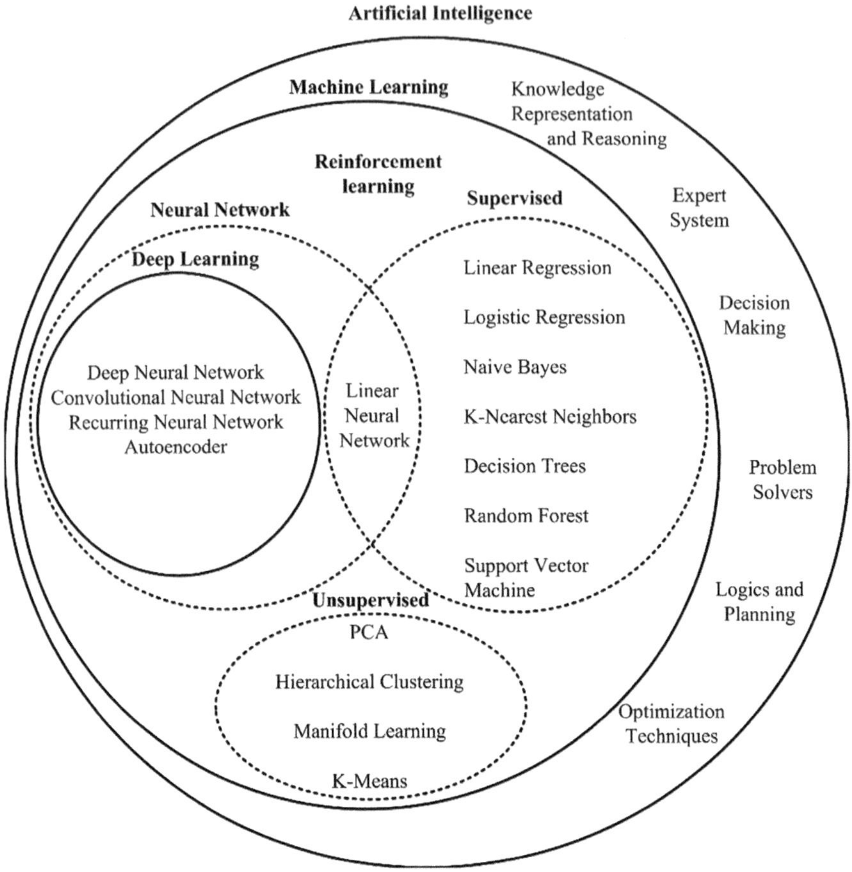

Fig. 1.3 Venn diagram illustrating the interrelationship between AI, ML and DL, showing common ML and DL algorithms

each connected to input features with corresponding weights. These neurons compute a weighted sum of the inputs, which are then processed through an activation function to produce an output. The connections of multiple perceptrons can form an artificial neural network (ANN) analogous to the neural system in the human brain [138]. DL constitutes a subset of ML that employs an ANN featuring multiple layers, hence the term "deep", to learn representations of data [139]. These neural networks are composed of interconnected nodes or neurons organized in layers, where each layer progressively extracts more abstract features from the input data. DL algorithms automatically learn these hierarchical representations from large amounts of labeled data without relying on handcrafted features. DL has been particularly transformative in areas where large amounts of data are available, driving significant advancements in AI research and applications. Speaking about the immense possibilities of AI,

Elon Mask stated, *"AI will be the best or worst thing ever for humanity"* [140]. The most common AI techniques for medical imaging are discussed in Sect. 3.4.

1.3.2 Historical Evolution of AI

In modern medical practice, AI has seen widespread utilization across healthcare, leveraging sophisticated algorithms and data analytics to enhance diagnostic capabilities, treatment strategizing, and the quality of patient care. This advancement has markedly heightened the efficiency and precision of medical decision-making protocols. Figure 1.4 presents the trajectory of the development of AI, marked by significant milestones and phases of heightened activity commonly known as 'AI booms'. The progression of AI integration in medical imaging can be divided into three distinct AI booms, representing periods of noteworthy advancement and increased interest in AI.

1.3.2.1 First AI Boom (1950–1970s)

This period witnessed the initial exploration and enthusiasm for AI research, marked by the development of foundational concepts, algorithms, and early AI applications.

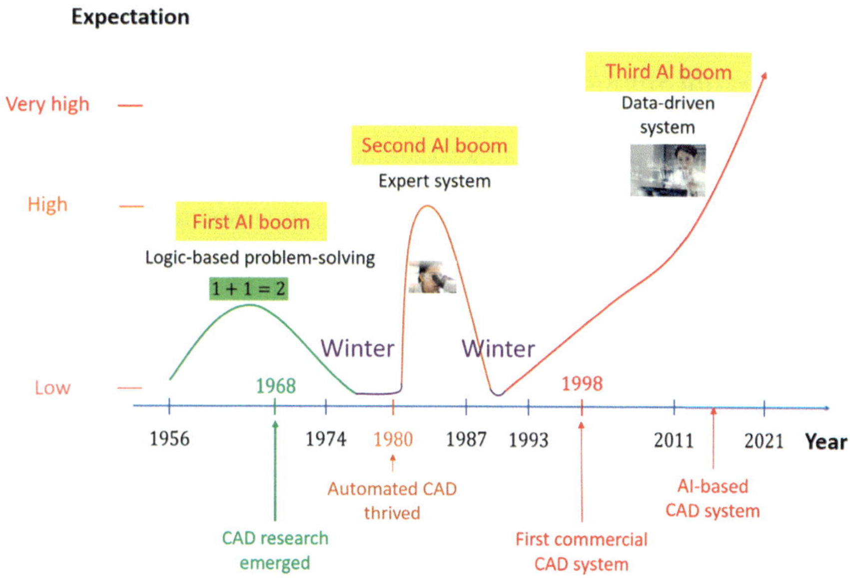

Fig. 1.4 Pictorial representation of the timeline of AI booms. The figure has been reproduced from Hoang-Thi et al. [141], Copyright © 2022 by the authors, Creative Commons Attribution (CC BY) License

In medical imaging, the first AI boom laid the groundwork for future developments by establishing basic principles of image processing and pattern recognition. The research on CAD started in this period [142–144].

1.3.2.2 Second AI Boom (1980–Early 2000s)

The second AI boom was characterized by renewed interest and investment in AI technologies, driven by advancements in computing power, algorithmic improvements, and the accessibility of larger sets of data. In medical imaging, this period saw the emergence of CAD systems, which began to assist radiologists in tasks such as mammography interpretation [142–144].

1.3.2.3 Third AI Boom (2000s–Present)

The third AI boom is ongoing and is fueled by breakthroughs in DL, increased computational capabilities, and the proliferation of big data. In medical imaging, this period has seen remarkable progress, with the widespread adoption of DL algorithms, particularly CNNs, for tasks such as image segmentation, classification, and disease detection. AI applications in medical imaging have expanded beyond mammography to various modalities. Additionally, regulatory approval for AI algorithms in medical imaging tasks has been obtained since 2010 [142–144].

1.3.3 Overall Workflow of AI-Assisted Biomedical Imaging

Figure 1.5 shows a representation of workflow in AI-assisted medical imaging. The process begins with acquiring medical images using various imaging modalities. These images capture detailed anatomical structures or physiological functions of the human body. Preprocessing steps are often performed on raw medical images to improve their quality and eliminate noise or artifacts. Relevant features are extracted from the pre-processed images. These features could be morphological, textural, or functional characteristics of the tissues or organs. After this step, the images are divided into two sections: the training dataset and the testing dataset. The training data are then fed to the ML/DL model for purposes such as image recognition, object detection, image segmentation, image enhancement, image restoration, image segmentation and registration, image analysis and interpretation, etc. These tasks are discussed in detail in the following sections. The trained models are validated and evaluated using the testing dataset. Once the ML/DL model has been trained and validated, it is integrated into the clinical workflow. This integration might involve deploying the model within existing medical imaging software or developing standalone applications for specific tasks.

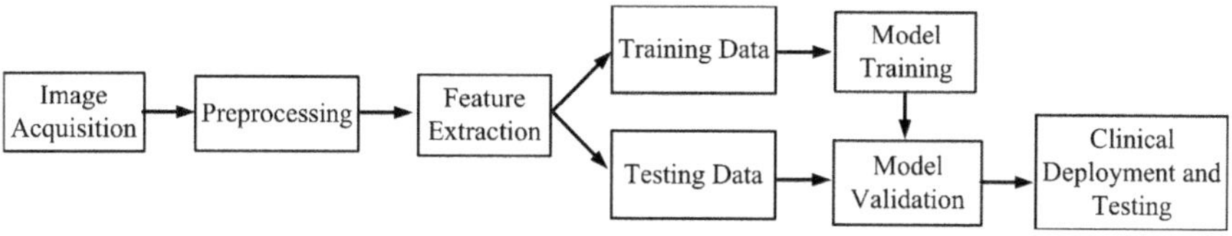

Fig. 1.5 Block diagram of the workflow of general AI-assisted biomedical imaging

1.3.3.1 Image Analysis and Interpretation

Image analysis involves processing digital images using algorithms to extract meaningful information, while image interpretation is the human cognitive process of understanding and deriving insights from visual data within images. AI-based algorithms can automatically detect and localize abnormalities, such as tumors, nodules, or fractures, in medical images. These algorithms assist radiologists in identifying and characterizing lesions more efficiently and accurately. For instance, AI algorithms are used in mammography to detect and classify breast lesions indicative of breast cancer. Automated breast cancer multi-classification from histopathological images holds clinical significance. A newly proposed DL model achieved 93.2% accuracy, offering an efficient tool for comprehensive breast cancer diagnosis and prognosis [145]. Another study compared a state-of-the-art DL method (CNN) with four classical ML methods for classifying mediastinal lymph node metastasis of NSCLC from 18F-FDG PET/CT images. They also assessed the discriminative power of PET/CT texture features against traditional diagnostic features. This study evaluated the sensitivity, specificity, ACC, and AUC via cross-validation on 1397 lymph nodes. Furthermore, this study revealed ideal feature sets for traditional approaches, which compare favorably to CNNs and doctors [146].

1.3.3.2 Image Segmentation and Registration

Image segmentation is a crucial process in computer vision that involves the division of an image into multiple segments based on criteria such as color, intensity, texture or shape. AI algorithms can segment medical images into distinct regions of interest, such as organs, tissues, or lesions. This segmentation is crucial for accurate diagnosis, treatment planning, and disease monitoring. DL has emerged as a potent and clinically relevant tool for segmenting head and neck anatomy in radiotherapy applications [147]. In this context, a pioneering algorithmic framework has been proposed for supervised multimodal image analysis. This framework shows enhanced performance in segmenting soft tissue sarcoma lesions by integrating information across different imaging modalities within a deep CNN. Such cross-modality fusion strategies provide valuable insights applicable to a range of multimodal image analysis tasks [148]. Moreover, a comparative study between standard UNet and dilated UNet architectures was conducted for automatic organ delineation using CT images. The findings indicate that the dilated UNet architecture surpasses the standard UNet architecture in accurately delineating soft tissues such as the esophagus and spinal cord, as assessed by spatial overlap metrics. However, in regard to lung delineation, there appears to be no significant difference in Dice scores between the two architectures [149].

In addition, studies on tumor segmentation have shown promising results. Accurate lymph node segmentation is crucial for disease assessment; a proposed DRL-LNS model leverages RECIST annotations for weakly supervised learning, achieving state-of-the-art performance with a mean Dice similarity coefficient (DSC) of 77.17% and a mean IoU of 64.78% on a public CT dataset, demonstrating superior bounding box prediction compared to label propagation [150]. An AI-based software solution for liver volumetric analysis showed excellent accuracy, with a mean absolute error of 44.3 mL (2.37% of averaged liver volumes), and significantly reduced processing time compared to manual segmentation (9.94 s vs 219.34 s per dataset), indicating its potential for efficient and precise clinical use [151].

1.3.3.3 Image Enhancement and Reconstruction

AI-based reconstruction techniques improve the quality and resolution of medical images, particularly in low-dose imaging or noisy environments. Current DL-based techniques for artifact reduction and motion correction can be divided into two categories: picture reconstruction and image postprocessing. Postprocessing is used to improve the quality of a previously reconstructed picture. A work proposes a novel approach for real-time dynamic medical image reconstruction from sparse samples, utilizing an autoencoder trained on representative data to achieve faster reconstructions than compressed sensing methods, albeit with slightly degraded image quality in the case of CT. Departing from standard autoencoder norms, the method employs a robust l1-norm and solves the resulting problem using the Split Bregman method, enabling efficient operation during testing while minimizing reconstruction time [152]. A reinforcement learning-based approach was proposed for automatically adjusting parameters in optimization problems, aiming to alleviate the tedious and impractical manual tuning process. By training a parameter-tuning policy network (PTPN), the system achieves similar or improved quality in optimization-based iterative CT reconstruction compared to manually tuned parameters [153].

Image reconstruction for conversion between image modalities may be useful for patients, such that one does not need to undergo multiple imaging processes. A novel deep CNN method is proposed to generate synthetic CT (sCT) images from MRI, achieving a mean absolute error below 85 HU for 13 out of 18 brain tumor patients. These results outperform an atlas-based approach and provide faster computation. Further validation of the dose computation accuracy and extension to multi-sequence MR images are suggested for future research [154]. A DL GAN approach efficiently generated accurate sCT images from MRI for brain radiation therapy, addressing alignment issues. Despite hours of initial training, the model quickly produced precise CT volumes, showing potential for clinical MRI-only treatment planning workflows [155].

1.3.3.4 Computer–Aided Diagnosis

CAD is a vital tool for assisting medical practitioners in image interpretation [156]. Basic research on CAD, including efforts in automatic diagnosis, commenced in the 1960s [142]. CAD is classified into four categories: second-reader, concurrent-reader, first-reader, and continuous learning CAD. In 1998, the first commercially available CAD device was for mammography. This was a second-reader CAD system, where a doctor sees pictures without CAD before verifying them with CAD results. Since then, several CADs have been produced. Simultaneous reader-type CAD, also known as concurrent reader CAD, was created as a way to include CAD data in the interpretation process from the outset. Riverain Technologies created the first concurrent-reader CAD system, ClearRead CT, for chest CTs. The first-reader CAD is a type of CAD in which the CAD first interprets the data. The doctor's interpretation is confined to the CAD-specified pictures. As a result, the interpretation period may have decreased. It needs to be used from a doctor's perspective, especially during mass screenings such as mammography. Continuous learning CAD is a CAD method that anticipates that learning will continue with new data and grow more intelligent as the CAD adapts.

Although there is plenty of recognition of the necessity for CAD in clinical settings, conventional CAD systems have many limitations, such as high development costs, workflow, cost-effectiveness challenges [157], low diagnostic accuracy of mammography [158], inconsistent effectiveness in clinical evaluation [142], etc. These limitations will likely be addressed and solved in the coming years using DL technologies. Figure 1.6 shows the fundamental difference between CAD in conventional and DL technologies.

1.3.3.5 Treatment Planning

AI-assisted treatment planning reduces human effort and improves accuracy. A novel PSIGAN framework was developed for unsupervised domain adaptation and multi-organ segmentation from MRI by employing a structure discriminator to model joint probability distributions between images and segmentations, facilitating cyclical optimization of generator and segmentation sub-networks. Through extensive experimentation on various MRI sequences totaling 257 scans, the method achieved significant segmentation accuracies, with an average DSC of 0.87 for abdominal organs on T1w and 0.90 on T2w, 0.82 for parotid glands on T2wFS, and 0.77 for lung tumors on T2w MRI [159]. A reinforcement learning-based planning bot was developed to streamline and optimize pancreas stereotactic body radiation therapy treatment planning, efficiently balancing organ-at-risk sparing and planning target volume

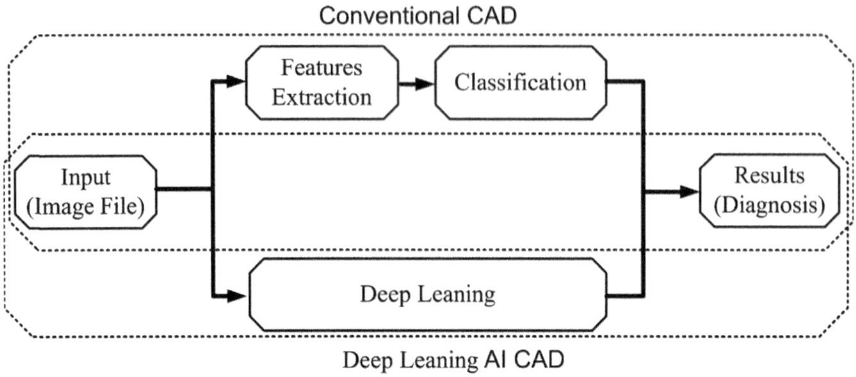

Fig. 1.6 Conventional CAD versus DL AI CAD

(PTV) coverage. The bot, trained on 48 plans and validated on 24 cases, consistently achieved similar PTV coverages as clinical plans while satisfying all clinical planning constraints, demonstrating its potential to reduce planning inefficiencies in the clinical workflow [160]. Over one million Americans suffer from Parkinson's disease, often managed with medications such as levodopa, but treatment plans rely on limited physician–patient interactions. Wearable sensors offer continuous symptom monitoring, prompting the development of a deep reinforcement learning model to personalize medication timing and dosage, significantly improving symptom management compared to static approaches, and suggesting promise for enhancing chronic disease treatment planning [161].

A treatment planning system was proposed for IMRT using DL-based 3D dose prediction and distribution algorithms. The results indicate clinically acceptable dose predictions and generated plans, suggesting promise for automated treatment planning in radiotherapy [162]. Innovation addresses critical challenges in achieving personalized high-quality plans, potentially enhancing tumor control and patient survival rates [163]. They proposed a DL-based dose estimate model (Hierarchically Densely Connected U-net) to streamline the complex treatment planning process for head and neck cancer. The results revealed significant improvements in accuracy and efficiency compared to those of conventional models, with reduced parameter complexity and faster prediction speed.

1.3.4 *Key Techniques and Algorithms in AI for Medical Imaging*

Various techniques and algorithms are employed to analyze, interpret, and extract useful information from medical images. Predictive models in AI focus on regression and classification tasks, aiming to predict continuous values or class probabilities based on input features. Regression is addressed through models such as linear regression or more complex universal approximators such as ANNs. ANNs consist of interconnected neurons with nonlinear activation functions, with capacity determined by the number of neurons and layers. With their hierarchical processing and feed-forward or recurrent architectures, deep neural networks have shown significant success, particularly in medical imaging applications.

Currently, attention has moved from traditional ML techniques such as SVMs [164, 165] and Random Forests [165] to DL techniques such as CNNs, Multilayer Perceptron (MLP), U-Net architecture, ResNet and Generative Adversarial Networks (GANs) [166]. This shift underscores the increasing recognition of the efficiency of DL approaches in medical imaging tasks. The DL techniques are detailed in the following sections.

1.3.4.1 Convolutional Neural Networks

CNNs are dominant in medical image processing due to their ability to detect hierarchical features; they consist of stacked layers of convolutions and down-sampling, followed by fully connected layers for output, as shown in Fig. 1.7. Fully Convolutional Networks (FCNs) efficiently handle image-to-image tasks such as segmentation, replacing fully connected layers with operations to restore resolution, as exemplified by U-net; FCNs are used in both supervised and unsupervised settings for tasks such as detection [167, 168], classification [169], and segmentation [170]. Recent advancements in CNN architecture include addressing class imbalance, adding squeeze-and-excitation blocks, and implementing attention mechanisms, leading to more robust models. CNNs have achieved success in various medical applications, such as classification, outcome prediction, treatment planning, and motion tracking, which are discussed in detail in Sect. 1.5.

1.3.4.2 Multilayer Perceptron

MLP is a network architecture, a type of ANN consisting of several layers of interconnected neurons. In an MLP, neurons within each layer process input data using an activation function, typically a nonlinear function such as the sigmoid or rectified linear unit (ReLU) function. The connections between neurons are associated with weights that determine the strength of influence between neurons. During training, these weights are adjusted iteratively through a process called backpropagation, where the

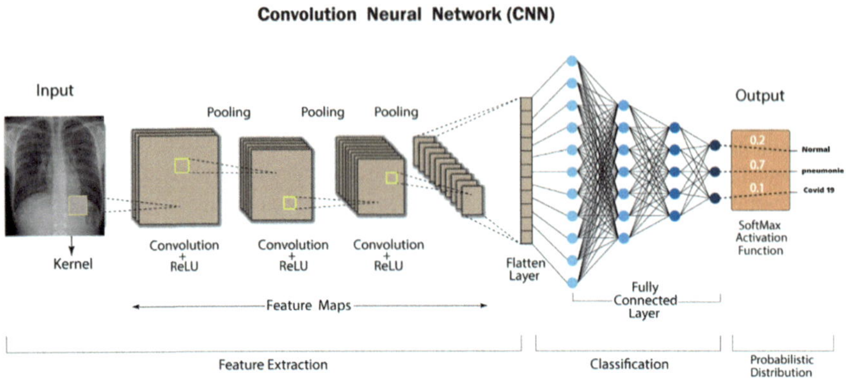

Fig. 1.7 Layered architecture of a convolutional neural network (CNN). The figure has been adapted from Rguibi et al. [171], Copyright © 2022 by the authors, Creative Commons Attribution (CC BY) License

error between the actual and predicted outputs is minimized using optimization techniques such as gradient descent. MLP has been extensively used in medical image analysis. It has been employed for tasks such as accelerated parallel MRI [172], emulating fast Fourier transform [173], low-resolution PET reconstruction [174], and enhancing maximum a posteriori reconstructed PET images, demonstrating its ability to map input data to desired outputs in various medical applications.

1.3.4.3 U-Net Architecture

It is another network architecture that has been extensively applied in various medical imaging tasks. The U-Net architecture is a CNN used for semantic segmentation tasks, particularly in medical image analysis. It consists of a contracting path (encoder) that captures context and high-level features through down-sampling, followed by an expansive path (decoder) that up-samples the feature maps to produce detailed segmentation masks. Skip connections are employed to preserve spatial information, and the final output layer generates pixel-wise probability maps for each class. The network is trained using loss functions such as cross-entropy or Dice coefficient loss, with backpropagation and data augmentation techniques. Overall, U-Net excels at accurate and detailed segmentation, especially in medical imaging applications (Fig. 1.8).

For MRI, U-Net-based methods, including deep residual learning and domain adaptation architectures, have successfully addressed challenges such as aliasing artifacts and undersampled k-space reconstruction [176]. Similarly, in CT reconstruction, U-Net variants have been effective in sparse-view reconstruction, noise suppression, artifact reduction, and super-resolution tasks [177]. Additionally, in

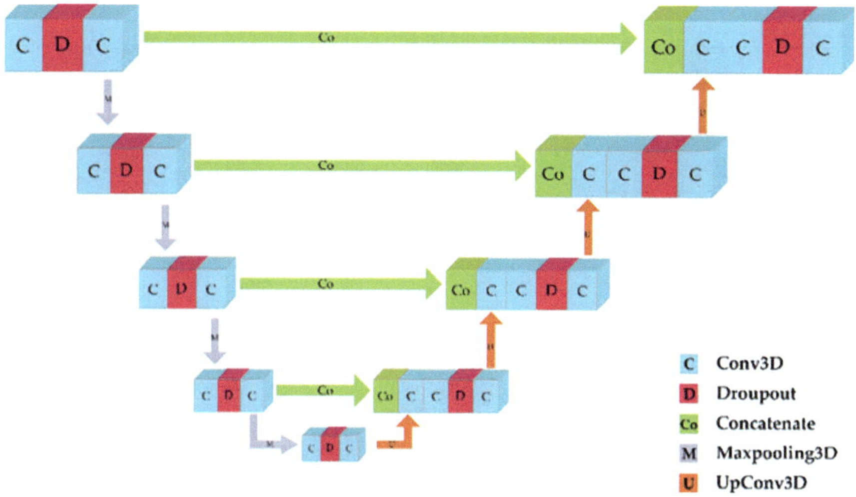

Fig. 1.8 Illustration of a three-dimensional U-Net architecture. The figure has been adapted from Asnawi et al. [175], Copyright © 2023 by the authors, Creative Commons Attribution (CC BY) License

PET reconstruction, U-Net frameworks, sometimes combined with other architectures such as residual networks, have demonstrated efficacy in denoising and integrating PET data from other modalities such as MRI and CT for improved diagnostic accuracy and image quality [178].

The 3D U-Net architecture is an extension of the popular U-Net architecture and is specifically designed for volumetric data such as 3D medical images (e.g., CT scans, MRI scans). It is widely used in tasks such as medical image segmentation, where the goal is to partition an image into different regions of interest [179].

1.3.4.4 ResNet Architecture

ResNet, or Residual Neural Network, uses residual blocks (as shown in Fig. 1.9) to train very deep networks effectively. It introduces skip connections to address the vanishing gradient problem by allowing the direct flow of information from earlier to later layers. By learning residual mappings instead of the entire mapping, ResNet can train deep networks more easily. Its architecture consists of multiple layers of residual blocks, varying in depth, and achieves state-of-the-art performance on tasks such as image classification and object detection through standard training techniques. Using ResNet, the computational time and resources may be reduced immensely. It is widely applied in medical imaging, including MRI and CT, for tasks such as image reconstruction, super-resolution, noise suppression, and artifact reduction [180].

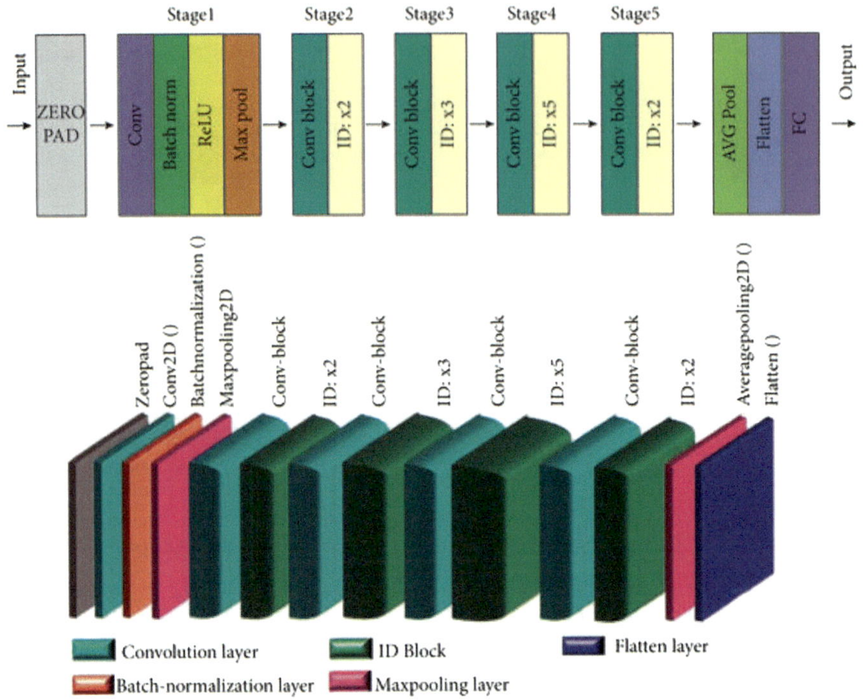

Fig. 1.9 Illustration of the ResNet-50 architecture. The figure has been adapted from Rezayi et al. [181], Copyright © 2021 Sorayya Rezayi et al., Creative Commons Attribution License

1.3.4.5 Generative Adversarial Networks

GAN consists of two neural networks, the generator and the discriminator, as presented in Fig. 1.10. The generator generates synthetic data samples from random noise to trick the discriminator, which differentiates between actual and false samples. During training, the generator and discriminator engage in a game in which the generator improves its ability to generate realistic samples, while the discriminator improves its ability to discriminate between real and false samples. GANs produce high-quality synthetic data via repeated training [182]. GANs use adversarial training to generate realistic data by minimizing the difference between generated and real data distributions, with applications ranging from artificial image generation for data augmentation to multi-modality image translation in medical imaging, leveraging advances such as conditional GANs and cycleGANs [183, 184]. GANs are increasingly used in tasks such as segmentation, radiotherapy dosage prediction, and artifact reduction, with the potential for quality assurance due to their discriminative ability, especially in radiation oncology.

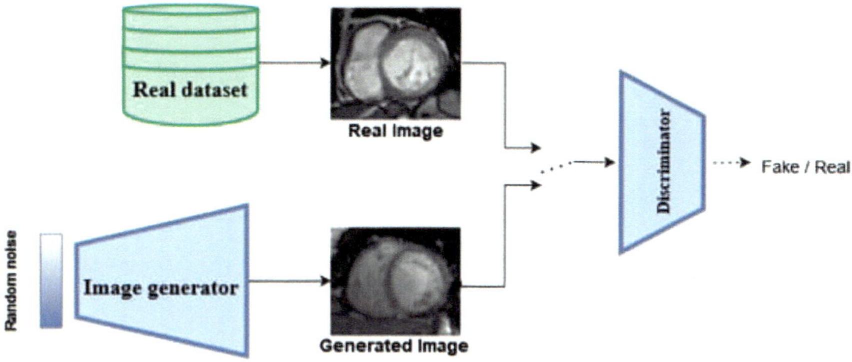

Fig. 1.10 A representative GAN architecture. The figure has been adapted from Skandarani et al. [33], Copyright © 2023 by the authors, Creative Commons Attribution (CC BY) License

1.4 Applications of AI in Medical Imaging

AI has been applied for automatic recognition of complex patterns in imaging data and for providing quantitative assessment of radiographic characteristics, making radiology a more objective science rather than a subjective perceptual skill. Compared with traditional methods, AI-based models achieve competitive results in classification tasks. The new era of DL-based methods has demonstrated substantial improvements in medical imaging research over traditional and ML algorithms. DL-based algorithms have become an automatic choice for different imaging modalities, such as CT, MRI, PET, ultrasonography (USG), etc., and different tasks, such as tumor detection, segmentation, disease prediction, etc. Table 1.1 lists some of the applications of different imaging modalities and the AI models and tasks involved.

1.4.1 Applications of AI in MRI

The complexity and variability of MRI pose a challenge for accurate classification, which has been effectively addressed using AI. The datasets of acquired MRI images are used for training AI algorithms such as CNN, RNN, U-Net, etc. The output accuracy of the AI methods increases based on the number of images in the dataset. In MRI, AI is used for image enhancement, image reconstruction, lesion segmentation, and analysis. A comparative study of liver MRI was carried out using a DL (ARDL) algorithm and without using the DL (NON-DL) to that of a conventional high-resolution acquisition (NAÏVE) [186]. The results showed that ARDL outperformed both the NAÏVE and NON-DL methods in the objective analysis of SSFSE T2 images, with higher SNR and CNR, as shown in Fig. 1.11. While DWI showed no differences in SNR, CNR was higher for ARDL than for NON-DL. Hence, ARDL is preferred for qualitative investigation of all sequences because of its better image quality, fewer

Table 1.1 Application of different AI tools and tasks in different medical imaging modalities

Modality	AI algorithm/network architecture	AI task	Application	Dataset	Evaluation metrics	References
MRI	DNN	Image reconstruction	Brain image reconstruction	MGH-USC HCP database, ImageNet database	SNR, RMSE	[185]
	ARDL	Image reconstruction and segmentation	Image reconstruction for liver images and reduction in scanning time	Liver MRI from 31 female	SNR, CNR	[186]
	CNN	Image reconstruction	Parallel imaging reconstruction framework	Brain and Knee image from 22 volunteers	PSNR, SSIM	[187]
	De-aliasing GAN and U-Net based generator	Image reconstruction	Image reconstruction for brain images	MICCAI 2013 grand challenge dataset	NMSE, PSNR	[188]
	U-Net	Segmentation	Lesion segmentation	Brain image from 320 women	Dice score	[189]
	CNN	Classification	Diagnosis of MCI	Brain image from 600 individuals	Accuracy, sensitivity, specificity, F-score, AUC-ROC	[190]
	CNN with GoogLeNet and ResNet	Classification	Diagnose Alzheimer's disease and MCI	38,024 brain images	Accuracy	[191]
	CNN	Classification	Diagnose Alzheimer's disease	ADNI database	Accuracy, sensitivity	[192]

(continued)

Table 1.1 (continued)

Modality	AI algorithm/network architecture	AI task	Application	Dataset	Evaluation metrics	References
	KNN, SVM, Naïve Bayes	Classification	Performance evaluation of ML algorithms in predicting fragility fractures	Fragility bone fractures image of 32 women	Sensitivity, specificity, precision, accuracy, F1-test	[193]
	SVM, ANN, Naïve Bayes, normal densities based linear classifier	Classification	MRI analysis	Abnormal brain images from Harvard Medical School website	Classification accuracy	[194]
	Variational model	Image reconstruction	Fast reconstruction of clinical accelerated multi-coil MRI	Five knee datasets	MSE, NRMSE, SSIM	[195]
	CNN	Image reconstruction	MRI reconstruction from incomplete k-space data	Cardiac MR data from 101 individual	MSE, PSNR, SSIM	[196]
	U-Net	Image reconstruction	MRI reconstruction with end-to-end motion-guided network	Cardiac MRI from 21 healthy people and 3 dyssynchrony disease patients	NMSE, SSIM, PSNR	[197]
CT	RetinaNet	Classification	Detection of abdominal hemorrhage lesions	Abd-CT-DB data	Sensitivity, Specificity	[198]
	CNN	Image reconstruction	Image restoration algorithm for coronary	Coronary CTA from 30 patients	CNR	[199]

(continued)

Table 1.1 (continued)

Modality	AI algorithm/network architecture	AI task	Application	Dataset	Evaluation metrics	References
	K-sparse autoencoder, GAN	Image reconstruction	Denoising image in low-dose CT	Low-dose CT Grand Challenge	SSIM, CNR, Haralick	[200]
	K-sparse autoencoder, GAN	Image reconstruction	Denoising image in low-dose CT	28 cardiac images	SNR, PSNR	[201]
	GAN	Image reconstruction	Denoising image in low-dose CT	2016 NIH-AAPM low dose CT grand challenge	PSNR, SSIM	[202]
	CNN	Image reconstruction	Denoising image in low-dose CT	2016 NIH-AAPM low-dose CT grand challenge	RMSE, PSNR, SSIM	[203]
	Sparse autoencoder	Classification	Breast density segmentation and mammographic risk scoring	(i) 493 mammograms of healthy women, 394 cancers, and 1182 healthy controls from the Dutch breast cancer screening program (ii) 668 mammograms from Mayo mammography health study	Dice coefficient, RMSE	[204]
	MBIR	Image reconstruction	Detection of subtle hypoattenuation in early cerebral infarction	64-slice multidetector CT	SNR	[205]

(continued)

Table 1.1 (continued)

Modality	AI algorithm/network architecture	AI task	Application	Dataset	Evaluation metrics	References
	CNN, LeNet	Classification	Distinguishing lung granulomas from adenocarcinomas	CT image from 290 patients	ROC AUC	[206]
	U-Net	Image reconstruction	Reduce the radiation dose using sparse projection views	AAPM low-dose CT challenge dataset	NMSE, PSNR, SSIM	[207]
	DD-Net	Image reconstruction	Fast data acquisition and radiation dose reduction	3059 two-dimensional images	Haralick, MSE, SSIM	[208]
	CNN	Image reconstruction	Super-resolution image generation	7670 image slices	NRMSE, PSNR	[209]
	GAN-Circle	Image reconstruction	Super-resolution image generation	(i) 25 micro-CT images from fresh-frozen cadaveric ankle specimens (ii) 2016 NIH-AAPM-Mayo clinic low dose CT grand challenge	PSNR, SSIM, IFC	[210]
	SADIR-Net	Image reconstruction	Super-resolution image generation	47 images from TCIA database	PSNR, SSIM, IFC, MTF	[211]
	CapsNet, FDNN	Classification	Intracerebral hemorrhage e-diagnosis	ICH dataset	Precision, Accuracy	[212]

(continued)

Table 1.1 (continued)

Modality	AI algorithm/network architecture	AI task	Application	Dataset	Evaluation metrics	References
PET	Deep convolutional encoder-decoder	Image reconstruction	PET image reconstruction from sinogram data	2,91,000 images	RMSE, PSNR, SSIM	[213]
	GAN	Image reconstruction	PET image reconstruction from sinogram data	Brain image from 9 patients	PSNR, SSIM	[214]
	GAN	Image reconstruction	PET image reconstruction from sinogram data	Simulated data	Bias, Variance	[215]
	DL	Image reconstruction	Attenuation correction	100 retrospective 3D FDG PET head images	Dice coefficient, MAE	[216]
	SVM, RF	Classification	Diagnose Parkinson's disease	(i) FDOPA PET in 44 patients and 14 controls (ii) FDG PET in 51 patients and 16 controls	AUC-ROC	[217]
	CNN	Classification	Diagnose Parkinson's disease	Images of 666 MCI patients	AUC-ROC	[218]
	CNN	Image reconstruction	Iterative PET image reconstruction	19 XCAT phantoms and 6 lung patient data	CR, STD	[219]
	CNN	Image reconstruction	Anatomically aided PET image reconstruction	Simulated data	CR, STD	[220]

(continued)

Table 1.1 (continued)

Modality	AI algorithm/network architecture	AI task	Application	Dataset	Evaluation metrics	References
	Logistic regression, Naïve Bayes, RF, SVM	Classification	Prediction for lung cancer	Image of 396 patients	AUC-ROC, accuracy, F1 score, precision, recall	[221]
	Res-UNets	Image reconstruction	Image reconstruction of low-dose PET	PET/CT scan of 10 patient	PSNR, SSIM	[222]
	U-Net	Image reconstruction	Full-count PET recovery from low-count image	Image from 35 subjects (~3500 slabs)	MAPE, PSNR, SSIM	[223]
	CycleGAN	Image reconstruction	Low-dose PET image denoising	Image of 18 patient	SSIM, PSNR, NRMSE	[183]
	CycleGAN	Image reconstruction	Convert images between amyloid-negative and amyloid-positive for dementia other than Alzheimer's disease	43 amyloid-negative and 45 positive images	…	[224]
	U-Net	Image reconstruction	Cerebral blood flow	Image from 32 subjects	SSIM, NRMSE,	[225]
USG imaging	Deep polynomial network	Classification	Tumor classification with a small USG dataset	200 pathology-proved breast ultrasound images	Accuracy	[226]
	SVM	Classification	Texture characterization of liver ultrasound images	51 ultrasound liver images with 54 malignant 'esions	Accuracy Sensitivity	[227]

(continued)

Table 1.1 (continued)

Modality	AI algorithm/network architecture	AI task	Application	Dataset	Evaluation metrics	References
	R-CNN	CAD	Biometric measurement during pregnancy	2,400 images of lateral ventricles	Precision, recall, sensitivity, specificity accuracy	[228]
	DGACNN	Classification	Diagnosis of fetal congenital heart disease	1000 video slices	F1-values, accuracy	[229]
	ML	Segmentation	Fetal head biometry	Image from 140 female	ICC	[230]
	Inception-v3	Image detection	Distinguish between papillary thyroid carcinomas and benign nodules	2,836 thyroid USG from 2,235 patients	Accuracy, sensitivity, specificity, A_Z value	[231]
	U^2-Net	Segmentation	Segmentation of nodules in thyroid	Thyroid images of 264 patients and 215 healthy volunteers	Accuracy	[232]
	CNN	Segmentation	Segmentation of nodules in thyroid	10,357 thyroid nodular lesions and 465 normal thyroids	Dice coefficient	[233]
	DSRU-Net	Segmentation	Segmentation of nodules in thyroid	5822 ultrasound images	Dice coefficient	[234]
Electrical impedance tomography (EIT)	CNN, DenseNet	Image reconstruction	EIT reconstruction	10,000 EIT simulation models generated from chest CT images of 60 patients	RMSE, ICC	[235]
	FC-UNet	Image reconstruction	EIT reconstruction	29,511 samples	SNR	[236]

(continued)

Table 1.1 (continued)

Modality	AI algorithm/network architecture	AI task	Application	Dataset	Evaluation metrics	References
	CNN, DenseNet	Image reconstruction	EIT reconstruction	Lung phantom test datasets	RMSE, SSIM, MAE, ICC	[237]
	U-Net	Segmentation	Separation of lung and heart bioimpedance images	Mixed EIT image lung-related and cardiac images	MAE, dice similarity coefficient	[238]
	ResNet-7 DenseNet-9	Classification	Predict metabolically vulnerable plaques	…	Accuracy	[239]
Optical microscopic imaging	MED-Net	Segmentation	Segmentation of cellular patterns in confocal images of melanocytic lesions	117 RCM mosaics collected at 6 clinics	Sensitivity, specificity, dice coefficient	[240]
	DNN	Image reconstruction	Real-time 3D histology-like imaging	SLAM microscope image	SNR	[241]
	DNN	Classification	Real-time intraoperative diagnosis	Laboratory-based image	AUC	[242]
	MICRA-Net	Segmentation	Detection, enumeration and segmentation from image-level annotations	F-actin and EM dataset	…	[243]
	CNN	Classification	Autonomous computational system to classify endoscopy	Normal and abnormal vocal fold images	Accuracy	[244]

(continued)

Table 1.1 (continued)

Modality	AI algorithm/network architecture	AI task	Application	Dataset	Evaluation metrics	References
Hybrid imaging modalities	DenseNet	Classification	Diagnosing and predicting Alzheimer's disease	Kaggle dataset containing 6200 Alzheimer's disease images	Accuracy, specificity, F1-score	[245]
	CNN	Segmentation, detection, classification	Detection of lung/ liver lesions	FDG PET-CT of 112 patients	Dice index	[246]
	CNN	Segmentation, detection	Detection of lung/ liver lesions	68 Ga-DOTATATE PET/ CT of 125 patients	PPV, F1 score, PR-AUC	[247]
	U-Net	Image reconstruction	Attenuation correction for brain PET imaging	Image Dataset: Brain image	Dice index	[248]
	deepCNN	Image reconstruction	Attenuation correction	PET/MR scan of 70 patients	F-score, p-value	[249]
	FCN, cGAN	Image reconstruction	Synthesize PET-like images from CT	60 PET/CT scans from Sheba Medical center	MAE, PSNR	[250]
	deepCNN	Image reconstruction	Attenuation correction	18F-FDG-PET/CT image of 164 patients	Average error	[251]
	CNN	Image reconstruction	Dose reduction	SPECT/CT image of 930 patients	Averaged correlation	[252]
	CNN	Detection, classification	Cancer detection	^{18}F-FDG-PET/CT image of 302 patients with lung cancer and 327 patients with lymphoma	AUC, respectively	[253]

(continued)

Table 1.1 (continued)

Modality	AI algorithm/network architecture	AI task	Application	Dataset	Evaluation metrics	References
	CNN	Detection, classification	Cancer detection	^{18}F-FDG-PET/CT image of 125 patients	SUV_{max}	[254]
	DNN	Image reconstruction	MRI based attenuation correction in brain PET/MR imaging	Image Dataset: Head image	Dice coefficient	[255]
	RF, k-NN, neural network, naive Bayes, logistic regression, SVM	Classification	Predict the prognosis of cancer	^{18}F-FDG-PET/CT image of 50 cervical cancer patients	AUC, accuracy, F1 score, precision, recall	[256]
Other imaging modalities						
Quantitative susceptibility mapping	CNN	Detection	Diagnose Parkinson's disease	Image data of 87 PD participants	AUC, BAC, sensitivity, specificity	[257]
SPECT	Multi vector pattern analysis	Classification	Discriminate Parkinson's disease and different atypical Parkinsonian syndromes	^{123}I-ioflupane SPECT image of 392 patients	MSA, PSP	[258]

(continued)

Table 1.1 (continued)

Modality	AI algorithm/network architecture	AI task	Application	Dataset	Evaluation metrics	References
Optical coherence tomography	Ensemble, SVM, kNN	Classification	Automated classification platform for the identification of otitis media	58 previously imaged subject datasets	SNR	[259]
Diffuse optical tomography	DOTnet 2.0	Image reconstruction	Image reconstruction to improve breast cancer	The dataset contains blood vessel structures from breast structures and artificially created vessels using the Lindenmayer-system algorithm	MS-SSIM	[260]

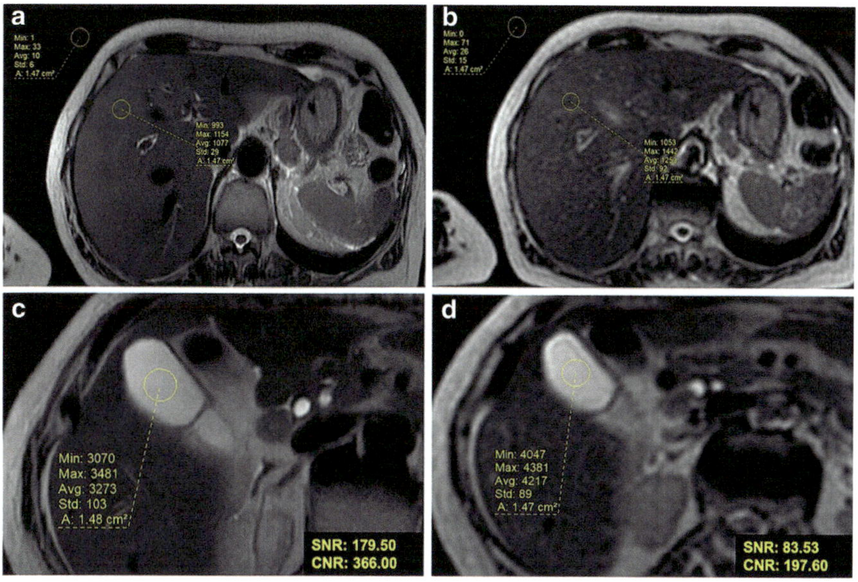

Fig. 1.11 AI-based quantitative image analysis in liver MRI. The figure has been adapted from Zerunian et al. [186], Copyright © 2022 The Author(s), Creative Commons Attribution 4.0 International License

truncation artifacts, higher sharpness, and contrast. ARDL sequences also had shorter scanning times than did NAÏVE sequences.

Accelerated imaging can be advantageous in neuroimaging when there is subject movement that causes image artifacts and distortion or when lengthier imaging techniques including many sequences dramatically increase the overall scan duration. The introduction of AI, particularly DL algorithms, revealed endless opportunities to increase MRI picture quality. DL has been shown to be advantageous in several stages of the medical imaging process, including data collection, scheduling, image reconstruction, enhancement correction, and interpretation [261–263]. Aggarwal et al. [264] suggested MoDL, a DL approach for picture reconstruction in the brain, and demonstrated that the algorithm had a better PSNR than compressed sensing across a wide range of acceleration factors.

The automated detection of stroke lesions in MRI images is critical for a variety of academic and therapeutic applications, including identifying impairments and predicting recovery. The availability of datasets such as ATLAS has considerably accelerated AI-based lesion segmentation research. The majority of research has employed traditional CNN designs, demonstrating a paucity of strong AI models capable of managing different lesion patterns.

MRI has revealed several structural brain alterations in patients with Alzheimer's disease and moderate cognitive impairment (MCI). There is a growing attempt to employ AI/DL models to evaluate brain MRI. To avoid model overfitting, most AI/DL models require a large number of training photos. However, excellent MRI training in general is difficult to obtain due to secrecy or other privacy concerns. Furthermore, varying MRI quality and contrasts owing to anatomical locations, sequence settings, and hardware configurations make it difficult to collect a significant volume of quality data [265]. Thus, a shortage of high-quality training datasets might considerably influence the performance of AI/DL models [266].

1.4.2 Applications of AI in CT

AI is an emerging technology in the field of CT image reconstruction that aims to minimize radiation exposure while enhancing image fidelity. Numerous studies have delved into the application of DL in CT reconstruction, particularly focusing on mitigating image noise. This is particularly relevant in ultra-high-resolution CT (U-HRCT), where finer detector elements and narrower tube focus sizes often lead to heightened noise levels. Nakamura et al. conducted a retrospective study [267] revealing that their DL Reconstruction (DLR) algorithm effectively reduced noise and improved the CNR surpassing hybrid-IR and MBIR models, especially at low doses, which is advantageous for low-dose CT examinations. Notably, images produced by the DLR algorithm from ultra-low dose chest CT scans exhibited superior task-based detectability [268]. This improvement in detectability addresses previous challenges associated with low-contrast detectability using IR algorithms at minimal doses. Figure 1.12a and b show a CT image generated by the DL noise reduction algorithm from an ultra-low dose chest CT image [268]. A stronger task-based detection was also observed in DLR images, which addresses earlier challenges with low-contrast detectability related to IR algorithms at extremely low doses. Consistent with the results of Kuo et al.'s 2016 study [269], DLR has emerged as a superior technique for noise reduction and SNR enhancement. Further validation through phantom studies supported the consensus that DLR outperforms other methods in terms of noise reduction and preserving image quality at low doses. The evaluation of image quality utilized objective metrics such as modulation transfer function (MTF), CNR, and SNR, complemented by expert assessments from radiologists and specialists. Despite variations in the definition of image quality across studies, consistent results underscore the efficacy of DLR in enhancing CT image reconstruction.

Researchers have created a variety of DL algorithms to evaluate TBI data from CT images. Jadon et al. [270] analyzed CNN architectures and developed algorithms for segmenting various types of injuries, resulting in high DSC. Kuo et al. [271] showed that training on random locations and assessing sliding windows may effectively identify cerebral bleeding. Guo et al. [272] devised a multiview CNN to detect small lesions, while Piçarra et al. [273] introduced BLAST-CT, a CNN model for segmenting and classifying four types of TBI injuries. Additionally, some studies focused on quantifying CT images, with Jain et al. [274] measuring cistern volume and midline shift and Yao et al. [275] automatically measuring hematoma volumes and predicting mortality.

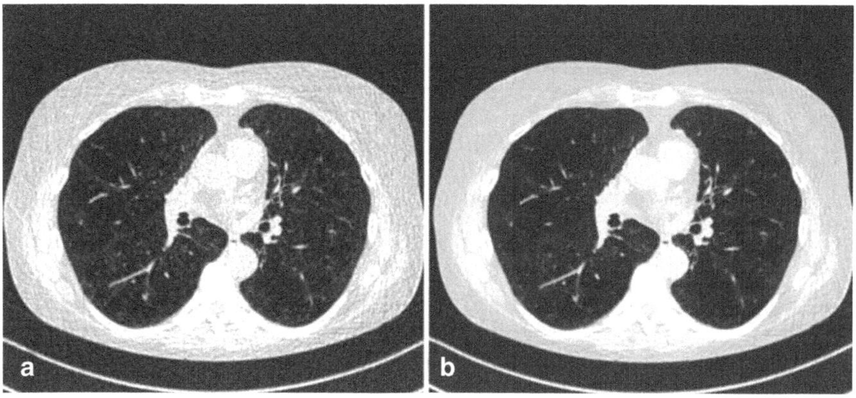

Fig. 1.12 Examples of **a** an ultra-low dose chest CT image and **b** a noise reduced chest CT image generated by the AI algorithm. The figure has been reproduced from Oever et al. [268], Copyright © 2020 The Authors, Creative CC BY License

1.4.3 Applications of AI in PET

While PET imaging presents unique challenges due to its cost and technical complexity, AI-driven solutions are increasingly being explored. Research in this area has focused on image processing and analysis techniques to standardize image acquisition and improve diagnostic accuracy. Recent studies have utilized AI for tasks such as lesion detection, with promising results in quickly and accurately identifying abnormalities. Additionally, efforts are underway to refine parameters such as the standardized uptake value (SUVmax) for better lesion localization, aiding in the development of AI training datasets.

In the field of oncology, AI has made significant strides in aiding diagnosis, differential diagnosis, and cancer staging. Studies have utilized texture analysis with PET/CT scans to diagnose breast cancer and differentiate spinal metastases, respectively, demonstrating the comparable or superior accuracy of AI-assisted methods compared to manual diagnosis [23, 24]. Similarly, a study demonstrated the potential of AI-driven radiomics models for predicting lymph node staging in large cell lung cancer patients, offering valuable insights for treatment planning.

In neurology, AI is aiding in distinguishing and studying neurodegenerative diseases. Recently, Mathotaarachchi et al. [reference] examined the effectiveness of amyloid PET as an early detection tool for individuals on the Alzheimer's disease trajectory. The group introduced a novel algorithm, RUS random forest (RUSRF), using the random under sampling (RUS) method to predict progression to Alzheimer's disease dementia within individuals with mild cognitive impairment (MCI) based on their Aβ PET measurements (Fig. 1.13) [276]. A new DL model, using rate-distortion theory and an extreme learning machine, to differentiate Alzheimer's disease, MCI, and normal control (NC) groups in 18F-AV45 PET/MR scans. This model outperformed previous models, demonstrating higher accuracy,

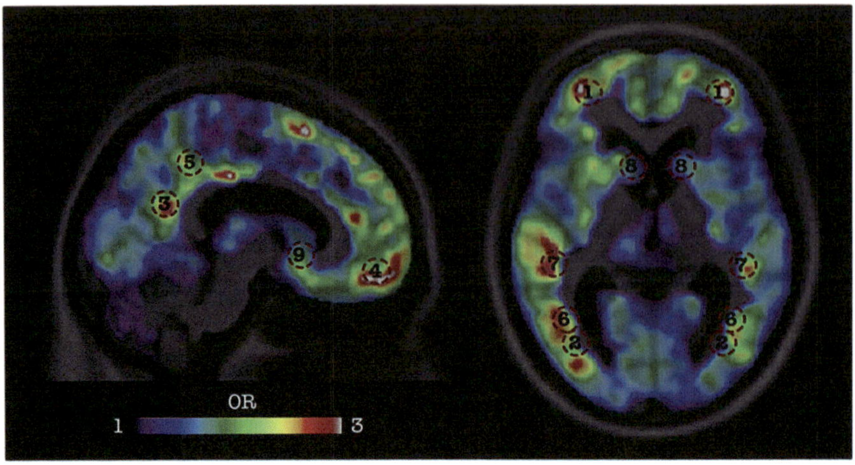

Fig. 1.13 Scaled odds ratio map highlighting regions with the highest values obtained from voxel-wise logistic regression analysis. The numbered regions: (1) mid frontal cortex, (2) angular gyrus, (3) posterior cingulate cortex, (4) orbitofrontal cortex, (5) precuneus, (6) temporal occipital junction, (7) mid-temporal cortex, (8) putamen, and (9) nucleus accumbens, are indicated. The figure has been adapted from Mathotaarachchi et al. [276], Copyright © 2017 The Author(s), CC BY-NC-ND license

specificity, sensitivity, and area under the curve (AUC) in separating Alzheimer's disease, MCI, and NC groups [207].

1.4.4 Applications of AI in USG Imaging

Prenatal ultrasonography plays a crucial role throughout pregnancy, although challenges such as fetal movement and inconsistencies among observers can limit the effectiveness of traditional ultrasound in clinical settings [277]. However, AI combined with obstetric ultrasound shows promise for improving examination efficiency, reducing physician workload, and enhancing diagnostic accuracy. The development of AI-based fetal ultrasound images entails several tasks categorized into image segmentation, classification, and object detection. Figure 1.14 represents various AI applications for fetal ultrasound and outlines three commonly employed AI models in medical imaging: CNN, U-Net, and RNN. AI aids in automatic fetal ultrasound biometric measurement, plane detection, and disease diagnosis, complementing radiation imaging methods. A novel approach proposed in a study proposed to automatically measure fetal head circumference (HC) in prenatal ultrasound scans, aiming to mitigate errors caused by human factors such as the skill and experience of sonographers [278]. The proposed method utilizes a one-stage network based on an anchor-free method, combining transformer and CNN structures for efficient feature extraction.

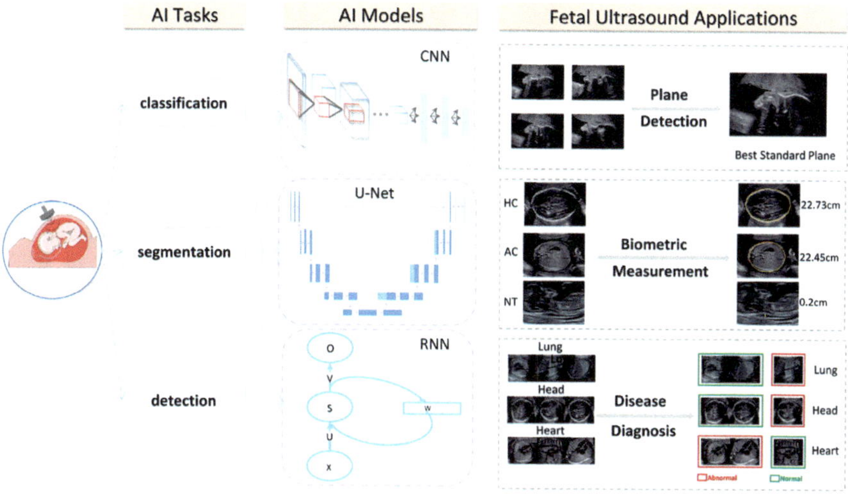

Fig. 1.14 The figure illustrates an overview of AI applications in fetal ultrasound. It outlines the computer vision tasks required for developing AI-based fetal ultrasound images, which are categorized into three main types: image classification, image segmentation, and object detection. Additionally, the figure provides a list of three commonly employed AI models in medical imaging: CNN, U-Net, and recurrent neural network (RNN). The figure has been adapted from Xiao et al. [280], Copyright © 2020 by the authors, Creative Commons Attribution (CC BY) license

In a case study involving a 58-year-old male experiencing exertional chest pain, quantitative plaque AI-based measurements were utilized [279]. The images revealed a diffuse calcified atherosclerotic plaque in the left circumflex artery, as well as various plaque components in the left anterior descending artery depicted in multiplanar reformatted images and cross-sectional views at different locations.

1.4.5 Applications of AI in EIT

EIT has been studied extensively for its cost-effectiveness in contrast to conventional imaging modalities such as MRI and CT scans, and it carries no discernible risks. Over the past three decades, EIT has garnered significant attention as an innovative imaging technique. The process involves the administration of a small electrical current to ascertain tissue electrical properties, followed by the reconstruction of tomographic images based on resultant voltage measurements. EIT is more cost effective than MRI and CT. Recent advancements have primarily revolved around enhancing technology and incorporating AI to address inherent challenges. From a clinical perspective, EIT has potential for monitoring lung disorders and mechanical ventilation, providing portability and non-radiological applicability over prolonged durations. It assists in the identification of pneumothoraxes, the quantification of pulmonary edema, and

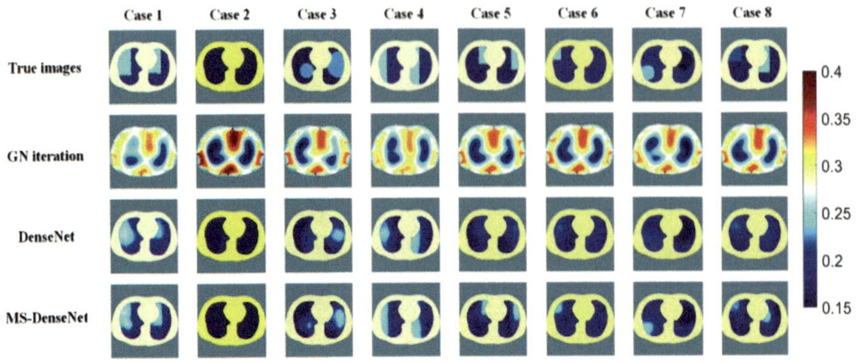

Fig. 1.15 Reconstructed images of eight lung cases on 20-dB noise datasets. The figure has been adapted from Yang et al. [237], Copyright © 2023 the Author(s), Creative Commons Attribution License

the optimization of ventilator parameters. Nevertheless, further research is necessary for its widespread adoption in intensive care units (ICUs). Furthermore, an ML-driven electrochemical impedance spectroscopy (EIS) platform has been devised to analyze atherosclerotic plaques for predicting vulnerability. This platform employs a DenseNet model, which achieves high accuracy in discerning metabolically active plaques, indicating potential clinical usefulness in plaque assessment [235, 239]. Figure 1.15 shows the same reconstructed images that show that the DL method improves the artifact and edge blur problems, achieves higher values on the image metrics and improves the EIT image quality [237].

1.4.6 Applications of AI in Optical Microscopy

Optical microscopy methodologies present advantages, including high contrast, affordability, and noninvasiveness, for visualizing both structural and functional aspects of tumor tissues. Diverse optical imaging techniques such as optical endoscopy, optical coherence tomography, and photoacoustic imaging are discussed alongside AI integration. A research study introduced a pioneering imaging technique named Simultaneous Label-free Autofluorescence Multiharmonic (SLAM) microscopy, which facilitates rapid and detailed visualization of tissue characteristics devoid of traditional histological processing requirements [241]. By amalgamating multiple nonlinear optical imaging modalities, such as second harmonic generation and fluorescence, the SLAM microscope generates images with substantial contrast, capturing crucial histological features such as collagen fibers and nuclei. DL algorithms are employed to segment nuclei and convert grayscale SLAM images into histology-like color images, closely resembling conventional stained histology. Comparison with gold-standard histology revealed a high degree of similarity. Furthermore, the technique enables real-time 3D imaging and quantification of nuclear-to-cytoplasmic ratios, potentially providing diagnostic insights into conditions such as cancer.

Figure 1.16 represents another study that introduces a DL framework adept at distinguishing between cancerous and healthy human breast tissue in real-time label-free virtual histology images [242]. The framework delivers great accuracy on concealed samples, with an AUC of 95% at the tile level and 100% at the slide level. Importantly, models trained on a laboratory dataset show significant generalizability to independent datasets obtained using portable intraoperative imaging technologies. Using classification activation maps and feature visualization, researchers have identified discriminatory patterns linked with cancer status, such as tumor cells and tumor-related vesicles. These findings suggest the possibility of reliable real-time diagnosis in clinical settings, hence improving point-of-procedure applications.

A study endeavored to develop an autonomous computational system capable of classifying endoscopic findings, with a focus on vocal fold images from a laryngology practice. They collected normal and abnormal images, including benign lesions such as nodules, papilloma, polyps, and malignant lesions such as carcinoma and leukoplakia. Utilizing an ANN, they devised two classifiers: one for benign lesions and another for distinguishing malignant-premalignant from benign lesions. The results demonstrated high precision and recall rates for both classifiers, with the binary classifier achieving 93% overall accuracy in identifying malignant-premalignant lesions [281].

Another article discusses the significance of colorectal cancer screening for the initial detection and elimination of precancerous adenomas, aiming to reduce mortality rates. It introduces "leave-in situ" and "resect-and-discard" approaches for the management of small polyps detected during colonoscopies, with a focus on cost savings and efficiency [282]. However, implementation faces challenges such as financial incentives and medical-legal risks. Although CAD adoption has yielded mixed results, it could improve colorectal polyp management. Various reviewed studies suggest that while AI does not consistently outperform expert endoscopists, it could serve as a valuable secondary reader, augmenting diagnostic accuracy and endoscopists' confidence.

1.4.7 Applications of AI in Hybrid Imaging Modalities

Alzheimer's disease is a common neurodegenerative disorder resulting in dementia, for which there is presently no effective treatment to stop its progression. Research indicates that successful interventions should focus on early stages, such as MCI, or preclinical phases. Diagnosis and prognosis during these stages are difficult but achievable using hybrid imaging methods. Recent studies have utilized statistical ML and AI to analyze these imaging data for diagnosing and predicting Alzheimer's disease. Although the utilization of multimodality imaging for Alzheimer's disease

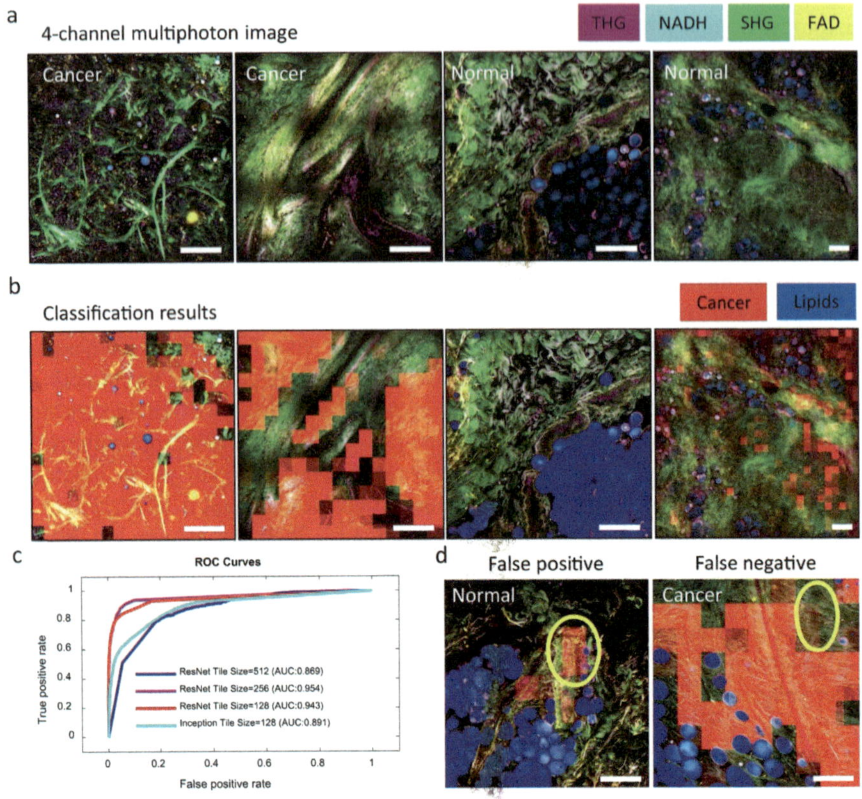

Fig. 1.16 Classification of human breast cancer using a DNN. **a** Two cancer and two normal slides depicted in 4-channel (THG, NADH, SHG, and FAD) multiphoton composite images. **b** Classification results corresponding to the multiphoton images in (**a**). The cancer probability map is coded in red and overlaid with the original multiphoton composite image. To mitigate bias from low-impact high-frequency content, lipid-filled adipocytes were segmented (shown in the blue mask) as a preprocessing step before classification. **c** AUC (area under the receiver operating curve) statistics on the test set for different model architectures and tile sizes. **d** Representative example of false positive (normal tiles classified as cancer, highlighted by the yellow oval) and false negative (cancer tiles classified as normal, highlighted by the yellow oval). Scale bar: 200 μm. The figure has been reproduced from You et al. [242], Copyright © The Author(s) 2019, Creative Commons Attribution 4.0 International License

diagnosis and prognosis in memory clinics primarily relies on the expertise of dementia specialists, researchers have devised and continued to develop diverse statistical models and ML algorithms for the quantitative analysis of imaging data to yield diagnostic and prognostic outcomes. This field of research is progressing rapidly. We anticipate that in the near future, AI will empower memory and aging centers through automated, computerized algorithms to support clinicians' decision-making. SPECT/CT imaging has shown promise in distinguishing between benign

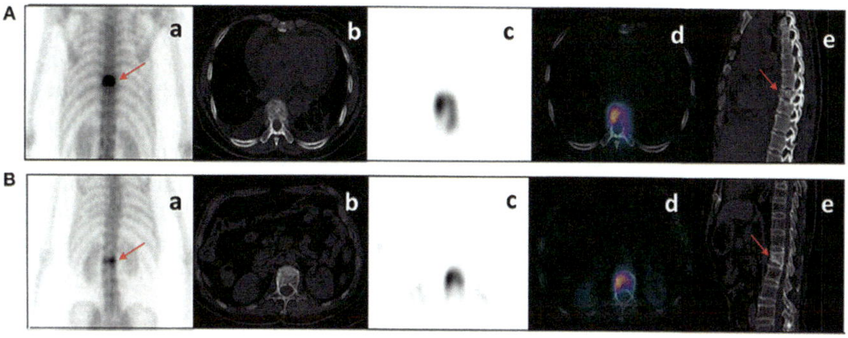

Fig. 1.17 Clinical cases of SPECT/CT images depicting bone metastases (**A**) and benign bone lesions (**B**) are shown in the figure. The images included whole-body scan (WBS), axial CT, SPECT, fusion image, and sagittal CT (**a**–**e**, respectively). In (**A**), bone metastases are observed in a 53-year-old female with adenocarcinoma of the left lung. Wedge-like changes are evident in the T8 vertebral body with an abnormal concentration of radioactive tracer (indicated by arrows). In (**B**), benign bone lesions are shown in a 68-year-old female with breast cancer. Wedge-like changes are observed in the L1 vertebral body with higher bone density and increased radioactive tracer distribution (indicated by arrows). Conventional images alone were insufficient for determining whether the lesions were metastatic. Subsequent imaging follow-up revealed lesion **A** to be a pathological fracture due to bone metastases confirmed by pathological examination, whereas lesion **B** was identified as a benign compression fracture based on imaging follow-up and clinical information. The figure and the caption have been adapted from Zin et al. [283], Copyright © 2022 Jin, Zhang, Wang, Tian, Zhang, Chen and Yu, Creative Commons Attribution License (CC BY)

and malignant tumors and represents a novel approach for disease staging and treatment strategizing. Figure 1.17 shows radiomics differentiation of bone metastases from benign bone lesions in tumor patients using SPECT and CT, as reported by Jin et al. [283].

Research has investigated the utilization of AI alongside 18F-FDG PET imaging for cancer detection and management. The results indicate that ML models can assist in various aspects of cancer imaging, including detection, differentiation, segmentation, staging, response evaluation, and prognosis determination [253, 254]. However, the routine clinical application of AI-based models and PET-derived radiomics encounters challenges due to the absence of standardized, generalizable, reproducible, and accurate methods.

While initial endeavors concentrated on disease classification, there is increasing potential to improve image quality itself. Hybrid imaging modalities such as PET/MRI and PET/CT are particularly suitable for these advancements [249, 251]. These enhancements enable shorter scanning durations, improved patient comfort, reduced motion artifacts, and the possibility of lower radiotracer doses. Moreover, AI-driven advances have enabled dual-tracer investigations, more frequent follow-up exams, and investigations of novel imaging indications.

1.5 Benefits and Implications of AI in Medical Imaging

AI is changing diagnostics from medical imaging by enabling increased accuracy and efficiency. The continuous learning process in AI is useful in the medical industry, as AI can adapt to changing clinical demands and stay up with the latest medical knowledge. These advantages are discussed further in the following subsections.

1.5.1 Improved Accuracy and Efficiency

AI systems, particularly DL models, have demonstrated high accuracy in recognizing and classifying anomalies in medical imaging, hence improving diagnostic precision. These algorithms, which are skilled at understanding nuanced patterns from large datasets, can detect irregularities that human observers miss, improving patient outcomes and reducing diagnostic mistakes [284, 285]. Furthermore, AI-powered systems excel in early illness identification by identifying modest imaging signals that indicate pathology, such as small tumors or nodules, allowing prompt interventions and treatment planning [159–161]. Furthermore, AI-driven technologies improve resource utilization in healthcare settings by optimizing imaging methods, eliminating needless tests, and reducing manual work. By highlighting clinically significant data, AI algorithms provide optimal resource allocation, improving patient care while lowering expenses and needless treatments.

1.5.2 Impact on Research and Development in Healthcare

The amount of research on AI in medical imaging has increased substantially in recent years. Researchers can greatly increase picture quality, resolution, and contrast by using AI for tasks such as image reconstruction, augmentation, and post-processing, resulting in more exact imaging of anatomical structures and disease characteristics. In addition, AI automates the interpretation and quantification of medical pictures, making it easier to extract quantitative imaging indicators such as lesion volume, density, and textural features. These biomarkers serve as objective measures of illness severity, progression, and therapeutic response, providing essential information on disease causes and therapeutic efficacy. Furthermore, AI-driven radiomics and imaging genomics approaches combine imaging data with genomic, proteomic, and clinical information to reveal hidden patterns and correlations between imaging phenotypes and disease molecular features. Such techniques enable researchers to find imaging biomarkers related to genetic mutations, molecular subtypes, and disease outcomes, accelerating progress in personalized medicine and precision oncology.

1.6 Challenges and Considerations

AI is still in its early stages, facing challenges such as reliability and user confidence. Challenges include ethics, validation, data sharing, regulations, and education. AI promises to automate tasks, improve disease characterization, and integrate multiple data sources. However, a sustainable development plan is needed to bridge theory and practice, considering contacts between professionals, technology, industry, government body, and society.

1.6.1 Ethical and Regulatory Considerations

The ethical concerns surrounding AI in medical imaging are varied and important. Patient consent, openness, accountability, and justice are critical considerations in this arena. Healthcare professionals must certify that patients are informed about the practice of AI algorithms in the treatment, including the possible effects on privacy, autonomy, and decision-making processes [286]. Regulatory monitoring is critical for maintaining the safety, effectiveness, and moral use of AI in medical imaging. Regulatory bodies must set thorough norms and standards for the development, validation, and application of AI algorithms, with necessities for data and algorithm transparency as well as rigorous clinical validation. Furthermore, ethical quandaries may arise around the ownership of medical imaging data, intellectual property rights, and economic interests [156]. As a result, healthcare institutions, researchers, and industry stakeholders must carefully negotiate these challenges while prioritizing patient rights, protecting intellectual property, and promoting fair access to AI-driven healthcare breakthroughs.

1.6.2 Data Privacy and Security

The incorporation of AI with medical imaging poses a significant issue in terms of data privacy and security [287]. This difficulty is exacerbated by the sensitive nature of patient health information, which necessitates strict measures against unauthorized access and data breaches. Compliance with strict standards such as HIPAA [288] in the United States and GDPR [289] in the European Union is critical for healthcare firms to maintain patient confidentiality and uphold privacy rights. To allow safe data interchange for research and therapeutic initiatives, secure data sharing frameworks and encryption procedures must be implemented. Emerging technologies such as blockchain and federated learning have the potential to provide decentralized, privacy-preserving solutions for exchanging medical imaging data in AI-driven research and clinical settings [290, 291].

1.6.3 *Integration with Clinical Workflow*

The addition of AI algorithms to clinical workflows involves a variety of issues in terms of usability, interoperability, and workflow effectiveness. Although studies have shown that the implementation of AI-based facilities has improved the workflow of doctors [292], to efficiently integrate AI technologies with current healthcare IT infrastructure and other clinical systems, flawless interoperability is required. Radiologists, physicians, and healthcare providers must accept and effectively use AI solutions, which require user-friendly interfaces and simple processes. Comprehensive training and education programs are required to familiarize users with AI algorithms, increase their ability to comprehend AI-generated insights and facilitate their incorporation into workflows. Interdisciplinary cooperation among radiologists, data scientists, software engineers, and healthcare administrators is critical for building and deploying AI solutions that are specific to clinical needs, improving diagnostic accuracy, and improving patient care outcomes. Continuous feedback loops and ongoing development of AI algorithms using real-world clinical experience are critical for addressing usability issues and enhancing workflow integration over time.

1.7 Conclusions

AI-driven technologies in medical imaging have the potential to increase diagnostic accuracy and optimize healthcare resource consumption, eventually transforming the way diseases are identified, treated, and controlled. By leveraging large-scale imaging datasets, AI algorithms can discover hidden patterns, predict patient outcomes, and personalize treatment strategies. DL enables researchers to extract actionable insights from heterogeneous data sources and uncover novel imaging biomarkers for disease diagnosis, prognosis, and treatment response prediction. The recent introduction of physics-enabled DL networks shows promise in refining the accuracy of AI-driven technologies in medical imaging. The benefits of AI can be increased with collaborative initiatives, such as multi-center studies, consortia, and data-sharing networks, facilitating the pooling of diverse imaging data to build robust AI models with broad clinical utility. Collaborative research networks and consortia bring together multidisciplinary teams of researchers, clinicians, data scientists, and industry stakeholders to tackle complex problems, such as data standardization, algorithm validation, and clinical implementation. Open-access datasets, benchmarking challenges, and shared repositories facilitate knowledge sharing, benchmarking, and reproducibility of AI research in medical imaging. These collaborative platforms foster transparency, rigor, and innovation, driving continuous improvement and advancement in the field.

Despite the tremendous advancement and promise of AI in medical imaging, there are significant challenges lie ahead. Addressing issues such as data quality, algorithm robustness, regulatory compliance, and ethical considerations is imperative to fully harnessing the potential of AI in clinical settings. The profound dependence of AI techniques on data underscores the necessity for efficient data storage infrastructures, potentially leading to substantial energy requirements in the future. Future research topics will involve improving AI algorithms for multimodal imaging analysis, longitudinal data modeling, and real-time picture interpretation.

Acknowledgements Financial support received under project No. GP/2021/RR/017, IHFC-IIT Delhi; project No. BBF/BITS/FY2023-24/ART-103, BITS BioCyTiH Foundation and DST-INSPIRE fellowship No. IF190773, Department of Science and Technology, Government of India is gratefully acknowledged. AG gratefully acknowledges the financial support received from the Science and Engineering Research Board (SERB), Department of Science and Technology (DST), India (Grant No. CRG/2019/004868). GYZ deeply appreciate the financial support received from the National Science and Technology Council (NSTC), Taiwan (Grant No. 110-2112-M-039-001-MY3).

References

1. J. Peter, Medical imaging modalities-An introduction. in *Advanced imaging in biology and medicine: technology, software environments, applications* (Springer Berlin Heidelberg, Berlin, Heidelberg), pp. 225–254
2. A. Elangovan, T. Jeyaseelan, February. Medical imaging modalities: a survey. in *2016 International Conference on emerging trends in engineering, technology and science* (2016), pp. 1–4
3. L. Kaufman, R.S. Hattner, December. Comparison of medical imaging modalities: Clinical realization and engineering potential of nuclear imaging. in *Recent and future developments in medical imaging II*, vol. 206 (1979), pp. 27–32
4. S.M.S. Islam, M.A.A. Nasim, I. Hossain, D.M.A. Ullah, D.K.D. Gupta, M.M.H. Bhuiyan, Introduction of medical imaging modalities. in *Data driven approaches on medical imaging* (2023), pp. 1–25
5. S. Dash, S.K. Shakyawar, M. Sharma, S. Kaushik, Big data in healthcare: management, analysis and future prospects. J. Big Data **6**(1), 1–25 (2019)
6. J.S. Duncan, M.F. Insana, N. Ayache, Biomedical imaging and analysis in the age of big data and deep learning [scanning the issue]. Proc. IEEE **108**(1), 3–10 (2019)
7. K. Wang, S. Li, X. Wang, J. Feng, Y. Xu, Intelligent analysis of biomedical imaging data for precision medicine. Front. Med. **9**, 1017751 (2022)
8. A.O.A. Deheyab, M.H. Alwan, I.K.A. Rezzaqe, O.A. Mahmood, Y.I. Hammadi, A.N. Kareem, M. Ibrahim, December. An overview of challenges in medical image processing. in *Proceedings of the 6th international conference on future networks & distributed systems* (2022), pp. 511–516
9. N. Hasani, F. Farhadi, M.A. Morris, M. Nikpanah, A. Rahmim, Y. Xu, A. Pariser, M.T. Collins, R.M. Summers, E. Jones, E. Siegel, Artificial intelligence in medical imaging and its impact on the rare disease community: threats, challenges and opportunities. PET clinics **17**(1), 13–29 (2022)
10. D. Shen, G. Wu, H.I. Suk, Deep learning in medical image analysis. Annu. Rev. Biomed. Eng. **19**, 221–248 (2017)
11. S.J. Adams, R.D. Henderson, X. Yi, P. Babyn, Artificial intelligence solutions for analysis of X-ray images. Can. Assoc. Radiol. J. **72**(1), 60–72 (2021)

12. M.M. Abuzaid, H.O. Tekin, M. Reza, I.R. Elhag, W. Elshami, Assessment of MRI technologists in acceptance and willingness to integrate artificial intelligence into practice. Radiography **27**, S83–S87 (2021)
13. T. Davenport, R. Kalakota, The potential for artificial intelligence in healthcare. Futur. Healthc. J. **6**(2), 94 (2019)
14. W. Yin, L. Li, F.X. Wu, Deep learning for brain disorder diagnosis based on fMRI images. Neurocomputing **469**, 332–345 (2022)
15. A. Kazeminejad, S. Golbabaei, H. Soltanian-Zadeh, Graph theoretical metrics and machine learning for diagnosis of Parkinson's disease using rs-fMRI. in *2017 Artificial Intelligence and Signal Processing Conference (AISP)* (IEEE), pp. 134–139
16. E. Helmy, A. Elnakib, Y. ElNakieb, M. Khudri, M. Abdelrahim, J. Yousaf, A. El-Baz, Role of artificial intelligence for autism diagnosis using DTI and fMRI: A survey. Biomedicines **11**(7), 1858 (2023)
17. M.N.I. Qureshi, J. Oh, B. Lee, 3D-CNN based discrimination of schizophrenia using resting-state fMRI. Artif. Intell. Med. **98**, 10–17 (2019)
18. C. Brossard, B. Lemasson, A. Attyé, J.A. De Busschère, J.F. Payen, E.L. Barbier, P. Bouzat, Contribution of CT-scan analysis by artificial intelligence to the clinical care of TBI patients. Front. Neurol. **12**, 666875 (2021)
19. S. Akram, M.Y. Javed, A. Hussain, F. Riaz, M. Usman Akram, Intensity-based statistical features for classification of lungs CT scan nodules using artificial intelligence techniques. J. Exp. Theor. Artif. Intell. **27**(6), 737–751 (2015)
20. R. Fusco, R. Grassi, V. Granata, S.V. Setola, F. Grassi, D. Cozzi, A. Petrillo, Artificial intelligence and COVID-19 using chest CT scan and chest X-ray images: machine learning and deep learning approaches for diagnosis and treatment. J. Pers. Med. **11**(10), 993 (2021)
21. Y.T. Shen, L. Chen, W.W. Yue, H.X. Xu, Artificial intelligence in ultrasound. Eur. J. Radiol. **139**, 109717 (2021)
22. M. Kawakami, K. Hirata, S. Furuya, K. Kobayashi, H. Sugimori, K. Magota, C. Katoh, Development of combination methods for detecting malignant uptakes based on physiological uptake detection using object detection with PET-CT MIP images. Front. Med. **7**, 616746 (2020)
23. Y. Satoh, K. Hirata, D. Tamada, S. Funayama, H. Onishi, Texture analysis in the diagnosis of primary breast cancer: comparison of high-resolution dedicated breast positron emission tomography (dbPET) and whole-body PET/CT. Front. Med. **7**, 603303 (2020)
24. X. Fan, H. Zhang, Y. Yin, J. Zhang, M. Yang, S. Qin, F. Yu, Texture analysis of 18F-FDG PET/CT for differential diagnosis spinal metastases. Front. Med. **7**, 605746 (2021)
25. M.B. Applegate, R.E. Istfan, S. Spink, A. Tank, D. Roblyer, Recent advances in high speed diffuse optical imaging in biomedicine. APL Photonics **5** (2020)
26. S. Sabir, S. Cho, Y. Kim, R. Pua, D. Heo, K.H. Kim, Y. Choi, S. Cho, Convolutional neural network-based approach to estimate bulk optical properties in diffuse optical tomography. Appl. Opt. **59**, 1461 (2020)
27. J. Yoo et al., Deep learning diffuse optical tomography. IEEE Trans. Med. Imaging **39**(4), 877–887 (2020). https://doi.org/10.1109/TMI.2019.2936522
28. Society for Imaging Informatics in Medicine Web site, "Imaging informatics." [Online]. Available http://www.siimweb.org/index.cfm. Accessed on: 22 Feb 2024.
29. C. Walsh, P. Tafforeau, W.L. Wagner, D.J. Jafree, A. Bellier, C. Werlein, P.D. Lee, *Multi-scale three-dimensional imaging of intact human organs down to the cellular scale using hierarchical phase-contrast tomography* (2021). biorxiv.
30. A. Gogoi, G. Keiser, F.J. Kao, A.S. Chiang, Multiscale and multimodal imaging for connectomics. Adv. Opt. Methods Brain Imaging, 3–45 (2019)
31. R.F. Betzel, D.S. Bassett, Multi-scale brain networks. Neuroimage **160**, 73–83 (2017)
32. A.J. Weinhaus, K.P. Roberts, *Anatomy of the human heart. Handbook of cardiac anatomy, physiology, and devices* (2005), pp. 51–79
33. Y. Skandarani, P.M. Jodoin, A. Lalande, Gans for medical image synthesis: An empirical study. J. Imaging **9**(3), 69 (2023)

34. P.C. Africa, R. Piersanti, M. Fedele, L. Dede, A. Quarteroni, Lifex-fiber: an open tool for myofibers generation in cardiac computational models. BMC Bioinform., **24**(1), 143 (2023)
35. R. Chabiniok, V.Y. Wang, M. Hadjicharalambous, L. Asner, J. Lee, M. Sermesant, D.A. Nordsletten, Multiphysics and multiscale modelling, data–model fusion and integration of organ physiology in the clinic: ventricular cardiac mechanics. Interface focus **6**(2), 20150083 (2016)
36. J.P. Singh, J.L. Young, The cardiac nanoenvironment: form and function at the nanoscale. Biophys. Rev. **13**(5), 625–636 (2021)
37. A.R. Sun, R.M. Hengst, J.L. Young, All the small things: Nanoscale matrix alterations in aging tissues. Curr. Opin. Cell Biol. **87**, 102322 (2024)
38. A.S. Panayides, A. Amini, N.D. Filipovic, A. Sharma, S.A. Tsaftaris, A. Young, C.S. Pattichis, AI in medical imaging informatics: current challenges and future directions. IEEE J. Biomed. Health Inform. **24**(7), 1837–1857 (2020)
39. R. Smith-Bindman, D.L. Miglioretti, E.B. Larson, Rising use of diagnostic medical imaging in a large integrated health system. Health Aff. **27**(6), 1491–1502 (2008)
40. P. Lambin, R.T. Leijenaar, T.M. Deist, J. Peerlings, E.E. De Jong, J. Van Timmeren, S. Walsh, Radiomics: the bridge between medical imaging and personalized medicine. Nat. Rev. Clin. Oncol. **14**(12), 749–762 (2017)
41. A. Quarteroni, T. Lassila, S. Rossi, R. Ruiz-Baier, Integrated heart—coupling multiscale and multiphysics models for the simulation of the cardiac function. Comput. Methods Appl. Mech. Eng. **314**, 345–407 (2017)
42. B. Asrican, G.J. Augustine, K. Berglund, S. Chen, N. Chow, K. Deisseroth, S. Zhao, Next-generation transgenic mice for optogenetic analysis of neural circuits. Front. Neural Circuits **7**, 160 (2013)
43. W. Denk, K. Svoboda, Photon upmanship: why multiphoton imaging is more than a gimmick. Neuron **18**(3), 351–357 (1997)
44. N. Korogod, C.C. Petersen, G.W. Knott, Ultrastructural analysis of adult mouse neocortex comparing aldehyde perfusion with cryo fixation. Elife, **4**, e05793 (2015)
45. B. Hammond, J. Kough, C. Herouet-Guicheney, J.M. Jez, ILSI international food biotechnology committee task force on the use of mammalian toxicology studies in the safety assessment of GM Foods, Toxicological evaluation of proteins introduced into food crops. Crit. Rev. Toxicol., **43**(sup2), 25–42 (2013)
46. S. Wang, G. Cao, Y. Wang, S. Liao, Q. Wang, J. Shi, D. Shen, Review and prospect: artificial intelligence in advanced medical imaging. Front. Radiol. **1**, 781868 (2021)
47. A.S. Panchbhai, Wilhelm Conrad Röntgen and the discovery of X-rays: Revisited after centennial. J. Indian Acad. Oral Med. Radiol. **27**(1), 90–95 (2015)
48. The Nobel Prize in Physics 1901, *NobelPrize.org* (n.d.). https://www.nobelprize.org/prizes/physics/1901/summary/
49. M.T. Mustapha, B. Uzun, D.U. Ozsahin, I. Ozsahin, A comparative study of X-ray based medical imaging devices. in Applications of multi-criteria decision-making theories in healthcare and biomedical engineering (Academic Press, 2021), pp. 163–180
50. P.J. Withers, C. Bouman, S. Carmignato, V. Cnudde, D. Grimaldi, C.K. Hagen, S.R. Stock, X-ray computed tomography. Nat. Rev. Methods Prim. **1**(1), 18 (2021)
51. R.A. Schulz, J.A. Stein, N.J. Pelc, How CT happened: the early development of medical computed tomography. J. Med. Imaging **8**(5), 052110–052110 (2021)
52. F. Bruno, V. Granata, F. Cobianchi Bellisari, F. Sgalambro, E. Tommasino, P. Palumbo, F. Arrigoni, D. Cozzi, F. Grassi, M.C. Brunese, S. Pradella, Advanced Magnetic Resonance Imaging (MRI) techniques: Technical principles and applications in nanomedicine. Cancers **14**(7), 1626 (2022)
53. P. Mahlknecht, A. Hotter, A. Hussl, R. Esterhammer, M. Schocke, K. Seppi, Significance of MRI in diagnosis and differential diagnosis of Parkinson's disease. Neurodegener. Dis. **7**(5), 300–318 (2010)
54. A. Tiwari, S. Srivastava, M. Pant, Brain tumor segmentation and classification from magnetic resonance images: Review of selected methods from 2014 to 2019. Pattern Recogn. Lett. **131**, 244–260 (2020)

55. E.T. Chou, J.A. Carrino, Magnetic resonance imaging. Pain Manag., 106–117 (2017). https://doi.org/10.1016/b978-0-7216-0334-6.50014-5
56. S. Hussain, I. Mubeen, N. Ullah, S.S.U.D. Shah, B.A. Khan, M. Zahoor, M.A. Sultan, Modern diagnostic imaging technique applications and risk factors in the medical field: A review. BioMed Res. Int., **2022**, (2022)
57. D. Le Bihan, M. Iima, Diffusion magnetic resonance imaging: what water tells us about biological tissues. PLoS Biol. **13**(7), e1002203 (2015)
58. B.A. Mueller, K.O. Lim, L. Hemmy, J. Camchong, Diffusion MRI and its role in neuropsychology. Neuropsychol. Rev. **25**(3), 250–271 (2015)
59. R.B. Buxton, The physics of functional magnetic resonance imaging (fMRI). Rep. Prog. Phys. **76**(9), 096601 (2013)
60. E. Düzel, M. Costagli, G. Donatelli, O. Speck, M. Cosottini, Studying Alzheimer disease, Parkinson disease, and amyotrophic lateral sclerosis with 7-T magnetic resonance. Eur. Radiol. Exp. **5**, 1–17 (2021)
61. J. Cohen-Adad, W. Zhao, B. Keil, E.M. Ratai, C. Triantafyllou, R. Lawson, C. Dheel, L.L. Wald, B.R. Rosen, M. Cudkowicz, N. Atassi, 7-T MRI of the spinal cord can detect lateral corticospinal tract abnormality in amyotrophic lateral sclerosis. Muscle Nerve **47**(5), 760–762 (2013)
62. S. Mayor, *Nobel prize in medicine awarded to MRI pioneers* (2003)
63. The Nobel Prize in Physiology or Medicine 2003. NobelPrize.org (n.d.). https://www.nobelprize.org/prizes/medicine/2003/press-release/
64. Y. Yao, Y. Pan, S. Liu, Power ultrasound and its applications: A state-of-the-art review. Ultrason. Sonochem. **62**, 104722 (2020)
65. A. Carovac, F. Smajlovic, D. Junuzovic, Application of ultrasound in medicine. Acta Informatica Medica **19**(3), 168 (2011)
66. L. Troxclair, D. Smetherman, E.I. Bluth, Shades of gray: a history of the development of diagnostic ultrasound in a large multispecialty clinic. Ochsner J. **11**(2), 151–155 (2011)
67. G. Crişan, N.S. Moldovean-Cioroianu, D.G. Timaru, G. Andrieş, C. Căinap, V. Chiş, Radiopharmaceuticals for PET and SPECT imaging: A literature review over the last decade. Int. J. Mol. Sci. **23**(9), 5023 (2022)
68. J. Czernin, M.E. Phelps, Positron emission tomography scanning: current and future applications. Annu. Rev. Med. **53**(1), 89–112 (2002)
69. A.K. Shukla, U. Kumar, Positron emission tomography: An overview. J. Med. Phys./Assoc. Med. Phys. India **31**(1), 13 (2006)
70. T.A. Holly, B.G. Abbott, M. Al-Mallah, D.A. Calnon, M.C. Cohen, F.P. DiFilippo, P. Soman, *Single photon-emission computed tomography* (2010)
71. D. Jain, R. Torres, R. Celli, J. Koelmel, G. Charkoftaki, V. Vasiliou, Evolution of the liver biopsy and its future. Transl. Gastroenterol. Hepatol., **6**, (2021)
72. N. Sun, Y. Jia, S. Bai, Q. Li, L. Dai, J. Li, The power of super-resolution microscopy in modern biomedical science. Adv. Colloid Interface Sci., 102880 (2023)
73. P.J. Rigby, R.G. Goldie, Confocal microscopy in biomedical research. Croat. Med. J. **40**, 346–352 (1999)
74. H. Blom, H. Brismar, STED microscopy: increased resolution for medical research? J. Intern. Med. **276**(6), 560–578 (2014)
75. F.J. Kao, G. Keiser, A. Gogoi (eds.), *Advanced optical methods for brain imaging, vol. 5* (Springer, Berlin, 2019)
76. H. Zeng, A. Weiss, R. Cline, C.E. MacAulay, Real-time endoscopic fluorescence imaging for early cancer detection in the gastrointestinal tract. Bioimaging **6**(4), 151–165 (1998)
77. M. Mujat, J.D. Akula, A.B. Fulton, R.D. Ferguson, N. Iftimia, Non-rigid registration for high-resolution retinal imaging. Diagnostics **13**(13), 2285 (2023)
78. H. Yoshikawa, T. Azuma, K.I. Kawabata, Precise vascular visualization in ultrasonic power Doppler imaging with motion compensation. Jpn. J. Appl. Phys., **49**(7S), 07HF13 (2010)
79. M. Pivette, J.E. Mueller, P. Crepey, A. Bar-Hen, Surveillance of gastrointestinal disease in France using drug sales data. Epidemics **8**, 1–8 (2014)

80. O. Elemento, C. Leslie, J. Lundin, G. Tourassi, Artificial intelligence in cancer research, diagnosis and therapy. Nat. Rev. Cancer **21**(12), 747–752 (2021)
81. B.J. Tromberg, Z. Zhang, A. Leproux, T.D. O'Sullivan, A.E. Cerussi, P.M. Carpenter, R.S. Mehta, D. Roblyer, W. Yang, K.D. Paulsen et al., Predicting responses to neoadjuvant chemotherapy in breast cancer: ACRIN 6691 trial of diffuse optical spectroscopic imaging. Cancer Res. **76**, 5933–5944 (2016)
82. The Nobel Prize in Chemistry 2014. NobelPrize.org. (n.d.). https://www.nobelprize.org/prizes/chemistry/2014/press-release/
83. H. Hricak, B.I. Choi, A.M. Scott, K. Sugimura, A. Muellner, G.K. von Schulthess, M.F. Reiser, M.M. Graham, N.R. Dunnick, S.M. Larson, Global Trends in hybrid imaging. Radiology, **257**(2), 498–506 (2010). https://doi.org/10.1148/radiol.10100579
84. R.J. Hicks, E.W. Lau, D.S. Binns, Hybrid imaging is the future of molecular imaging. Biomed. Imaging Interv. J., **3**(3), (2007)
85. S. Walrand, M. Hesse, F. Jamar, SPECT/CT, PET/CT and PET/MR principles. Diagn. Ther. Nucl. Med. Neuroendocr. Tumors, 163–200 (2017)
86. L. Papp, C.P. Spielvogel, I. Rausch, M. Hacker, T. Beyer, Personalizing medicine through hybrid imaging and medical big data analysis. Front. Phys. **6**, 51 (2018)
87. P. Parlanti, V. Cappello, Microscopes, tools, probes, and protocols: A guide in the route of correlative microscopy for biomedical investigation. Micron **152**, 103182 (2022)
88. J. Caplan, M. Niethammer, R.M. Taylor II., K.J. Czymmek, The power of correlative microscopy: multi-modal, multi-scale, multi-dimensional. Curr. Opin. Struct. Biol. **21**(5), 686–693 (2011)
89. K.A. Jahn, D.A. Barton, K. Kobayashi, K.R. Ratinac, R.L. Overall, F. Braet, Correlative microscopy: providing new understanding in the biomedical and plant sciences. Micron **43**(5), 565–582 (2012)
90. M.L. Scarpelli, D.R. Healey, S. Mehta, V.D. Kodibagkar, C.C. Quarles, A practical method for multimodal registration and assessment of whole-brain disease burden using PET, MRI, and optical imaging. Sci. Rep. **10**(1), 17324 (2020)
91. R. LoGullo, J. Horvat, J. Reiner, K. Pinker, Multimodal, multiparametric and genetic breast imaging. Radiologe **61**, 183–191 (2021)
92. M.A. Jacobs, A.C. Wolff, K.J. Macura, V. Stearns, R. Ouwerkerk, R. El Khouli, D.A. Bluemke, R. Wahl, Multiparametric and multimodality functional radiological imaging for breast cancer diagnosis and early treatment response assessment. J. Natl. Cancer Inst. Monogr. **2015**(51), 40–46 (2015)
93. A. Bockisch, L.S. Freudenberg, D. Schmidt, T. Kuwert, Hybrid imaging by SPECT/CT and PET/CT: proven outcomes in cancer imaging. in *Seminars in nuclear medicine*, vol. 39, no. 4 (WB Saunders, 2009)
94. M. Daniel, P. Andrzejewski, A. Sturdza, K. Majercakova, P. Baltzer, K. Pinker, D. Georg, Impact of hybrid PET/MR technology on multiparametric imaging and treatment response assessment of cervix cancer. Radiother. Oncol. **125**(3), 420–425 (2017)
95. V. Goh, A. Engledow, M. Rodriguez-Justo, M. Shastry, J. Peck, G. Blackman, A.M. Groves, The flow-metabolic phenotype of primary colorectal cancer: assessment by integrated 18F-FDG PET/perfusion CT with histopathologic correlation. J. Nucl. Med. **53**(5), 687–692 (2012)
96. D. Singh, K. Miles, Multiparametric PET/CT in oncology. Cancer Imaging **12**(2), 336 (2012)
97. C. Eichhorn, S. Greulich, C. Bucciarelli-Ducci, R. Sznitman, R.Y. Kwong, C. Gräni, Multiparametric cardiovascular magnetic resonance approach in diagnosing, monitoring, and prognostication of myocarditis. Cardiovasc. Imaging **15**(7), 1325–1338 (2022)
98. A.R. Padhani, K.A. Miles, Multiparametric imaging of tumor response to therapy. Radiology **256**(2), 348–364 (2010)
99. A. Radjenovic, A.G. Christodoulou, Simultaneous multiparametric and multidimensional cardiovascular magnetic resonance imaging. Front. Cardiovasc. Med. **10**, 1205994 (2023)
100. A. Maier, Y.C. Toner, J. Munitz, N.A. Sullivan, K. Sakurai, A.E. Meerwaldt, M.M. van Leent, Multiparametric Immunoimaging maps inflammatory signatures in murine myocardial infarction models.JACC: Basic Transl. Sci, (2023)

101. G. Wang, L. Nardo, M. Parikh, Y.G. Abdelhafez, E. Li, B.A. Spencer, R.D. Badawi, Total-body PET multiparametric imaging of cancer using a voxelwise strategy of compartmental modeling. J. Nucl. Med. **63**(8), 1274–1281 (2022)
102. L.S. Chow, R. Paramesran, Review of medical image quality assessment. Biomed. Signal Process. Control **27**, 145–154 (2016)
103. I. Domingues, G. Pereira, P. Martins, H. Duarte, J. Santos, P.H. Abreu, Using deep learning techniques in medical imaging: a systematic review of applications on CT and PET. Artif. Intell. Rev. **53**, 4093–4160 (2020)
104. M. Lehne, J. Sass, A. Essenwanger, J. Schepers, S. Thun, Why digital medicine depends on interoperability. NPJ Digit. Med. **2**(1), 79 (2019)
105. L. Lévêque, M. Outtas, H. Liu, L. Zhang, Comparative study of the methodologies used for subjective medical image quality assessment. Phys. Med. & Biol., **66**(15), 15TR02 (2021)
106. L.K.S. Sundar, O. Muzik, I. Buvat, L. Bidaut, T. Beyer, Potentials and caveats of AI in hybrid imaging. Methods **188**, 4–19 (2021)
107. M. Li, Y. Zhang, Medical image analysis using deep learning algorithms. Front. Public Health **11**, 1273253 (2023)
108. K. Doi, Computer-aided diagnosis in medical imaging: historical review, current status and future potential. Comput. Med. Imaging Graph. **31**(4–5), 198–211 (2007)
109. S. Legg, M. Hutter, A collection of definitions of intelligence. Frontiers Artificial Intelligence Appl. **157**, 17 (2007)
110. v. Bush, As we may think. The Atlantic Monthly (1945)
111. S.H. Chalasani, J. Syed, M. Ramesh, V. Patil, T.P. Kumar, Artificial intelligence in the field of pharmacy practice: A literature review. Explor. Res. Clin. Soc. Pharm. **12**, 100346 (2023)
112. A. Aadhityan, A novel method for implementing artificial Intelligence, Cloud and Internet of Things in Robots. in 2015 *International Conference on Innovations in Information, Embedded and Communication Systems* (ICIIECS) (pp. 1–4). IEEE.
113. N.J. Nilsson, *John McCarthy* (National Academy of Sciences, 2012), pp. 1–27
114. J. McCarthy, M.L. Minsky, N. Rochester, C.E. Shannon, A proposal for the dartmouth summer research project on artificial intelligence, august 31, 1955. AI Mag. **27**(4), 12–12 (2006)
115. V. Rajaraman, John McCarthy—Father of artificial intelligence. Resonance **19**, 198–207 (2014)
116. R. Cordeschi, AI turns fifty: revisiting its origins. Appl. Artif. Intell. **21**(4–5), 259–279 (2007)
117. J. Moor, The dartmouth college artificial intelligence conference: The next fifty years. AI Mag. **27**(4), 87–87 (2006)
118. G. Solomonoff, The meeting of the minds that launched AI. IEEE Spectrum., (2023). https://spectrum.ieee.org/dartmouth-ai-workshop
119. S.L. Andresen, John McCarthy: father of AI. IEEE Intell. Syst. **17**(5), 84–85 (2002)
120. P. Hamet, J. Tremblay, Artificial intelligence in medicine. Metabolism **69**, S36–S40 (2017)
121. A.N. Ramesh, C. Kambhampati, J.R. Monson, P.J. Drew, Artificial intelligence in medicine. Ann. R. Coll. Surg. Engl. **86**(5), 334 (2004)
122. A. Konar, *Artificial intelligence and soft computing: behavioral and cognitive modeling of the human brain* (CRC Press, 2018)
123. L. Zhou, S. Pan, J. Wang, A.V. Vasilakos, Machine learning on big data: Opportunities and challenges. Neurocomputing **237**, 350–361 (2017)
124. M.W. Berry, A. Mohamed, B.W. Yap (Eds.), Supervised and unsupervised learning for data science (Springer Nature, 2019)
125. X. Su, X. Yan, C.L. Tsai, Linear regression. Wiley Interdisciplinary Reviews: Computational Statistics **4**(3), 275–294 (2012)
126. T.G. Nick, K.M. Campbell, Logistic regression. Top. Biostat., 273–301 (2007)
127. I. Wickramasinghe, H. Kalutarage, Naive Bayes: applications, variations and vulnerabilities: a review of literature with code snippets for implementation. Soft. Comput. **25**(3), 2277–2293 (2021)
128. O. Kramer, O. Kramer, *K-nearest neighbors. Dimensionality reduction with unsupervised nearest neighbors* (2013), pp. 13–23.

129. C. Kingsford, S.L. Salzberg, What are decision trees? Nat. Biotechnol. **26**(9), 1011–1013 (2008)
130. G. Biau, E. Scornet, A random forest guided tour. TEST **25**, 197–227 (2016)
131. D.A. Pisner, D.M. Schnyer, Support vector machine. in *Machine learning* (Academic Press, 2020), pp. 101–121
132. S. Wold, K. Esbensen, P. Geladi, Principal component analysis. Chemom. Intell. Lab. Syst. **2**(1–3), 37–52 (1987)
133. F. Nielsen, F. Nielsen, *Hierarchical clustering. Introduction to HPC with MPI for Data Science* (2016), pp. 195–211
134. A.J. Izenman, Introduction to manifold learning. Wiley Interdiscip. Rev.: Comput. Stat. **4**(5), 439–446 (2012)
135. M. Ahmed, R. Seraj, S.M.S. Islam, The k-means algorithm: A comprehensive survey and performance evaluation. Electronics **9**(8), 1295 (2020)
136. K. Arulkumaran, M.P. Deisenroth, M. Brundage, A.A. Bharath, Deep reinforcement learning: A brief survey. IEEE Signal Process. Mag. **34**(6), 26–38 (2017)
137. F. Rosenblatt, The perceptron: A probabilistic model for information storage and organization (1958) (2021)
138. J.K. Mai, G. Paxinos (Eds.), The human nervous system (Academic press, 2011)
139. I.H. Sarker, Deep learning: a comprehensive overview on techniques, taxonomy, applications and research directions. SN Computer Science **2**(6), 420 (2021)
140. C. Clifford, *Elon Musk says 'A.I. will be the best or worst thing ever for humanity,' recommends a book on the topic* (CNBC, 2018). https://www.cnbc.com/2017/08/29/elon-musk-recommends-a-book-on-the-future-of-artificial-intelligence.html
141. T.N. Hoang-Thi, G. Chassagnon, H.D. Tran, N.N. Le-Dong, A.T. Dinh-Xuan, M.P. Revel, How artificial intelligence in imaging can better serve patients with bronchial and parenchymal lung diseases? J. Pers. Med. **12**(9), 1429 (2022)
142. H. Fujita, AI-based computer-aided diagnosis (AI-CAD): the latest review to read first. Radiol. Phys. Technol. **13**(1), 6–19 (2020)
143. P. Ghosh,. *AI – Past, Present and Future. ThirdEye Data* (2022). https://thirdeyedata.ai/ai-past-present-and-future/
144. Broken Promises & Empty Threats: The Evolution of AI in the USA, 1956–1996—Technology's Stories. (n.d.). https://www.technologystories.org/ai-evolution/
145. Z. Han, B. Wei, Y. Zheng, Y. Yin, K. Li, S. Li, Breast cancer multi-classification from histopathological images with structured deep learning model. Sci. Rep. **7**(1), 4172 (2017)
146. H. Wang, Z. Zhou, Y. Li, Z. Chen, P. Lu, W. Wang, L. Yu, Comparison of machine learning methods for classifying mediastinal lymph node metastasis of non-small cell lung cancer from 18 F-FDG PET/CT images. EJNMMI Res. **7**, 1–11 (2017)
147. S. Nikolov, S. Blackwell, A. Zverovitch, R. Mendes, M. Livne, J. De Fauw, O. Ronneberger, Clinically applicable segmentation of head and neck anatomy for radiotherapy: deep learning algorithm development and validation study. J. Med. Internet Res. **23**(7), e26151 (2021)
148. Z. Guo, X. Li, H. Huang, N. Guo, Q. Li, Deep learning-based image segmentation on multimodal medical imaging. IEEE Trans. Radiat. Plasma Med. Sci. **3**(2), 162–169 (2019)
149. U. Javaid, D. Dasnoy, J.A. Lee, Multi-organ segmentation of chest CT images in radiation oncology: comparison of standard and dilated UNet. in *Advanced concepts for intelligent vision systems: 19th international conference, ACIVS 2018, Poitiers, France, September 24–27, 2018, Proceedings 19* (Springer International Publishing, 2018), pp. 188–199
150. Z. Li, Y. Xia, Deep reinforcement learning for weakly-supervised lymph node segmentation in CT images. IEEE J. Biomed. Health Inform. **25**(3), 774–783 (2020)
151. D.J. Winkel, T.J. Weikert, H.C. Breit, G. Chabin, E. Gibson, T.J. Heye, D.T. Boll, Validation of a fully automated liver segmentation algorithm using multi-scale deep reinforcement learning and comparison versus manual segmentation. Eur. J. Radiol. **126**, 108918 (2020)
152. J. Mehta, A. Majumdar, Rodeo: robust de-aliasing autoencoder for real-time medical image reconstruction. Pattern Recogn. **63**, 499–510 (2017)

153. C. Shen, Y. Gonzalez, L. Chen, S.B. Jiang, X. Jia, Intelligent parameter tuning in optimization-based iterative CT reconstruction via deep reinforcement learning. IEEE Trans. Med. Imaging **37**(6), 1430–1439 (2018)
154. X. Han, MR-based synthetic CT generation using a deep convolutional neural network method. Med. Phys. **44**(4), 1408–1419 (2017)
155. S. Kazemifar, S. McGuire, R. Timmerman, Z. Wardak, D. Nguyen, Y. Park, A. Owrangi, MRI-only brain radiotherapy: Assessing the dosimetric accuracy of synthetic CT images generated using a deep learning approach. Radiother. Oncol. **136**, 56–63 (2019)
156. J.H. Rodríguez, F.J.C. Fraile, M.J.R. Conde, P.L.G. Llorente, Computer aided detection and diagnosis in medical imaging: a review of clinical and educational applications. in *Proceedings of the fourth international conference on technological ecosystems for enhancing multiculturality* (2016), pp. 517–524
157. A. Kohli, S. Jha, Why CAD failed in mammography. J Am Coll Radiol, **15**(3 Pt B), 535–537(2018)
158. C.D. Lehman, R.D. Wellman, D.S. Buist, K. Kerlikowske, A.N. Tosteson, D.L. Miglioretti Breast Cancer Surveillance Consortium.. Diagnostic accuracy of digital screening mammography with and without computer-aided detection. JAMA Intern. Med., **175**(11), 1828–1837 (2015)
159. J. Jiang, Y.C. Hu, N. Tyagi, A. Rimner, N. Lee, J.O. Deasy, H. Veeraraghavan, PSIGAN: Joint probabilistic segmentation and image distribution matching for unpaired cross-modality adaptation-based MRI segmentation. IEEE Trans. Med. Imaging **39**(12), 4071–4084 (2020)
160. J. Zhang, C. Wang, Y. Sheng, M. Palta, B. Czito, C. Willett, Q.J. Wu, An interpretable planning bot for pancreas stereotactic body radiation therapy. Int. J. Radiat. Oncology* Biology* Phys., **109**(4), 1076–1085 (2021)
161. J. Watts, A. Khojandi, R. Vasudevan, R. Ramdhani, Optimizing individualized treatment planning for Parkinson's disease using deep reinforcement learning. in *2020 42nd annual international conference of the IEEE engineering in medicine & biology society (EMBC)* (IEEE, 2020), pp. 5406–5409
162. J. Fan, J. Wang, Z. Chen, C. Hu, Z. Zhang, W. Hu, Automatic treatment planning based on three-dimensional dose distribution predicted from deep learning technique. Med. Phys. **46**(1), 370–381 (2019)
163. D. Nguyen, X. Jia, D. Sher, M.H. Lin, Z. Iqbal, H. Liu, S. Jiang, 3D radiotherapy dose prediction on head and neck cancer patients with a hierarchically densely connected U-net deep learning architecture. Phys. Med. Biol. **64**(6), 065020 (2019)
164. B. Chen, L. Zhang, H. Chen, K. Liang, X. Chen, A novel extended Kalman filter with support vector machine based method for the automatic diagnosis and segmentation of brain tumors. Comput. Methods Programs Biomed. **200**, 105797 (2021)
165. A. Subudhi, M. Dash, S. Sabut, Automated segmentation and classification of brain stroke using expectation-maximization and random forest classifier. Biocybern. Biomed. Eng. **40**(1), 277–289 (2020)
166. M. Torres-Velázquez, W.J. Chen, X. Li, A.B. McMillan, Application and construction of deep learning networks in medical imaging. IEEE Trans. Radiat. Plasma Med. Sci. **5**(2), 137–159 (2020)
167. D. Wang, Y. Zhang, K. Zhang, L. Wang, Focalmix: Semi-supervised learning for 3d medical image detection. in *Proceedings of the IEEE/CVF conference on computer vision and pattern recognition* (2020), pp. 3951–3960
168. G. Dhiman, S. Juneja, W. Viriyasitavat, H. Mohafez, M. Hadizadeh, M.A. Islam, K. Gulati, A novel machine-learning-based hybrid CNN model for tumor identification in medical image processing. Sustainability **14**(3), 1447 (2022)
169. R. Ashraf, M.A. Habib, M. Akram, M.A. Latif, M.S.A. Malik, M. Awais, Z. Abbas, Deep convolution neural network for big data medical image classification. IEEE Access **8**, 105659–105670 (2020)
170. L. Liu, A.I. Avilés-Rivero, C.B. Schönlieb, Contrastive registration for unsupervised medical image segmentation. IEEE Trans. Neural Netw. Learn. Syst., (2023)

171. Z. Rguibi, A. Hajami, D. Zitouni, A. Elqaraoui, A. Bedraoui, Cxai: Explaining convolutional neural networks for medical imaging diagnostic. Electronics **11**(11), 1775 (2022)
172. E.Z. Chen, C. Zhang, X. Chen, Y. Liu, T. Chen, S. Sun, Computationally efficient 3D MRI reconstruction with adaptive MLP. in *International conference on medical image computing and computer-assisted intervention* (Springer Nature Switzerland, 2020), pp. 195–205
173. C. Keleşoğlu, H. KŘšŘk, M. DemetgŘl, Fault diagnosis of bevel gears using neural pattern recognition and MLP neural network algorithms. Int. J. Precis. Eng. Manuf., **21**, 843–856 (2020)
174. V.S.S. Kandarpa, Tomographic image reconstruction with direct neural network approaches. Doctoral dissertation, Université de Bretagne occidentale-Brest, 2020
175. M.H. Asnawi, A.A. Pravitasari, G. Darmawan, T. Hendrawati, I.N. Yulita, J. Suprijadi, F.A.L. Nugraha, Lung and infection CT-scan-based segmentation with 3D UNet architecture and its modification. in *Healthcare*, vol. 11, no. 2. (MDPI, 2023), p. 213
176. A. Aghabiglou, E.M. Eksioglu, Projection-based cascaded U-net model for MR image reconstruction. Comput. Methods Programs Biomed. **207**, 106151 (2021)
177. Y. Han, J.C. Ye, Framing U-Net via deep convolutional framelets: Application to sparse-view CT. IEEE Trans. Med. Imaging **37**(6), 1418–1429 (2018)
178. S. Kaviani, A. Sanaat, M. Mokri, C. Cohalan, J.F. Carrier, Image reconstruction using UNET-transformer network for fast and low-dose PET scans. Comput. Med. Imaging Graph. **110**, 102315 (2023)
179. L.M. Hsu, S. Wang, L. Walton, T.W.W. Wang, S.H. Lee, Y.Y.I. Shih, 3D U-net improves automatic brain extraction for isotropic rat brain magnetic resonance imaging data. Front. Neurosci. **15**, 801008 (2021)
180. T. Lu, X. Zhang, Y. Huang, D. Guo, F. Huang, Q. Xu, X. Qu, PFISTA-SENSE-ResNet for parallel MRI reconstruction. J. Magn. Reson. **318**, 106790 (2020)
181. S. Rezayi, N. Mohammadzadeh, H. Bouraghi, S. Saeedi, A. Mohammadpour, Timely diagnosis of acute lymphoblastic leukemia using artificial intelligence-oriented deep learning methods. Comput. Intell. Neurosci., (2021)
182. Y. Ma, J. Liu, Y. Liu, H. Fu, Y. Hu, J. Cheng, Y. Zhao, Structure and illumination constrained GAN for medical image enhancement. IEEE Trans. Med. Imaging **40**(12), 3955–3967 (2021)
183. L. Zhou, J.D. Schaefferkoetter, I.W. Tham, G. Huang, J. Yan, Supervised learning with cyclegan for low-dose FDG PET image denoising. Med. Image Anal. **65**, 101770 (2020)
184. K. Zhao, L. Zhou, S. Gao, X. Wang, Y. Wang, X. Zhao, H. Ye, Study of low-dose PET image recovery using supervised learning with CycleGAN. PLoS ONE **15**(9), e0238455 (2020)
185. B. Zhu, J.Z. Liu, S.F. Cauley, B.R. Rosen, M.S. Rosen, Image reconstruction by domain-transform manifold learning. Nature **555**(7697), 487–492 (2018)
186. M. Zerunian, F. Pucciarelli, D. Caruso, M. Polici, B. Masci, G. Guido, A. Laghi, Artificial intelligence based image quality enhancement in liver MRI: A quantitative and qualitative evaluation. Radiol. Med. (Torino) **127**(10), 1098–1105 (2022)
187. S. Wang, H. Cheng, L. Ying, T. Xiao, Z. Ke, H. Zheng, D. Liang, DeepcomplexMRI: Exploiting deep residual network for fast parallel MR imaging with complex convolution. Magn. Reson. Imaging **68**, 136–147 (2020)
188. G. Yang, S. Yu, H. Dong et al., DAGAN: Deep de-aliasing generative adversarial networks for fast compressed sensing MRI reconstruction. IEEE Trans. Med. Imaging **37**(6), 1310–1321 (2018)
189. A.M. Rauschecker, T.J. Gleason, P. Nedelec, M.T. Duong, D.A. Weiss, E. Calabrese, C.P. Hess, Interinstitutional portability of a deep learning brain MRI lesion segmentation algorithm. Radiol.: Artif. Intell., **4**(1), e200152 (2021)
190. H. Taheri Gorji, N. Kaabouch, A deep learning approach for diagnosis of mild cognitive impairment based on MRI images. Brain Sci. **9**(9), 217 (2019)
191. A. Farooq, S. Anwar, M. Awais, M. Alnowami, Artificial intelligence based smart diagnosis of Alzheimer's disease and mild cognitive impairment. in *2017 International Smart cities conference (ISC2)* (IEEE, 2017), pp. 1–4

192. X. Bi, W. Liu, H. Liu, Q. Shang, Artificial Intelligence-based MRI images for brain in prediction of Alzheimer's disease. J. Healthc. Eng., (2021)
193. U. Ferizi, H. Besser, P. Hysi, J. Jacobs, C.S. Rajapakse, C. Chen, G. Chang, Artificial intelligence applied to osteoporosis: a performance comparison of machine learning algorithms in predicting fragility fractures from MRI data. J. Magn. Reson. Imaging **49**(4), 1029–1038 (2019)
194. F. Wahid, R. Ghazali, M. Fayaz, A.S. Shah, Using probabilistic classification technique and statistical features for brain magnetic resonance imaging (MRI) classification: an application of AI technique in bio-science. Int. J. Bio-Sci. Bio-Technol **8**(6), 93–106 (2017)
195. K. Hammernik, T. Klatzer, E. Kobler, M.P. Recht, D.K. Sodickson, T. Pock, F. Knoll, Learning a variational network for reconstruction of accelerated MRI data. Magn. Reson. Med. **79**(6), 3055–3071 (2018)
196. S. Wang, Z. Ke, H. Cheng, S. Jia, L. Ying, H. Zheng, D. Liang, DIMENSION: dynamic MR imaging with both k-space and spatial prior knowledge obtained via multi-supervised network training. NMR Biomed. **35**(4), e4131 (2022)
197. Q. Huang, Y. Xian, D. Yang, H. Qu, J. Yi, P. Wu, D.N. Metaxas, Dynamic MRI reconstruction with end-to-end motion-guided network. Med. Image Anal. **68**, 101901 (2021)
198. Y.J. Park, H.S. Cho, M.N. Kim, AI model for detection of abdominal hemorrhage lesions in abdominal CT images. Bioengineering **10**(4), 502 (2023)
199. F. Tatsugami, T. Higaki, Y. Nakamura, Z. Yu, J. Zhou, Y. Lu, K. Awai, Deep learning–based image restoration algorithm for coronary CT angiography. Eur. Radiol. **29**, 5322–5329 (2019)
200. D. Wu, K. Kim, G. El Fakhri, Q. Li, Iterative low-dose CT reconstruction with priors trained by artificial neural network. IEEE Trans. Med. Imaging **36**(12), 2479–2486 (2017)
201. J.M. Wolterink, T. Leiner, M.A. Viergever, I. Išgum, Generative adversarial networks for noise reduction in low-dose CT. IEEE Trans. Med. Imaging **36**(12), 2536–2545 (2017)
202. Q. Yang, P. Yan, Y. Zhang, H. Yu, Y. Shi, X. Mou, G. Wang, Low-dose CT image denoising using a generative adversarial network with Wasserstein distance and perceptual loss. IEEE Trans. Med. Imaging **37**(6), 1348–1357 (2018)
203. E. Kang, W. Chang, J. Yoo, J.C. Ye, Deep convolutional framelet denosing for low-dose CT via wavelet residual network. IEEE Trans. Med. Imaging **37**(6), 1358–1369 (2018)
204. M. Kallenberg, K. Petersen, M. Nielsen, A.Y. Ng, P. Diao, C. Igel, M. Lillholm, Unsupervised deep learning applied to breast density segmentation and mammographic risk scoring. IEEE Trans. Med. Imaging **35**(5), 1322–1331 (2016)
205. M. Nishizawa, H. Tanaka, Y. Watanabe, Y. Kunitomi, A. Tsukabe, N. Tomiyama, Model-based iterative reconstruction for detection of subtle hypoattenuation in early cerebral infarction: a phantom study. Jpn. J. Radiol. **33**, 26–32 (2015)
206. M. Alilou, M. Orooji, N. Beig et al., Quantitative vessel tortuosity: A potential CT imaging biomarker for distinguishing lung granulomas from adenocarcinomas. Sci. Rep. **8**, 15290 (2018). https://doi.org/10.1038/s41598-018-33473-0
207. P. Zhou, S. Jiang, L. Yu, Y. Feng, C. Chen, F. Li, Z. Huang, Use of a sparse-response deep belief network and extreme learning machine to discriminate Alzheimer's disease, mild cognitive impairment, and normal controls based on amyloid PET/MRI images. Front. Med. **7**, 621204 (2021)
208. Z. Zhang, X. Liang, X. Dong, Y. Xie, G. Cao, A sparse-view CT reconstruction method based on combination of DenseNet and deconvolution. IEEE Trans. Med. Imaging **37**(6), 1407–1417 (2018)
209. J. Park, D. Hwang, K.Y. Kim, S.K. Kang, Y.K. Kim, J.S. Lee, Computed tomography super-resolution using deep convolutional neural network. Phys. Med. Biol. **63**(14), 145011 (2018)
210. C.L. *Ensemble, CT Super-resolution GAN constrained by the identical, residual, and cycle learning ensemble*
211. M.Z. Zhang, S. Yu, W. Qin, X. Liang, Y. Xie, G. Cao, Self-supervised CT super-resolution with hybrid model. Comput. Biol. Med., **138**, 104775 (2021)
212. R.F. Mansour, J. Escorcia-Gutierrez, M. Gamarra, V.G. Díaz, D. Gupta, S. Kumar, Artificial intelligence with big data analytics-based brain intracranial hemorrhage e-diagnosis using CT images. Neural Comput. Appl. **35**(22), 16037–16049 (2023)

213. I. Haggstrom, C.R. Schmidtlein, G. Campanella, T.J. Fuchs, DeepPET: a deep encoder-decoder network for directly solving the PET image reconstruction inverse problem. Med. Image Anal. **54**, 253–262 (2019)
214. Q. Feng, H. Liu, Rethinking PET image reconstruction: ultra-low-dose, sinogram and deep learning. in *Medical Image Computing and Computer Assisted Intervention–MICCAI 2020: 23rd International Conference, Lima, Peru, October 4–8, 2020, Proceedings, Part VII 23* (Springer International Publishing, 2020), pp. 783–792
215. Z. Liu, H. Chen, H. Liu, Deep learning based framework for direct reconstruction of PET images. in *medical image computing and computer assisted intervention–MICCAI 2019: 22nd international conference, Shenzhen, China, October 13–17, 2019, Proceedings, Part III 22* (Springer International Publishing, 2019), pp. 48–56
216. F. Liu, H. Jang, R. Kijowski, G. Zhao, T. Bradshaw, A.B. McMillan, A deep learning approach for (18)F-FDG PET attenuation correction. Eur J Nucl Med Mol Imaging Phys **5**(1), 24 (2018)
217. E. Glaab, J.P. Trezzi, A. Greuel, C. Jäger, Z. Hodak, A. Drzezga, L. Timmermann, M. Tittgemeyer, N.J. Diederich, C. Eggers, Integrative analysis of blood metabolomics and PET brain neuroimaging data for Parkinson's disease. Neurobiol. Dis. **124**, 555–562 (2019)
218. H. Choi, Y.K. Kim, E.J. Yoon, J.Y. Lee, D.S. Lee, Alzheimer's disease neuroimaging initiative. Cognitive signature of brain FDG PET based on deep learning: domain transfer from Alzheimer's disease to Parkinson's disease. Eur. J. Nucl. Med Mol. Imaging. **47**(2), 403–412 (2020)
219. Z. Xie, T. Li, X. Zhang, W. Qi, E. Asma, J. Qi, Anatomically aided PET image reconstruction using deep neural networks. Med. Phys. **48**(9), 5244–5258 (2021)
220. K. Gong, J. Guan, K. Kim, X. Zhang, J. Yang, Y. Seo, Q. Li, Iterative PET image reconstruction using convolutional neural network representation. IEEE Trans. Med. Imaging **38**(3), 675–685 (2018)
221. S.H. Hyun, M.S. Ahn, Y.W. Koh et al., A machine-learning approach using PET-based radiomics to predict the histological subtypes of lung cancer. Clin. Nucl. Med. **44**, 956–960 (2019)
222. L. Xiang, L. Wang, E. Gong, G. Zaharchuk, T. Zhang, Noise-aware standard-dose PET reconstruction using general and adaptive robust loss. in *Machine learning in medical imaging: 11th International Workshop, MLMI 2020, Held in Conjunction with MICCAI 2020, Lima, Peru, October 4, 2020, Proceedings* 11 (Springer International Publishing, 2020), pp. 654–662
223. K. Spuhler, M. Serrano-Sosa, R. Cattell, C. DeLorenzo, C. Huang, Full-count PET recovery from low-count image using a dilated convolutional neural network. Med. Phys. **47**(10), 4928–4938 (2020)
224. Y. Kimura, A. Watanabe, T. Yamada, S. Watanabe, T. Nagaoka, M. Nemoto et al., AI approach of cycle-consistent generative adversarial networks to synthesize PET images to train computer-aided diagnosis algorithm for dementia. Ann. Nucl. Med. **34**(7), 512–515 (2020)
225. J. Guo, E. Gong, A.P. Fan, M. Goubran, M.M. Khalighi, G. Zaharchuk, Predicting (15)O-Water PET cerebral blood flow maps from multi-contrast MRI using a deep convolutional neural network with evaluation of training cohort bias. J. Cereb. Blood Flow Metab. **40**(11), 2240–2253 (2020)
226. J. Shi, S. Zhou, X. Liu, Q. Zhang, M. Lu, T. Wang, Stacked deep polynomial network based representation learning for tumor classification with small ultrasound image dataset. Neurocomputing **194**, 87–94 (2016)
227. J. Virmani, V. Kumar, N. Kalra, N. Khandelwal, Characterization of primary and secondary malignant liver lesions from B-mode ultrasound. J. Digit. Imaging **26**, 1058–1070 (2013)
228. X. Chen, M. He, T. Dan, N. Wang, M. Lin, L. Zhang, H. Xie, Automatic measurements of fetal lateral ventricles in 2D ultrasound images using deep learning. Front. Neurol. **11**, 526 (2020)
229. Y. Gong, Y. Zhang, H. Zhu, J. Lv, Q. Cheng, H. Zhang, S. Wang, Fetal congenital heart disease echocardiogram screening based on DGACNN: adversarial one-class classification combined with video transfer learning. IEEE Trans. Med. Imaging **39**(4), 1206–1222 (2019)

230. I.D. Pluym, Y. Afshar, K. Holliman, L. Kwan, A. Bolagani, T. Mok, L.D. Platt, Accuracy of automated three-dimensional ultrasound imaging technique for fetal head biometry. Ultrasound Obstet. Gynecol. **57**(5), 798–803 (2021)
231. Q. Guan, Y. Wang, J. Du, Y. Qin, H. Lu, J. Xiang, F. Wang, Deep learning based classification of ultrasound images for thyroid nodules: a large scale of pilot study. Ann. Transl. Med., **7**(7), (2019)
232. Q. Liu, Y. Li, Z.X. Zhai, H.Y. Jia, L.P. Liu, An improved method for thyroid nodule ultrasound image segmentation based on U2-Net. Int. J. Imaging Syst. Technol. **33**(6), 2118–2127 (2023)
233. J. Ma, F. Wu, T.A. Jiang, Q. Zhao, D. Kong, Ultrasound image-based thyroid nodule automatic segmentation using convolutional neural networks. Int. J. Comput. Assist. Radiol. Surg. **12**, 1895–1910 (2017)
234. T. Zheng, H. Qin, Y. Cui, R. Wang, W. Zhao, S. Zhang, L. Zhao, Segmentation of thyroid glands and nodules in ultrasound images using the improved U-Net architecture. BMC Med. Imaging **23**(1), 56 (2023)
235. Y. Wu, B. Chen, K. Liu, C. Zhu, H. Pan, J. Jia et al., Shape reconstruction with multiphase conductivity for electrical impedance tomography using improved convolutional neural network method. IEEE Sens. J. **21**(7), 9277–9287 (2021)
236. Z. Chen, Y. Yang, J. Jia, P. Bagnaninchi, Deep learning based cell imaging with electrical impedance tomography, in *2020 IEEE International Instrumentation and Measurement Technology Conference (I2MTC), Dubrovnik, Croatia, 25–28 May 2020* (2020b)
237. D. Yang, S. Li, Y. Zhao, B. Xu, W. Tian, An EIT image reconstruction method based on DenseNet with multi-scale convolution. Math. Biosci. Eng. **20**(4), 7633–7660 (2023)
238. Y.F. Ko, K.S. Cheng, Semi-siamese U-net for separation of lung and heart bioimpedance images: A simulation study of thorax EIT. PLoS ONE **16**(2), e0246071 (2021)
239. J. Chen, S. Wang, K. Wang, P. Abiri, Z.Y. Huang, J. Yin, T.K. Hsiai, Machine learning-directed electrical impedance tomography to predict metabolically vulnerable plaques. Bioeng. & Transl. Med. **9**(1), e10616 (2024)
240. K. Kose, A. Bozkurt, C. Alessi-Fox, M. Gill, C. Longo, G. Pellacani, M. Rajadhyaksha, Segmentation of cellular patterns in confocal images of melanocytic lesions in vivo via a multiscale encoder-decoder network (MED-Net). Med. Image Anal. **67**, 101841 (2021)
241. Y. Sun, S. You, X. Du, A. Spaulding, Z.G. Liu, E.J. Chaney, S.A. Boppart, Real-time three-dimensional histology-like imaging by label-free nonlinear optical microscopy. Quant. Imaging Med. Surg. **10**(11), 2177 (2020)
242. S. You, Y. Sun, L. Yang, J. Park, H. Tu, M. Marjanovic, S.A. Boppart, Real-time intraoperative diagnosis by deep neural network driven multiphoton virtual histology. NPJ Precis. Oncol. **3**(1), 33 (2019)
243. A. Bilodeau, C.V. Delmas, M. Parent, P. De Koninck, A. Durand, F. Lavoie-Cardinal, Microscopy analysis neural network to solve detection, enumeration and segmentation from image-level annotations. Nat. Mach. Intell. **4**(5), 455–466 (2022)
244. M.E. Dunham, K.A. Kong, A.J. McWhorter, L.K. Adkins, Optical biopsy: automated classification of airway endoscopic findings using a convolutional neural network. Laryngoscope **132**, S1–S8 (2022)
245. S. Sharma, S. Gupta, D. Gupta, A. Altameem, A.K.J. Saudagar, R.C. Poonia, S.R. Nayak, HTLML: Hybrid AI based model for detection of Alzheimer's disease. Diagnostics **12**(8), 1833 (2022)
246. P. Borrelli, J. Ly, R. Kaboteh, J. Ulén, O. Enqvist, E. Trägårdh, L. Edenbrandt, AI-based detection of lung lesions in [18F] FDG PET-CT from lung cancer patients. EJNMMI Phy **8**(1), 1–11 (2021)
247. J. Wehrend et al., Automated liver lesion detection in (68)Ga DOTATATE PET/CT using a deep fully convolutional neural network. EJNMMI Res. **11**(01), 98 (2021)
248. K. Gong, J. Yang, K. Kim, G. El Fakhri, Y. Seo, Q. Li, Attenuation correction for brain PET imaging using deep neural network based on Dixon and ZTE MR images. Phys. Med. Biol. **63**(12), 125011 (2018)

249. P. Blanc-Durand, M. Khalife, B. Sgard et al., Attenuation correction using 3D deep convolutional neural network for brain 18F-FDG PET/MR: Comparison with atlas, ZTE and CT based attenuation correction. PLoS ONE **14**(10), e0223141 (2019)
250. A. Ben-Cohen, E. Klang, S.P. Raskin, S. Soffer, S. Ben-Haim, E. Konen et al., Cross-modality synthesis from CT to PET using FCN and GAN networks for improved automated lesion detection. Eng. Appl. Artif. Intell. **78**, 186–194 (2019)
251. C.N. Ladefoged, D. Benoit, I. Law et al., Region specific optimization of continuous linear attenuation coefficients based on UTE (RESOLUTE): Application to PET/MR brain imaging. Phys. Med. Biol. **60**(20), 8047–8065 (2015)
252. A.J. Ramon, Y. Yang, P.H. Pretorius, K.L. Johnson, M.A. King, M.N. Wernick, Initial investigation of low-dose SPECT-MPI via deep learning. in *2018 IEEE nuclear science symposium and medical imaging conference proceedings (NSS/MIC)* (IEEE, 2018), pp. 1–3
253. L. Sibille, R. Seifert, N. Avramovic, T. Vehren, B. Spottiswoode, S. Zuehlsdorff, M. Schäfers, 18F-FDG PET/CT uptake classification in lymphoma and lung cancer by using deep convolutional neural networks. Radiology **294**(2), 445–452 (2020)
254. Y.H. Qu et al., The correlation of (18)F-FDG PET/CT metabolic parameters, clinicopathological factors, and prognosis in breast cancer. Clin. Transl. Oncol. **23**(03), 620–627 (2021)
255. F. Liu, H. Jang, R. Kijowski, T. Bradshaw, A.B. McMillan, Deep learning MR imaging-based attenuation correction for PET/MR imaging. Radiology **286**(2), 676–684 (2018)
256. M. Nakajo, M. Jinguji, A. Tani, E. Yano, C.K. Hoo, D. Hirahara, T. Yoshiura, Machine learning based evaluation of clinical and pretreatment 18 F-FDG-PET/CT radiomic features to predict prognosis of cervical cancer patients. Abdom. Radiol., 1–10 (2022)
257. B. Xiao, N. He, Q. Wang, F. Shi, Z. Cheng, E.M. Haacke, D. Shen, Stability of AI-enabled diagnosis of Parkinson's disease: A study targeting substantia nigra in quantitative susceptibility mapping imaging. Front. Neurosci. **15**, 760975 (2021)
258. S. Badoud, D. Van De Ville, N. Nicastro, V. Garibotto, P.R. Burkhard, S. Haller, Discriminating among degenerative Parkinsonisms using advanced (123)I-ioflupane SPECT analyses. Neuroimage Clin **12**, 234–240 (2016)
259. G.L. Monroy, J. Won, R. Dsouza, P. Pande, M.C. Hill, R.G. Porter, S.A. Boppart, Automated classification platform for the identification of otitis media using optical coherence tomography. NPJ Digital Med **2**(1), 22 (2019)
260. Z.Y.G. Ko, Y. Li, J. Liu, H. Ji, A. Qiu, N. Chen, DOTnet 2.0: Deep learning network for diffuse optical tomography image reconstruction. Intell.-Based Med., 100133 (2023)
261. A. Selvikvag Lundervold, A. Lundervold, An overview of deep learning in medical imaging focusing on MRI. Z. Med. Phys. **29**(2), 102–127 (2018). https://doi.org/10.1016/j.zemedi.2018.11.002
262. D. Liang, J. Cheng, Z. Ke, L. Ying, Deep magnetic resonance image reconstruction: Inverse problems meet neural networks. IEEE Signal Process. Mag. **37**(1), 141–151 (2020). https://doi.org/10.1109/MSP.2019.2950557
263. F. Knoll, K. Hammernik, C. Zhang, S. Moeller, T. Pock, D.K. Sodickson, M. Akcakaya, Deep-learning methods for parallel magnetic resonance imaging reconstruction: a survey of the current approaches, trends, and issues. IEEE Signal Process. Mag. **37**(1), 128–140 (2020). https://doi.org/10.1109/MSP.2019.2950640
264. H.K. Aggarwal, M.P. Mani, M. Jacob, MoDL: Model-based deep learning architecture for inverse problems. IEEE Trans. Med. Imaging **38**(2), 394–405 (2018)
265. A.D. Gilmore, N.J. Buser, J.L. Hanson, Variations in structural MRI quality significantly impact commonly used measures of brain anatomy. Brain Inform. **8**(1), 7 (2021). https://doi.org/10.1186/s40708-021-00128-2
266. C. Shorten, T.M. Khoshgoftaar, A survey on image data augmentation for deep learning. J Big Data. **6**(1), 60 (2019). https://doi.org/10.1186/s40537-019-0197-0
267. Y. Nakamura, K. Narita, T. Higaki, M. Akagi, Y. Honda, K. Awai, Diagnostic value of deep learning reconstruction for radiation dose reduction at abdominal ultra-high-resolution CT. Eur. Radiol. **31**, 4700–4709 (2021)

268. L.B. Van den Oever, M. Vonder, M. Van Assen, P.M.A. van Ooijen, G.H. de Bock, X.Q. Xie, R. Vliegenthart, Application of artificial intelligence in cardiac CT: From basics to clinical practice. Eur. J. Radiol. **128**, 108969 (2020)
269. Y. Kuo, Y.Y. Lin, R.C. Lee, C.J. Lin, Y.Y. Chiou, W.Y. Guo, Comparison of image quality from filtered back projection, statistical iterative reconstruction, and model-based iterative reconstruction algorithms in abdominal computed tomography. Medicine, **95**(31), (2016)
270. S. Jadon, O.P. Leary, I. Pan, T.J. Harder, D.W. Wright, L.H. Merck, D.L. Merck, A comparative study of 2D image segmentation algorithms for traumatic brain lesions using CT data from the ProTECTIII multicenter clinical trial. in *Medical imaging 2020: Imaging informatics for healthcare, research, and applications*, vol. 11318 (SPIE, 2020), pp. 195–203
271. W. Kuo, C. Häne, E. Yuh, P. Mukherjee, J. Malik, *PatchFCN for intracranial hemorrhage detection* (2018). arXiv preprint arXiv:1806.03265
272. Y. Guo, Y. Wang, H. Yang, J. Zhang, Q. Sun, Dual-attention EfficientNet based on multi-view feature fusion for cervical squamous intraepithelial lesions diagnosis. Biocybern. Biomed. Eng. **42**(2), 529–542 (2022)
273. C. Piçarra, S. Winzeck, M. Monteiro, F. Mathieu, V.F. Newcombe, D.K. Menon, B. Glocker, Automatic localisation and per-region quantification of traumatic brain injury on head CT using atlas mapping. Eur. J. Radiol. Open **10**, 100491 (2023)
274. S. Jain, T.V. Vyvere, V. Terzopoulos, D.M. Sima, E. Roura, A. Maas, J. Verheyden, Automatic quantification of computed tomography features in acute traumatic brain injury. J. Neurotrauma **36**(11), 1794–1803 (2019)
275. H. Yao, C. Williamson, J. Gryak, K. Najarian, Automated hematoma segmentation and outcome prediction for patients with traumatic brain injury. Artif. Intell. Med. **107**, 101910 (2020)
276. S. Mathotaarachchi, T.A. Pascoal, M. Shin, A.L. Benedet, M.S. Kang, T. Beaudry, V.S. Fonov, S. Gauthier, P. Rosa-Neto, A.D.N. Initiative, Identifying incipient dementia individuals using machine learning and amyloid imaging. Neurobiol. Aging **59**, 80–90 (2017)
277. M.O. Thompson, *Ultrasound in labour and delivery. Contemporary obstetrics and gynecology for developing countries* (2021), pp. 77–89
278. C. Yang, S. Liao, Z. Yang, J. Guo, Z. Zhang, Y. Yang, Y. Kang, RDHCformer: Fusing ResDCN and transformers for fetal head circumference automatic measurement in 2D ultrasound images. Front. Med. **9**, 848904 (2022)
279. G. Muscogiuri, V. Volpato, R. Cau, M. Chiesa, L. Saba, M. Guglielmo, S. Sironi, Application of AI in cardiovascular multimodality imaging. Heliyon, **8**(10), (2022)
280. S. Xiao, J. Zhang, Y. Zhu, Z. Zhang, H. Cao, M. Xie, L. Zhang, Application and progress of artificial intelligence in fetal ultrasound. J. Clin. Med. **12**(9), 3298 (2023)
281. A. Paderno, F. Gennarini, A. Sordi, C. Montenegro, D. Lancini, F.P. Villani, C. Piazza, Artificial intelligence in clinical endoscopy: Insights in the field of videomics. Front. Surg. **9**, 933297 (2022)
282. A.Z. Gimeno-García, A. Hernández-Pérez, D. Nicolás-Pérez, M. Hernández-Guerra, Artificial intelligence applied to colonoscopy: is it time to take a step forward? Cancers **15**(8), 2193 (2023)
283. Z. Jin, F. Zhang, Y. Wang, A. Tian, J. Zhang, M. Chen, J. Yu, Single-photon emission computed tomography/computed tomography image-based radiomics for discriminating vertebral bone metastases from benign bone lesions in patients with tumors. Front. Med. **8**, 792581 (2022)
284. O. Adelaja, H. Alkattan, Operating artificial intelligence to assist physicians diagnose medical images: A narrative review. Mesop. J. Artif. Intell. Healthc. **2023**, 45–51 (2023)
285. H. Higham, C. Vincent, *Human error and patient safety. Textbook of patient safety and clinical risk management* (2021), pp. 29–44
286. T. Araujo, N. Helberger, S. Kruikemeier, C.H. De Vreese, In AI we trust? Perceptions about automated decision-making by artificial intelligence. AI & Soc. **35**, 611–623 (2020)
287. G.A. Kaissis, M.R. Makowski, D. Rückert, R.F. Braren, Secure, privacy-preserving and federated machine learning in medical imaging. Nat. Mach. Intell. **2**(6), 305–311 (2020)

288. HIPAA Home [online]. Available https://www.hhs.gov/hipaa/index.html. Accessed on: 22 Feb 2024
289. General Data Protection Regulation (GDPR)—Official Legal Text [online]. Available https://gdpr-info.eu/. Accessed on 22 Feb 2024
290. A. Rauniyar, D.H. Hagos, D. Jha, J.E. Håkegård, U. Bagci, D.B. Rawat, V. Vlassov, Federated learning for medical applications: A taxonomy, current trends, challenges, and future research directions. IEEE Internet Things J., (2023)
291. P. Tagde, S. Tagde, T. Bhattacharya, P. Tagde, H. Chopra, R. Akter, M.H. Rahman, Blockchain and artificial intelligence technology in e-Health. Environ. Sci. Pollut. Res. **28**, 52810–52831 (2021)
292. E. Ranschaert, L. Topff, O. Pianykh, Optimization of radiology workflow with artificial intelligence. Radiol. Clin. **59**(6), 955–966 (2021)

Chapter 2
From Pixels to Predictions: Exploring the Role of Artificial Intelligence in Radiology

M. J. Akshit Aiyappa and B. Suresh Kumar Shetty

Abstract Among the current generation of radiologists, the terms "artificial intelligence," "deep learning," and "machine learning" are used most often. Artificial intelligence (AI) is quickly making headway in the field of radiology. AI, when used effectively, can be beneficial in the education of future doctors. Technological developments frequently cause changes in the field of radiology. Therefore, a new storage device, image distribution system, or scanner won't be the next "game-changer" in radiology; rather, it will be a new technology that will enhance the interpretation of image data. With speech recognition software and image archiving and communication software (PACS), machine learning applications are already in use. A great many radiologists are concerned that extensive study and advancements in AI could result in the extinction of their field. There have been stories in the past warning that machines using deep learning-based software will soon take over their field. Corroborating evidence suggested that several algorithms could perform numerous tasks far more effectively than the typical radiologist. As previously said, by reason of recent events, radiology trainees feel vulnerable and uncertain about their future in this sector. Understanding the advantages of AI is our role, not stoking these ambiguities. Additionally, its urgent ethical and professional implications should be acknowledged. We will discuss the origins, uses, and present day developments of AI in radiology while keeping in mind the collaborative intelligence between humans and AI. It is crucial for working radiologists to comprehend how AI will develop in our area in the future. It is claimed that AI can improve both patient outcomes and radiologists' quality of life. We will therefore discuss data that demonstrates how radiologists and AI may complement each other's abilities, particularly in diagnostic imaging.

M. J. Akshit Aiyappa
Department of Radiodiagnosis, AJ Institute of Medical Sciences and Research Centre, Mangalore, India

B. S. K. Shetty (✉)
Department of Forensic Medicine, Kasturba Medical College, Manipal Academy of Higher Education, Mangalore, Manipal, Karnataka, India
e-mail: suresh.shetty@manipal.edu

A. Gogoi and N. Mazumder (eds.), *Biomedical Imaging*, Biological and Medical Physics, Biomedical Engineering, https://doi.org/10.1007/978-981-97-5345-1_2

2.1 Introduction

Modern imaging techniques such as multi-slice computed tomography (CT), multi-parametric magnetic resonance imaging (MRI), multi-dimensional ultrasonography (US), and multi-modal or hybrid PET/CT or PET/MRI have superseded the era of conventional radiography and fluoroscopic treatments [1]. For diagnosis, treatment planning, follow-up, and research, the extraction of pertinent, important, and quantifiable information from medical images obtained using these modalities can be extremely difficult and a serious bottleneck. Manual measurements by specialists, especially radiologists, could be a long-drawn-out, subject to inter-observer or intra-observer variability, thereby undermining the clinical importance of the data derived from the images. In order to enable computer-aided imaging analysis in routine clinical practice, it is necessary to develop more effective, reliable, and corroborative procedures, either automated or semi-automated, for these medical images. It goes without saying that the information, that must be extracted from the medical images, depends on the application. Quantification of specific geometric aspects of the object of interest, such as shape, size, symmetry, volume, extent, location, etc., is the main focus of medical image analysis. It also deals with assessment of any anatomical (such as growth, atrophy, etc.) or morphological changes over time. The availability of large amounts of data provides a much better opportunities for group-wise analysis within a particular subject group, such as normal versus abnormal variations between subjects, early discovery of markers of certain disease onset and progression, ideal treatment selection and prediction of treatment outcome. This computer-aided image analysis is essential for separating out and analysing pertinent information from the imaging data, supplementing it with other data, and afterwards supporting the interpretation. The majority of image analysis issues are a mixture of a few basic tasks due to the wide range of distinctive imaging modalities [2].

Image segmentation constitutes the perception of the borders of the image's objects of interest, such as distinguishing between the image voxels that belong to that specific object and those that belong to other objects [3]. Following image acquisition, image segmentation aids in estimating the object's geometrical characteristics. This can be done in a variety of ways, that includes boundary-wise, which involves identifying the borders of the object in 2D or 3D image slices; voxel-wise, which involves assigning each voxel in the image slice to a specific object or tissue of interest; and region-wise, which involves grouping the voxels that are most likely to belong to the specific object or tissue. Precise 3D segmentation of complex-shaped objects in medical images are complicated due to the limited resolution of images and partial volume artifacts causing detail and contrast distortion. Therefore, interpolation is necessary to fill in the missing information of the image data. Segmentation is commonly accomplished by computerised complex algorithms, that can vary from selecting pixels of similar values to those that are based on machine learning [4].

Image registration determines the structural relationship between different images, based on the image content [5]. Images that are obtained at various time intervals, such as before and after therapy, or using other modalities, such as CT and

MRI, can be combined and studied since they include additional information. To make use of the good resolution of images, the image data can be joined at the same voxel level. Image registration is performed to compensate for any differences, in comparison to the previous data, anatomical variations or to look for any tissue distortion between different time points. After registration, the image data can be grouped in a particular geometric space, or spatial correspondence, and are fused, which can help in the joint analysis of the images. Hybrid imaging needs image registration and helps in generating information by combining anatomical information from various modalities.

Image visualization is termed as the most optimal visual representation of information extracted from the images, to support the diagnosis and for treatment planning. Instead of using 2D multiplanar image visualisation techniques such as surface or volume rendering, true 3D visualisation methodologies are necessary to evaluate 3D medical images. Surface rendering is a 3D segmentation of objects of interest and furnishes them within a given 3D position, by assigning material characteristics to each surface element. Volume rendering renders the image voxels directly by specifying specific transfer functions, which assign each voxel a colour and opacity based on their intensity [6]. Prior segmentation is not necessary for volume rendering, however it is applied such that the transfer functions can be made structurally dependant and object specific, which can allow to differentiate between voxels belonging to different objects. For certain applications like image-guided preoperative navigation, additional tools are required to manipulate objects in that particular 3D location, add or fuse virtual objects to the scene. Such elevated presentation of all the available information is what is needed in arriving at a final diagnosis and for further treatment planning.

Image segmentation, registration and visualization cannot be addressed independently (Fig. 2.1). They are entwined and an optimal image is ready for interpretation when these tasks are performed together. For example, prior segmentation of similar structures that are already available, can benefit image registration as they can provide clues to lead the registration process. Suitable additional manipulation tools can be sorted for better visualization of the information in certain situations like image-guided therapy, intra-operative position of instruments, etc. Despite all the novel imaging techniques, the analysis of medical images is still intricate in terms of data, objects of interest and validation. Medical images are based on several physical principles. The quantification of data is hindered by obscurity brought on by restrictions during the image acquiring process, limited resolution, noise, lack of contrast, and artifacts [7]. The objects of interests in medical images are mostly anatomical structures, that can depict a complex shape, that cannot be easily expressed by a mathematical template. Few anatomical structures can show shape variability, such as bowel peristalsis, breathing motion, that can lead to inter-subject variability. Similar structures may appear on several images, but they may differ significantly in terms of shape and intensity. Assessment of precise accuracy of images is usually impossible in most applications, due to lack of ground truth [8]. In our present clinical practice, the ground truth is laid down by manual analysis performed by a human expert, which once again leads to inter-observer and intra-observer variability.

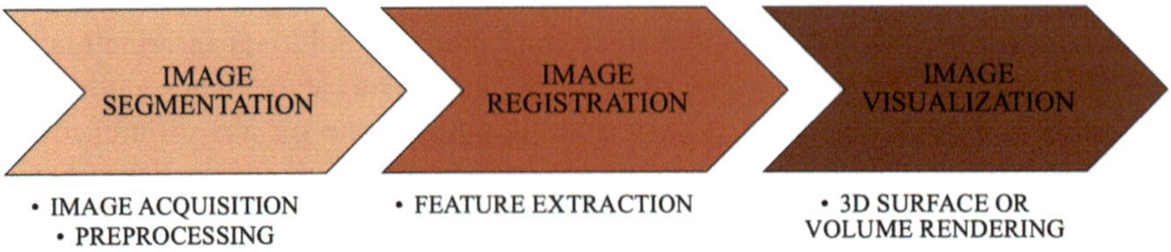

Fig. 2.1 Basic tasks in medical image analysis for interpretation

2.2 Medical Image Processing

Medical image processing or computing is a field of scientific computing that lies at the crossroads of machine learning, computer vision and medical imaging. It is a high-powered tool for disease diagnosis and prognosis. The main aim is to develop computational programs for medical image analysis in a certain way, where these strategies can manage the complexity of the medical imaging data. These programmes mostly use mathematical models that already take into account the past knowledge, about the object of interest in images. The prior knowledge can include texture, contrast, intensity, geometric properties, such as shape, position; and the relation of the object of interest with adjacent structures [(9)]. Model-based imaging processing and analysis includes an accurate specific representation for the model; assessing the integrity of the fit of the model to the data by deriving its objective function; and selecting the perfect optimization strategy for finding the optimal parameters of the model, that best fits the image data (Fig. 2.2). These models need to be adequately flexible to account for various imaging appearances. The flexibility of the model is determined by a particular representation that has been chosen for the model, its configuration and limitations imposed on its parameters. Model-based imaging provided integration of functional and structural information obtained from imaging [10]. The more simpler models are not suitable for medical image computing, in general, as they are not strong enough to capture the integrity of the problem. As a result, more sophisticated methods, that incorporate application-specific information, are needed for image analysis. A more suitable strategy would be to construct an ideal model from the data itself, and these models could ensure that the degrees of freedom of the model are adequately tuned to relevant variations within the data of each particular application. By modifying the correlations in the data, application-specific models can be built with fewer relevant model parameters. More supervised methods can be used to draw conclusions about the relationship between a vector of specific features derived from the data and the intended result, rather than speculating about a particular analytical form for a statistical model. Consequently, different machine learning and deep learning techniques might be applied for a superior result [11].

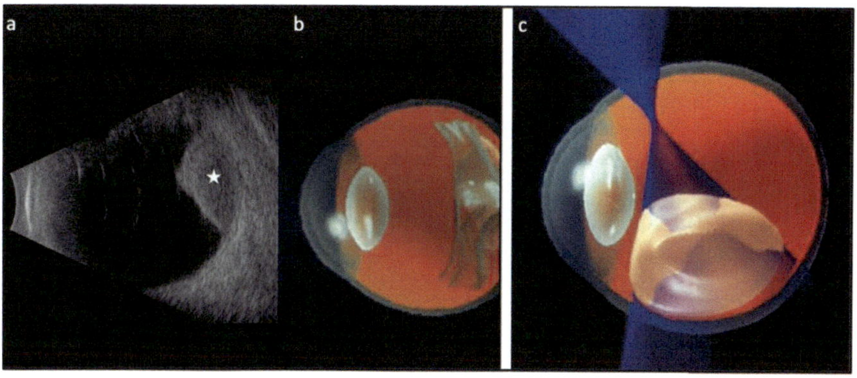

Fig. 2.2 Realistic 3D rendering of the eye using ultrasonic data (**a**) and conceptual model of ocular anatomy **b**. Image **c** shows a 3D reconstruction of the eye with malignant melanoma. This image was used for treatment planning [11]

2.3 Artificial Intelligence

The field of computer science that studies the reconstruction of intelligent behaviour in computers is referred to as artificial intelligence (AI) [12]. However, the definition garners controversy amongst various subject experts. It is a branch that is dedicated to generating algorithms that can puzzle out problems. It symbolises the capacity of machines to mimic human cognitive processes. Artificial intelligence can be further broken down into artificial narrow intelligence, where a computer can perform a specific task better than a human, and artificial general intelligence, where a computer performs more complex amalgams, that are beyond its current capabilities [13]. The Turing test was developed in the year 1950 by a British computer scientist, Alan Turing. It states that a computer passes the test when a human investigator is unable to determine whether the written responses to a series of written questions were generated by a person or a computer [14]. However, an American computer scientist named John McCarthy originated the phrase "artificial intelligence" in 1956, during a summer workshop at Dartmouth College. He received the 1971 Turing award for his contributions towards AI [15]. The number of research articles on the practical applications of AI in radiology has been increasing rapidly during the last several years. Due to AI's potential to advance the discipline, it is evident that its role in radiology is expanding. Many radiologists were worried if AI is going to replace them, but they have failed to recognize that AI can augment radiologist efficiency, increase diagnostic confidence, reduce their workload, spot urgent and difficult cases, know the patient prognosis and therapy planning (Fig. 2.3).

The three basic phases of detection, characterization, and monitoring can be used by AI to redesign the work of a radiologist [16] (Fig. 2.4). *Detection* is the process of identifying the location or boundaries of a specific area on an image which could contain the anomaly or lesion [17]. The current technology that helps in identifying the area of interest is Computer Aided Detection (CAD). CAD points specific areas

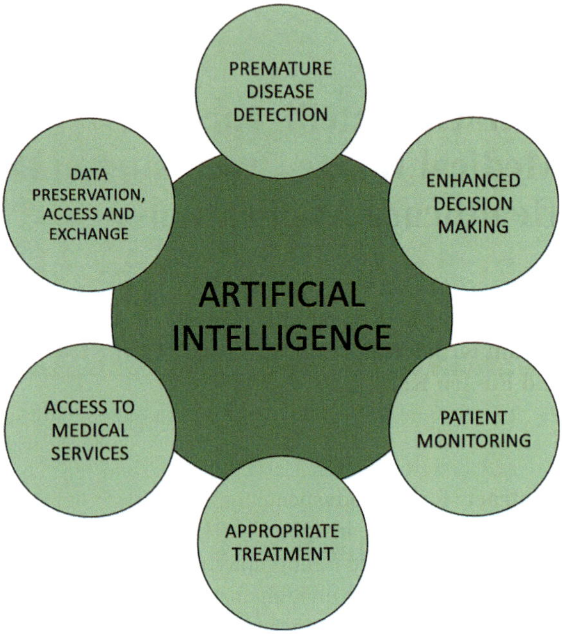

Fig. 2.3 Artificial intelligence in medical imaging

of images which could be abnormal and are made in such a way where the risk of missing pathologies are reduced [18]. Limitations of CAD includes higher rate of false-positives, higher labour requirements and each CAD is task-specific, not for generalised diseases [16]. AI-based CAD systems utilises pattern recognition tools to recognize large volumes of images in an appropriate manner. Any subtle areas, that are suspicious or missed by the radiologist, can be easily identified by AI based detection tools and presented to the reporting personnel [19].

Characterisation defines the particular qualities of the abnormal finding, such as size, shape, extent, internal architecture, etc. These qualities can help in differentiating between benign and malignant diseases. Computer Aided Diagnosis (CADx) is the term that is used in characterising the disease process and helps the radiologist is classifying and assessing the pathologies in medical images [17]. Limitations of CADx includes significant irregularity amongst readers and lack of coordination. For example, since benign and malignant lung nodules often have similar features, it is arduous for readers to identify high-risk nodules. Also, if the radiologist manually draws the border of object of interest, that would omit the subclinical disease as well. AI can take care of these limitations through machine learning or deep learning algorithms. AI-based CADx systems have been proven to predict the tumour response to various therapy options and can also help in deciding the correct management without the need for invasive procedures [16].

Monitoring mainly deals with the follow-ups of the already detected pathologies, assess the changes over time, pre-treatment or post-treatment, using certain

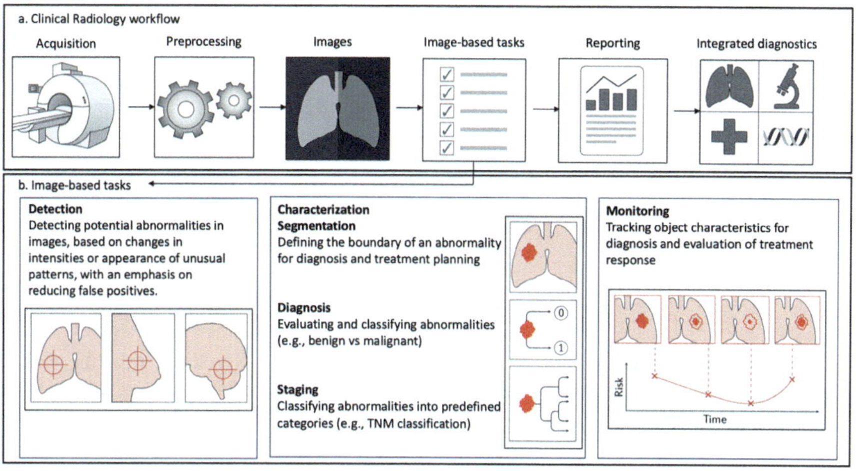

Fig. 2.4 Image depicting the tasks in radiology where AI implementation will have a large impact

protocols. AI-based monitoring systems could augment these protocols by gathering a large number of discriminative data that are unseen by the human expert. AI-based monitoring systems provide a clearer image for any subtle characteristics. Machine learning algorithms have been proven to distinguish between the pathological tumour versus the treatment-related changes [19]. Certain algorithms can help surgeons decide the area of tumour resection and anticipate areas of future recurrence after tumour resection.

2.3.1 *Machine Learning*

Machine learning (ML) is the field of study that enables computers to evolve without being explicitly taught, according to AI pioneer Arthur Samuel [20]. Technically speaking, it is a specific practical application of computer science that enables computers to automatically extract data from patterns that have been identified. This has been used in a number of applications to analyse complex data and find any links or patterns therein [21]. Machine learning algorithms adapt when they are exposed to more data, builds linear and non-linear predictive models that identify and categorise patterns, gets better with practise, and learns to provide precise responses by analysing the data [22] (Fig. 2.5). Machine learning comes in many forms.

If the outcome is unknown, the *unsupervised learning* algorithm finds patterns in the data. In *reinforcement learning*, the algorithm picks up new skills from a string of rewards or penalties. *Supervised learning* compares the accuracy of machine learning to the real world by using a data set that has already been categorised with the ground truth. If only few teacher's output have been provided to the algorithm, that is termed as *semi-supervised learning* [23]. There are various machine learning

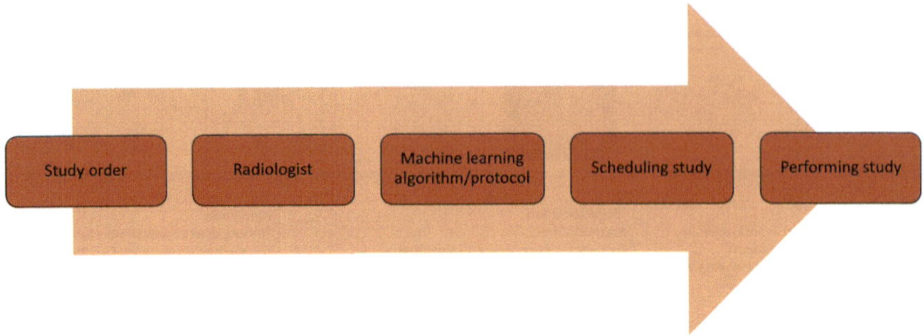

Fig. 2.5 A sample machine learning workflow for developing a study protocol. After the study is ordered, the radiologist creates a protocol for certain studies using machine learning software, and then the study is performed

algorithms, that include artificial neural networks (ANNs), k-nearest neighbours, random forest and support vector machines [24]. Beyond image interpretation, these algorithms have the potential to solve numerous problems in radiology. Some of its applications are in Fig. 2.6. The current machine learning models have narrow AI, which when combined with general human intelligence, can improve the patient management and radiology practices. After image acquisition, the region of interest is identified, relevant characteristics are extracted, and further categorization of the disease process are performed. These algorithms can increase radiologists' output, aid in the interpretation of results, lower the risk of error, and enhance patient care and satisfaction.

Fig. 2.6 Applications of machine learning in radiology

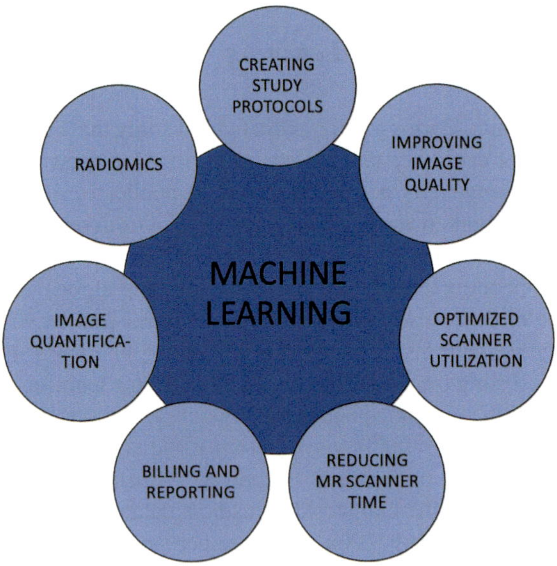

2.3.2 *Deep Learning*

Artificial neural networks (ANNs) are influenced by the parallel architecture of the brain and are built on the performance of biological neural networks. They hold a network of mathematical equations that are used to replicate certain biological processes such as memory and learning [25]. They have *links,* connecting millions of *nodes* or units. These linkages enable activation between units, and each link activation is assigned a numerical value that determines the connection's strength. A *perceptron* is the component that receives the activation and is linked via the connections. Perceptron is the building block for ANNs. It is single-layer neural network, which comprises of four main parameters—input nodes or input layer, which is the primary component that accepts the initial data for further processing; weight parameter, that constitutes the strength of the connection between nodes; activation function, which helps to determine whether the neural network will fire out; and finally the desired output [26]. A feed-forward network, in which connections only run in one direction, or a recurrent network, where the output is fed into its own inputs like a loop, is created by a perceptron. The nodes of the feed-forward network are organized in multiple layers. There are three main layers—input, hidden and output layers (Fig. 2.7). The two main ways on how these feed-forward networks function are, the input units are determined by the application or algorithm, and the information does not loop; it just moves from input to output. As the inputs are processed through these levels, the algorithm is created to increase the proportion of correct responses [27, 28]. The use of deep or multi-layered ANNs is broadly termed as *deep learning.*

Deep learning is a subset of machine learning, often referred to as deep structured learning or deep machine learning, and it serves as the foundation for AI detection technologies that aid in visual analysis and interpretation. The term "deep learning" refers to a particular class of neural networks, ideally suited for use in image processing, where the computer has numerous layers of interconnected algorithms. These layers gather information from the input layers, and once the AI absorbs new characteristics from the information, it produces an output [28]. Deep learning requires deep neural networks that can directly process input data and form algorithms that can attempt to shape high-level formulations from the data [29]. Deep learning algorithms can be of typical networks, where they take one-dimensional inputs, or of convolutional neural networks (CNNs), which receives inputs in two or three dimensions. For medial image analysis, CNNs have gained precedence, especially for feature extraction from the images [30]. CNNs are specialized architectures that are designed for multidimensional array signals. One input layer, several hidden layers, and one output layer make up CNNs (Fig. 2.8). The hidden layers comprise of multiple layers, including convolutional and pooling layers. These layers retrieve the needed information by filtering the input data. Convolution is done by finding patterns or filtering the signals from the input data. Kernels are filter components found in the convolutional layer. The pooling layer lessens the spatial size of the

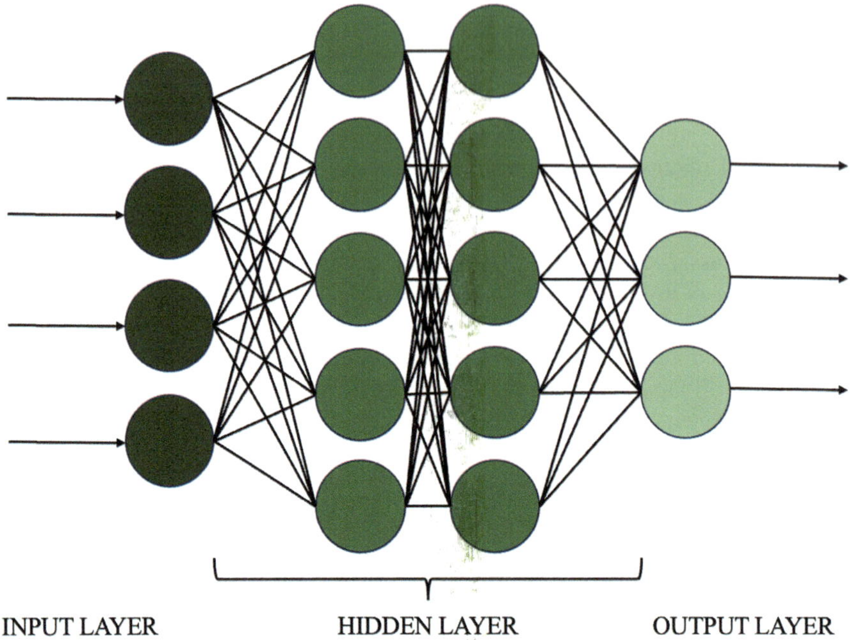

Fig. 2.7 A feed-forward artificial neural network. It is possible for mathematical algorithms to produce outputs by using multiple hidden layers with nodes

illustrative data to reduce the amount of parameters and computation. CNNs in radiology commonly undergo supervised learning. The training process of such deep networks takes a huge amount of time and the weights of the connections need to be modified by a process called backpropagation. Backpropagation is performed to calculate the error function within CNNs and estimates how the parameters should change to optimise work [31]. Once the trained data is ready, the network is verified with the test data and then deployed as the electronic expert, which is a faster process [32].

Deep learning is best suited for complicated data and huge datasets. It would be really helpful to interpret medical images and to arrive at a faster diagnosis. The deep learning algorithms learn on their own, therefore repeated coding is not required. A large number of data should be given into the algorithm, so that it could learn to classify [33].

2.3.3 Development of AI-Based Applications in Radiology

The use and expanding rise of CNNs, which are created specifically to analyse unstructured bidimensional data from the image dataset, is the primary AI turning

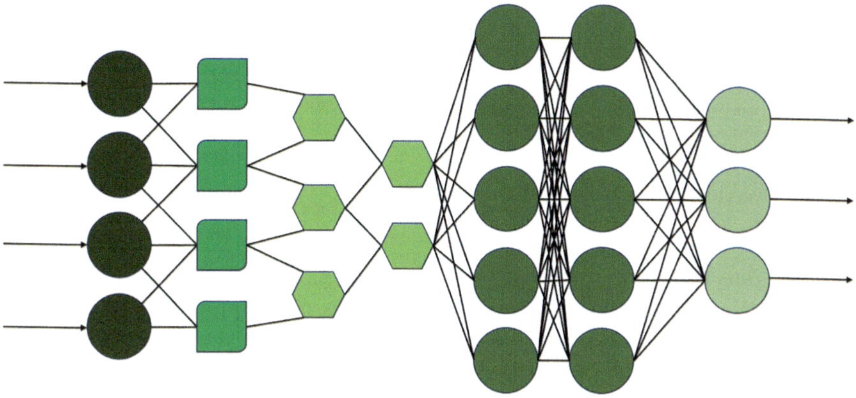

Fig. 2.8 Convolutional neural network, comprising of input, hidden and output layers

point in the discipline of radiology. Convolutional layer in CNNs includes complicated hidden layers that apply filters to the input data, distinguishing them from traditional ANNs. As a result, the network's number of free parameters can be significantly reduced without the need to extract previously hand-crafted features. At the moment, deep learning is the most popular emerging technology [34]. ML-based models go beyond CNNs architecture, as they use different techniques to learn new patterns from the data [35]. ML and DL-based models have proven to have strong problem-solving abilities, display human-like performance, and be accepted as an adjunct to radiologists rather than a replacement. Radiology is a field that, not only is for image recognition, but also contains other contextual information such as patient details, clinical status, previous imaging or surgical findings, etc. The future of AI in radiology hinges on how quickly the technology can be implemented in clinical settings, how much time is saved during image interpretation, how easily complex imaging studies can be analysed, and how many repetitive, time-consuming jobs can be eliminated.

Beyond automated image interpretation, AI applications in radiology span from image acquisition to management to population imaging, to improve daily practice productivity [36] (Fig. 2.9). *Image acquisition* encompasses the development of research techniques, optimisation, and image quality evaluation. The clinical justification for the imaging process determines the study methods. The patient's history and data, particularly earlier imaging data, are significant. AI develops the necessary protocols based on the disease guidelines, differential diagnosis, and also can be synchronized with the machine to assign the scheduling the image acquisition time. For optimal image quality, CT machines require image filters and radiation dose reduction modulations, whereas MRI requires techniques to reduce imaging time and enhance signal homogeneity. With the use of AI, it is possible to automatically extract the signs of good image quality and detect any potential anomalies in scanners.

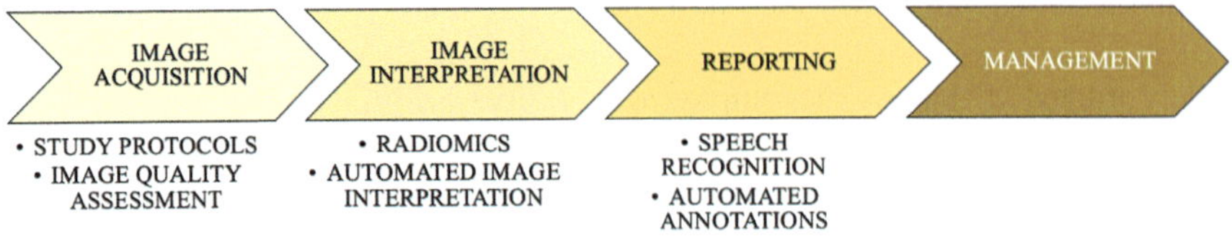

Fig. 2.9 Overview of developing applications in radiology

Image interpretation includes automated hanging protocols, radiomics and automated image interpretation. To prepare the images for interpretation, radiologists invest a lot of time. For quicker interpretation, AI can assist in loading the pertinent images or slices from the imaging dataset onto the picture archiving and communication system (PACS). One of the main steps in the analysis of imaging data is segmentation of lesion, organ or region of interest. Compared to manual performance, segmentations made possible by AI are extremely accurate. In the subject of radiomics, ML approaches are used to process vast amounts of quantitative data that go beyond simple descriptors. To accomplish this, techniques for automated patient "radiomics signature" or "imaging fingerprint" assembly are used, allowing the detection of image-based characterization, that may be connected to diagnosis or treatment. A large amount of data is needed for AI-based applications to perform better when interpreting images. Following image interpretation, reporting is performed with speech recognition, text translation and automated annotation software. Systems for speech recognition rely on dynamic temporal warping and hidden Markov models [37]. AI can help minimizing transcription error rates in reports and provide structured reporting. AI can also allow translation of the reports to various languages. The proper labelling of the imaging examination will be possible, thanks to automated translation of clinical data and radiology reports into keywords from MeSH (Medical Subject Headings) to RadLex (Radiology Lexicon), which will be helpful for future image interpretation. PACS and radiology information systems (RIS) store radiology reports, and natural language processing (NLP) tools can be used to automatically annotate cases with keywords [38]. The European Society of Radiology has developed software platforms that enable clinical data to be annotated, stored, and analysed alongside images [39]. Such applications are bound to help in better management of patients. AI enables improved administration by optimising the use of radiological equipment and scheduling staff and examinations.

Once the clinical indication is evident, AI-based applications must be developed, and ML technology should be determined for image interpretation [40]. After proper design of AI technology, imaging dataset is collected and preserved. Next step is the data annotation, which could pose a problem since PACS systems archives the images and not manage appropriate case annotations [41]. More AI-friendly PACS platforms, that enable the labelling of the tests, could be developed to address this issue. This can be done by using structured reporting and automatically connecting the case and report to the database's predefined fields that hold the findings. Retrospective annotations would also be helpful, to extract information from the pre-existing image datasets from PACS. Training the AI software is also another relevant step. AI models must be trained, using more than 80% of the image collection. The input data and annotations will be provided to the load of the model. Weights and configurations cannot be changed after training is complete. The new unseen images will be used to generate the final model. The next stage takes less time to complete because the test images have already been examined. The training process takes higher performance computing power than the testing phase, which could take days or months together. The required hardware and software must be available to support the specialists' work, in order to create radiology-specific AI applications.

2.3.4 AI Imaging Applications in Radiology

Artificial intelligence, deep learning, and machine learning have made their way into every aspect of medicine. These techniques, from image reconstruction to image processing and automated image interpretation, have advanced to state-of-the-art in radiology. AI applications using deep learning algorithms have extended beyond ultrasound, CT and MRI image evaluation, to provide accurate diagnosis and prognostic prediction. The growing momentum of artificial intelligence improves the workflow, diagnosis and enhances the quantitative imaging techniques. Let us explore the role and advantages of artificial intelligence in various aspects of radiology and how artificial intelligence can be complemented with the work of a radiologist.

2.3.4.1 Applications in Neuroradiology

Numerous AI-based clinical applications have been created in the area of neuroradiology over the last few years. The aim of these AI-based applications have been divided into four main categories—disease detection, prediction of outcome, disease status, etc.; improving quality of image and workflow optimization (Fig. 2.10). The first thought when considering AI in medical imaging is the detection of abnormalities. Hence, the primary focus would be to identify certain urgent findings such as intracranial haemorrhage, acute infarcts, vessel occlusion, aneurysm detection, and traumatic brain injury. AI can help predict the outcome of status of the disease process, such as stroke, Alzheimer's disease and brain tumours. It can also help in monitoring the treatment response of brain tumours. The quality of the medical images can be improved with the help of AI, by way of image reconstruction using image co-registration, dose reduction, perfusion or dynamic contrast MRI. Detection of critical findings using AI, can improve the clinical workflow and is proven to be successful [42]. DL-based models are shown to have a greater performance and in the past few years, dedicated DL algorithms have been created in neuroradiology for better outcomes [43].

One significant deep learning application in neuroradiology has been the identification of intracerebral haemorrhage [44-47]. This has been useful in creating an

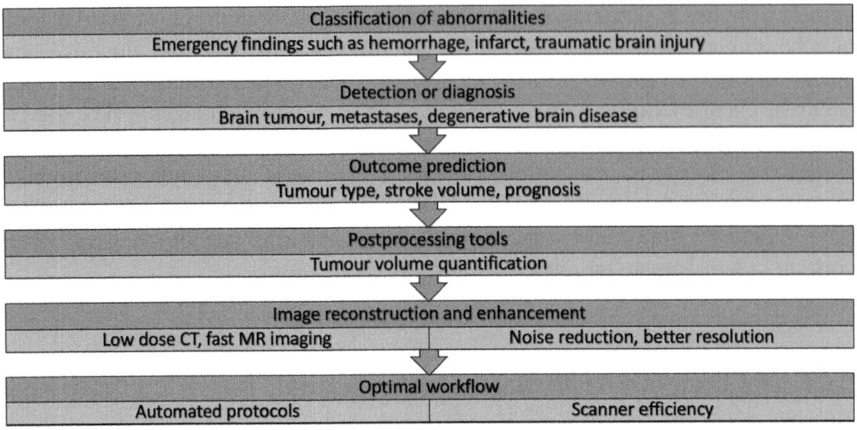

Fig. 2.10 Applications of AI in neuroradiological imaging

early warning system, classifying the haemorrhage types and in improving diagnostic confidence [48–50]. A 3D CNN model was created by Ker et al. [50], to detect four types of intracranial haemorrhage: subarachnoid haemorrhage, polytrauma haemorrhage, intraparenchymal haematoma, and acute subdural haemorrhage. To get better outcomes, they experimented with data augmentation techniques, using a data set of 399 CT scans, that were performed locally. By achieving ROC-AUC values ranging from 0.919 to 0.952, they evaluated the performance as a binary comparison between the normal and one of the haemorrhage types. Another turning point came with the RSNA 2019 Brain CT Hemorrhage Challenge, where 25,312 brain CT scans were analysed using logarithmic loss, and the top models produced the best results [51]. Aidoc software, on the other hand, is the best commercially available and authorised programme for haemorrhage detection [49]. Ginat et al. evaluated 2011 non-contrast CT heads for haemorrhage detection using the Aidoc programme, and they found that the programme had a sensitivity and specificity of 88.7 and 94.2%, respectively [52].

Deep learning in imaging of stroke has been actively researched [53]. Middle cerebral artery region infarct is a strong early predictor, thanks to the Alberta Stroke Programme Early CT Score (ASPECTS) [54]. There are various commercial applications that performs automated ASPECTS evaluation, such as e-ASPECTS (Brainomix), Frontier ASPECTS and RAPID ASPECTS (iSchemaView) [55–57]. The accuracy of these software depends on the chosen ground truth and the image reconstruction algorithms. There are also several algorithms for detection of large vessel occlusion. In non-contrast CT brain studies, You et al. created an AI-based model to recognise dense MCA sign, which demonstrated a sensitivity and specificity of 68.4 and 61.4%, respectively [58]. Additionally, deep learning has produced prognostication tools for patients who are most likely to benefit from treatment and assisted in determining infarct volume [59] (Fig. 2.11).

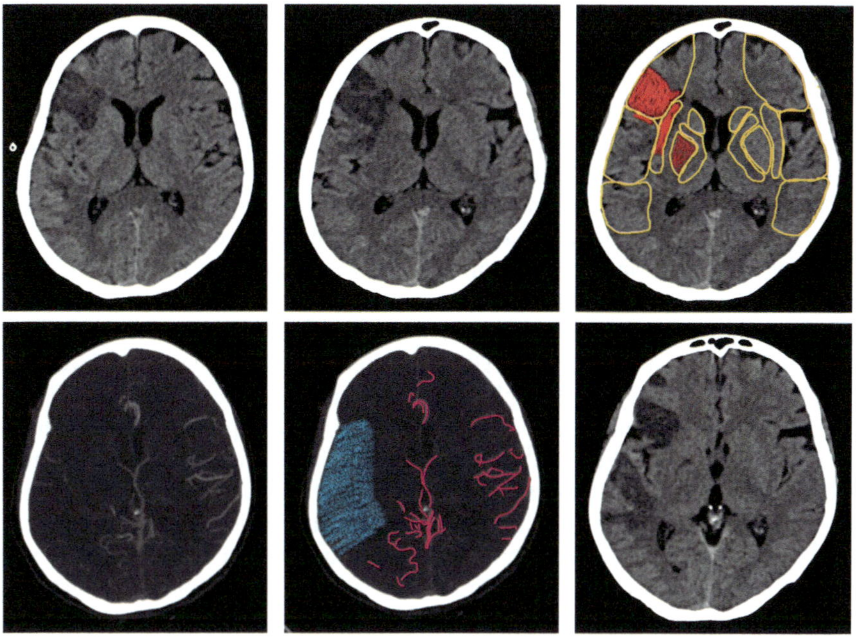

Fig. 2.11 Computer-aided analysis showing progression of acute ischemic stroke

Aneurysms are responsible for almost 85% of non-traumatic SAH. The best preferred methods for detecting aneurysms are CT angiography (CTA) and Time-of-flight MR angiography (TOF-MRA). AI has proven to be useful in detection of aneurysms, that can be missed by a human eye. Computer-assisted diagnosis (CAD) systems, based on DL, have been used to increase radiologists' reading precision [60, 61]. It has been helpful in morphological characterization, rupture risk stratification and outcome prediction [62].

Multiple AI-based applications have been created in early detection of degenerative brain diseases and has been trained using long-term data sets like the ADNI (Alzheimer's Disease Neuroimaging Initiative). AI can assist in distinguishing various types of dementia, Alzheimer's disease versus mild cognitive impairment, Parkinson's dementia and Lewy body disease. CNNs have shown impressive results in detecting AD [63]. For the purpose of forecasting when mild cognitive impairment can progress to Alzheimer's disease, DL can be combined with MRI, APOE genotype, and neurocognitive data. Combining AI applications in MRI and amyloid PET, can improve the diagnostic accuracy and management of Alzheimer's disease. Multi-approaches are required to improve the risk stratification, mortality prediction, treatment monitoring, assessment of cognition, clinical care and management [64]. Using AI to diagnose epilepsy may improve its diagnostic potential as it can overlap with other neurological diseases. Neuroimaging helps in diagnosis, follow-up and prognosis of patients with epilepsy. The two most prevalent causes of epilepsy are mesial temporal sclerosis and focal cortical dysplasia, which can sometimes go undetected on conventional MRI [65]. In individuals with hippocampal sclerosis and temporal lobe epilepsy, machine learning algorithms have been shown to lateralize the diseases in the hippocampus [66]. Using diffusion MRI and structural MRI, a recent ENIGMA-Epilepsy study examined the performance of ML and DL algorithms and found that both models had comparable accuracy [67]. AI can also combine clinical and imaging data, for example, volumetric MRI measurements, to predict the clinical outcomes of patients with epilepsy. MRI plays an important role in diagnosis of multiple sclerosis. DL-based tools have been used to estimate the disease burden and activity. MS patients and control groups could be distinguished using CNN applications on T2-weighted, fluid attenuated inversion recovery (FLAIR) or susceptibility weighted images [68]. AI also helped in identifying multiple sclerosis mimics, such as acute disseminated encephalomyelitis, neurosarcoidosis, neuromyelitis optica spectrum, etc. ML-based approaches have also been helpful in assessing the disease prognosis and treatment monitoring [68]. Radiomics in brain tumour imaging extracts information from the medical images [68]. The basic steps of radiomics includes, image preprocessing, lesion segmentation and extraction of quantitative features such as size, shape, intensity, etc. (Fig. 2.12). Various ML and DL-based approaches can then be applied, to analyse important information about the tumour such as infiltration, molecular markers and prognosis [70]. The majority of neuro-oncology research has been concentrated on gliomas, which make up 50% of primary brain tumours and have a poor prognosis and an aggressive course [71]. The best method for identifying and characterising tumours is MRI, including diffusion-weighted and perfusion imaging. Diffuse gliomas harbours more than 60 genetic alterations, which is essential in improving the diagnosis and providing targeted therapies. Imaging properties are combined with genetic and mutational trends in radiogenomics [72]. It helps in characterizing the phenotypes of various genetic alterations, such as IDH mutation, MGMT methylation, EGFR splice variant, and 1p/19q codeletion. AI helps in predicting the outcome of these patients, based on clinical, imaging and genetic variables. Response Assessment for Neuro-oncology (RANO) criteria, which integrate imaging characteristics with clinical data, have been described for evaluating treatment response [73]. Following treatment, the differentiation between recurrence, pseudo-progression and true progression can be well delineated by AI [74]. The margins of infiltrative tissue in preoperative MRI images have been identified with the use of machine learning techniques and this information may now be used to guide the planning of radiation, biopsies, and surgical resections [75]. These AI-based algorithms can assist in identifying individuals for clinical trials and guiding the appropriate treatment pathway, even while surgical excision is currently guided by enhancing tumour alone on imaging. AI-based methods have also been applied to other tumours, such as metastases and CNS lymphoma. There are various machine learning approaches that help in differentiating glioblastoma, metastases and primary CNS lymphoma, based on their molecular pattern and differences [76]. Multiparametric imaging data, clinical data, and genetic markers would all be appropriate components of an AI-based neuro-oncology diagnostic system, that can accurately

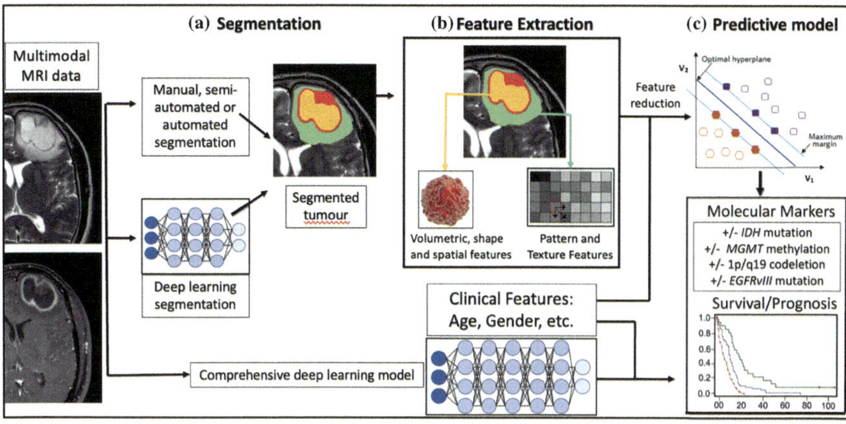

Fig. 2.12 Radiomics in neuro-oncology

predict the relevant subtype of the disease process, treatment strategy, and prognosis outcome [77].

2.3.4.2 Applications in Respiratory Imaging

More than forty AI applications have been introduced in chest radiology [78]. These applications mainly deal with the analysis of conventional chest radiographs and chest CT. AI has shown promising performance for various thoracic diseases and could address clinical problems that remained unsolved due to technical limitations. The first set of AI-based applications for conventional chest radiographs were designed to detect lung nodules. The most extensively studied application for conventional radiography is automated lung nodule detection. The first study was conducted in the year 1999, which showed a sensitivity of 35%, however by 2014, the sensitivity improved to 75% [79, 80]. Another study conducted by Nam et al. used DL based models, that were trained for 40,000 chest radiographs and showed that AI performed significantly better than the participating radiologists [81]. These AI applications are helpful in the early detection of lung malignancy. Numerous investigations have been done to identify consolidation on conventional radiography, after the discovery of lung nodules. The detection of pneumonia and tuberculosis would benefit from automated examination of chest radiographs. The Stanford Machine Learning Group tested CheXNeXt, a CNN model, that could identify 14 diseases on antero-posterior and postero-anterior radiographs. The model achieved same-level performance as the reporting radiologist [82]. There is a recent AI algorithm known as CAD4TB, that is used in the detection of tuberculosis in children younger than 4 years of age [83]. Few commercial software have been developed to detect pneumothorax and pleural effusions on chest radiographs [84, 85]. Position of lines and tubes on chest radiographs is important to be identified, and AI can do a better task at identifying and

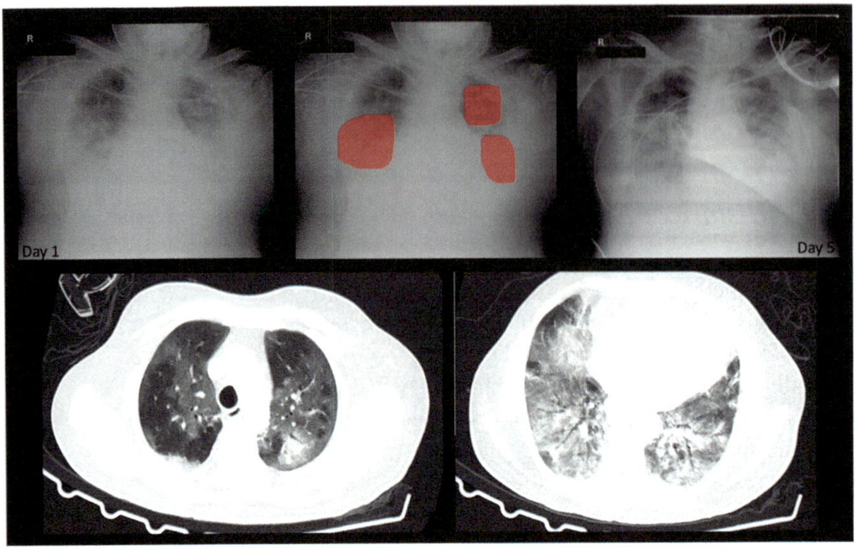

Fig. 2.13 Identification of COVID pneumonia on chest radiograph using AI-based model

visualising its correct positions. This would help in avoidance of repeated radiographs and reduce patient irradiation [86]. It would also be useful in identifying umbilical catheters amongst paediatric population [87].

Most AI-based applications in chest CT are employed in the detection of lung malignancy and further characterization. A CNN model performed exceptionally well in a recent study by Baldwin et al., for the risk assessment of tiny nodules in CT [88]. With the use of AI, respiratory conditions such obstructive lung disease, asthma, and cystic fibrosis can also be identified rapidly on chest CT. This can help in monitoring the disease process and can detect minute attenuation differences [89]. Only few applications are available in detection of pulmonary thromboembolism and interstitial lung diseases. Since the COVID emergence in 2019, there have been millions of deaths and healthcare workers have been devoted to COVID-related medical duties [90]. The researchers have been driven and rushed to develop AI models to detect COVID lung changes on conventional radiographs and chest CT. AI can help in automated diagnosis and screening of COVID with the use of image analysis from conventional radiographs, chest CT and lung ultrasonography [91] (Fig. 2.13).

2.3.4.3 Applications in Cardiovascular Imaging

The present cardiac AI applications are designed to further assist the radiologists, reduce the workflow and improve patient care. AI in cardiovascular imaging bridges the gap between newer technologies and the large data. The AI algorithms used in cardiovascular imaging are entirely data-driven and have achieved outstanding results in a number of time-consuming activities, including image segmentation, anomaly identification, and patient selection. The performance of ML and DL-based models will improve exponentially as the complexity of the data rises [92].

The most common imaging method utilised in cardiac evaluation is echocardiography. It aids in determining different pathologies and evaluating heart function [93]. The gold standard in echocardiographic cardiac evaluation is ejection fraction. AI algorithms can be used to evaluate cardiac strain in order to evaluate cardiac function, beyond ejection fraction [94]. Also, DL algorithms help in assessing valvular heart diseases and echocardiographic assessment of cardiac chambers. In certain investigations, CNN models have been used to assess regional wall motion problems using echocardiography [95].AI is used in cardiac CT imaging in a variety of ways, from diagnosis to prognosis. AI is applicable to both low-contrast and high-contrast images [96]. Coronary calcium scoring is the best predictor for coronary artery disease outcome and helps in prognostic stratification [97]. The application of AI in coronary calcium scoring has shown to be effective in rapidly determining the likelihood and risk of coronary artery disease [98]. Isgum et al. employed a pool of ECG-gated non-contrast images in one of the first methods that utilised AI for coronary calcium scoring. The AI application correctly detected the patient's risk score in 93.4% of cases and coronary calcification in 73.8% of cases [98]. There are also studies where image evaluation can be done without ECG-gating, which would still give good and positive results [99]. One of the other imaging applications of AI in cardiac CT is identification of coronary artery stenosis and plaque analysis (Fig. 2.14). Presently, CAD-RADS is used for classifying coronary artery stenosis, which provides useful information regarding the patient prognosis [100]. When reporting coronary CT angiography, CAD-RADS score can be provided promptly and automatically using DL-based algorithms. Muscogiuri et al., who constructed three CAD-RADS models, conducted a study in which they established that the AI-based method generated a CAD-RADS score faster and with higher diagnostic accuracy than human experts [101]. AI has also proven its ability in analysing the morphology and quantification of coronary atherosclerotic plaques in cardiac CT [102, 103]. In addition, AI has also created algorithms for assessment of ischaemic risk scores on coronary angiographic imaging. Fractional flow reserve CT is a helpful tool in evaluating the ischaemia caused by the plaques in coronary vessels. Machine learning algorithms have been useful in evaluating these ischaemic lesions [104].

From image acquisition to image interpretation, artificial intelligence is used in cardiac MR imaging [95]. The foremost entities to be evaluated on cardiac MRI include cardiac function and tissue characterization. In a reduced time and with less interobserver variability, fully automated AI-based software can identify biventricular cardiac volumes and cardiac activity from cine images [95]. The capability of cardiac MRI to assess and characterise cardiac tissue is another additional benefit. T1-weighted black blood sequences and T2-weighted sequences are used for evaluation of fibrofatty infiltration and myocardial oedema respectively. Late gadolinium enhancement sequences can depict reduced gadolinium washout in cases of myocardial fibrosis. Few studies have shown that AI-based algorithms are used to identify

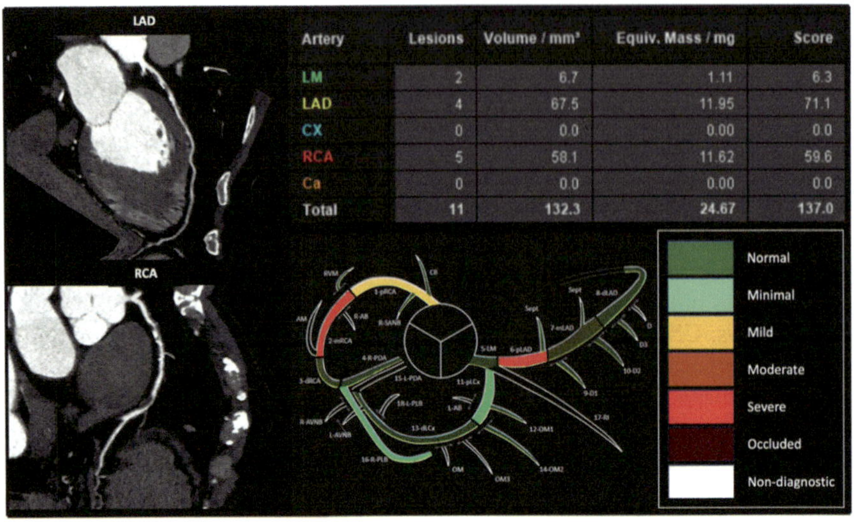

Fig. 2.14 AI analysis of coronary calcium score and stenosis in cardiac CT images

late gadolinium enhancement in cases of myocardial fibrosis and can be used as a future application. Also, AI algorithms in cardiac MRI have been useful in predicting prognosis, adverse events and patient mortality [105].

2.3.4.4 Applications in Hepatobiliary and Gastrointestinal Imaging

Segmentation is the main pre-processing step performed in AI algorithms for assessment of liver pathologies. It is commonly performed to assess the liver function and to differentiate between liver tumours. Manual processing is time-consuming, so hence automated technologies have been introduced, where the ground truth is derived directly from the image data itself. For example, an expert radiologist diagnoses a liver mass on a CT dataset and the AI algorithm is assigned to replicate the diagnosis in a supervised training test. There are also AI-based models that can predict prognosis from the imaging dataset. Accurate prognostication can help the treating doctor in adjusting follow-up intervals and adjust the treatment plan accordingly. Segmentation studies mainly outlines the region of interest and there are various applications that are approved for clinical use in hepatology [106, 107]. However, due to the complex nature of hepatology, the clinical implementation of DL methods lags far behind as compared to other systems. Multiple studies have gathered data from patients with non-alcoholic steatohepatitis (NASH), non-alcoholic fatty liver disease (NAFLD), and hepatocellular carcinoma (HCC). The studies focused on quantification of steatosis, morphological patterns of NAFLD and HCC, and staging of liver fibrosis [108–110]. Many studies created support vector machine algorithm to quantify macrosteatosis, biliary ducts [111, 112] and liver fibrosis [113]. There models showed excellent results in quantification and segmentation of imaging data in patients with NAFLD and NASH. CNN-based classification models can help in grading of the severity of NAFLD, replacing liver biopsy [114]. Patients with HCC undergo multiple imaging studies to establish a diagnosis, preoperative planning and monitoring treatment response. Even though LI-RADS gives the most standard guidelines for diagnosis of HCC, there has been inconsistencies that could hinder the treatment planning. Therefore, to transform the information of imaging data into clinically accessible information, AI has developed two categories for better imaging evaluation—radiomics and DL systems. Many DL-based models have been created to detect and the morphological patterns of HCC, and these models were able to distinguish between HCC and normal tissue [115–117]. Segmentation is performed to outline the margins of the tumour (Fig. 2.15). These models could identify the risk of HCC recurrence after resection or liver transplantation [106, 115–118]. Due to high rates of recurrence after surgical resection of HCC, microvascular invasion has been the main focus in radiomics nomogram [119], which helps in providing additional information in terms of treatment planning. AI has also helps in predicting the response after transarterial chemoembolization (TACE), by assessing MRI imaging fractures and clinical variables [120]. The non-invasive visualisation of bile ducts is another significant use of AI in hepatobiliary imaging. The biliary system is currently evaluated by imaging modalities like endoscopic ultrasonography (EUS) and magnetic resonance cholangiopancreatography (MRCP). Researchers have used deep learning algorithms to achieve more sensitive lesion detection in biliary system and pancreas [121].

The applications of AI in gastrointestinal radiology is so vast in literature. There are multiple ML and DL methods used for image interpretation and diagnosis of pathologies, starting from oesophagus up to rectum. Oesophageal malignancy is frequently identified on CT and PET scans, although therapy prediction and response assessment have had sporadic success. The image dataset has been used by DL and radiomics techniques to forecast prognosis [122, 123]. Combined model shows a better accuracy and prediction performance, than the model using radiomics alone [124]. This combined analysis can also be used to enhance stomach tumour characterisation and identification. CT radiomics nomogram can identify lymph node metastases and a study showed significant correlation with patient survival [125].

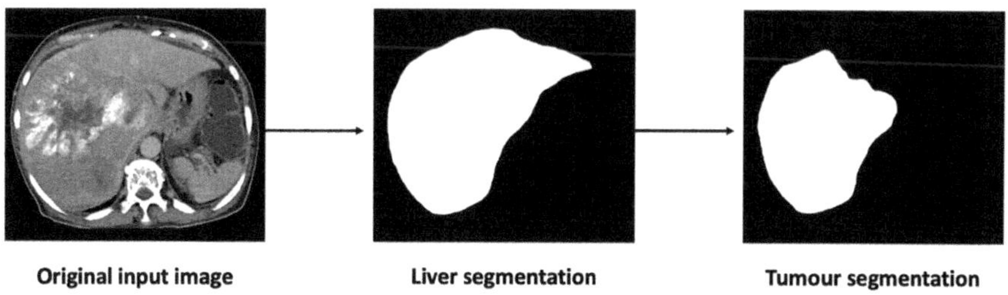

Fig. 2.15 Liver tumour detection and segmentation

In addition, characterization and staging of colon and rectal malignancies can be performed using AI [126, 127]. For patients with colorectal cancer, radiomic analysis of CT, MRI, and PET images is performed to forecast therapy outcomes and direct treatment planning [128]. AI in endoscopic imaging improves patient outcomes, creates better treatment plans, and enhances healthcare performance. Various AI applications have been introduced in endoscopic imaging, to see and interpret the visual data. Computer-aided polyp detection and diagnosis can be performed using AI [129], which helps in early diagnosis and surveillance monitoring of the disease process. AI tools have also been created in detecting gastrointestinal bleed. One of the earliest studies showed that CAD of bleeding from video endoscopic images could distinguish between bleeding and non-bleeding areas, with the aid of colour and feature extraction [130]. Automated recognition endoscopy software has been introduced, that gives better results and characterization [131]. The most recent applications of AI involves better identification of mucosal healing in inflammatory bowel disease [132]. Identifying these pathologies by AI algorithms, before the clinical onset of symptoms appear, would be quite helpful in treatment planning and improving patient prognosis.

2.3.4.5 Applications in Musculoskeletal Imaging

Imaging continues to be crucial for diagnosing musculoskeletal disorders, and its value has led to a rise in the use of MRI [133]. Every stage of image interpretation can be assisted by AI. In the current early stages of the introduction of AI into the field of radiology, numerous AI applications in musculoskeletal MRI have been introduced and have shown excellent results.

Participants in the Radiological Society of North America Bone Age Machine Learning Challenge were invited to create AI-based algorithms for bone age prediction [134]. The goal of the challenge was to identify the most effective MI-based methods for precisely estimating bone age. The model developed by Drs. Alexander Bilbily and Mark Cicero, who used Inception V3 architecture for pixel information and data augmentation, took the first place. Currently, BoneXpert is the commercially available software for automatically determining bone age [135]. Using 13 bones and traditional machine learning approaches, this software determines the bone age by analysing their shape, intensity, and texture. Early diagnosis of decreased bone mineral density through medical care can lower the risk of fracture and enhance quality of life [136]. First few CAD systems were created to assess the bone fragility on dental panoramic radiographs [137]. On CT scan, osteoporosis can also be screened. Bone mineral density measurements were done on low dose chest CTs for the purpose of screening for lung cancer using a DL-based system, reported by Pan et al. [138]. Additionally, CNN models have also been developed to automatically segment the proximal femur and evaluate the bone quality on MRI [139]. Combining trabecular bone microarchitecture and ML techniques allows for the precise prediction of the biomechanical strength of trabecular bone. Further algorithmic development is necessary for automated assessment of osteoporosis, disease progression, and treatment response. Other applications includes fracture detection, assessment of bone tumours and detection of osteoarthritis. AI has a very important role in automatic fracture detection. Deep CNNs has given excellent results in detecting fractures involving humerus, wrist and ankle on conventional radiographs [140, 141]. Additionally, ML may recognise compression fractures of the vertebral body and fractures involving the posterior elements [142, 143]. There are multiple studies that have used AI for detection of fractures in both axial and appendicular skeleton, on conventional radiographs and CT images. ML-based algorithms can be used in assessment of bone lesions, especially bone metastases. In the early 1960s, Lodwick et al. created AI models in early detection of primary bone tumours [144]. AI can also help in diagnosing tumour recurrence and soft tissue malignancies. Soft tissue sarcomas on MRI can have their post-treatment alterations evaluated using ML techniques [145]. AI can also be used to grade osteoarthritis on conventional radiographs and standardise their interpretation [146]. AI can improve the quantitative and qualitative evaluation of cartilage on MRI, and numerous deep learning systems have been developed for the automated detection of cartilage lesions on MRI [147]. There have also been studies where meniscal and ligament injuries of the knee joint can be identified using AI [148].

2.3.4.6 Applications in Breast Imaging

The most frequent malignant tumour found in women of all ages is breast cancer. Early detection of malignancy by means of imaging, such as ultrasonography, mammography and MRI, can help in treatment planning and better prognosis. AI has shown incredible performance in image analysis and is being used in breast cancer screening. Ultrasonography is the easiest modality available and requires no radiation. To avoid missing any findings, AI has developed an algorithm to detect breast lesions on ultrasound images [149]. AI is primarily used to identify regions of interest, extract features, and categorise lesions into benign and malignant types. To accurately detect the breast lesions by means of ultrasound, automated lesion segmentation and margin identification are performed from the background [150]. By segmenting lesions in breast ultrasound pictures, sonographers can diagnose the illness much more faster. Then, the morphological features are taken into account, such as size, shape, orientation, calcifications, etc. BIRADS is then used to assess the degree of cancer probability. There are few studies where morphological and texture features can be extracted from the image data and classified into benign and malignant disease. CAD systems are helpful in defining the lesion characteristics and planning the treatment protocol [151]. Presence of axillary lymphadenopathy, classification and diagnosis of breast diseases can speed up the diagnosis process and increase diagnostic precision.

Mammography is a non-invasive test performed for breast cancer screening. Mammography has means of identifying lesions that can be missed on ultrasound and can differentiate between benign and malignant breast diseases [152]. The main role of AI in mammography includes detection and classification of lesions, presence of microcalcifications, lesion segmentation, density and risk assessment, and improvement of image quality. Detection of mass lesion is the first step in CAD algorithm [153]. Due to variations in shape or margins of the lesion, it may be difficult to identify lesions on mammography. There are few DL-based models that are available, such as You-Only-Look-Once (YOLO) and full resolution convolutional network (FrCN), and have shown good diagnostic accuracy [154, 155]. Presence of microcalcifications usually indicates early signs of breast cancer and these can be detected early with the help of AI [156]. CNNs can identify mass lesions, detect microcalcifications automatically, give the referring physician information, and increase radiologists' productivity. The usage of automatic image segmentation would help in precise disease classification and diagnosis [157]. Furthermore, research has shown that women with higher breast densities are more likely to acquire breast cancer than women with lower densities [158]. AI can help in evaluating the breast density on mammographic images, using CNN models [159]. It can lessen inter-radiologist variances and improve breast cancer risk prediction. The application of AI to the prediction of breast cancer risk has been extensively studied. Studies have demonstrated that DL-based models can analyse mammography results, risk variables (such as family history, reproductive factors, lifestyle changes, etc.), and clinical evidence to determine the likelihood of breast cancer and determine its risk [160]. This breast risk assessment can help the referring doctors in conducting the appropriate interventions and treatment planning (Fig. 2.16).

MRI is a more sensitive modality in detection of breast cancer than mammography [161]. MRI can extensively identify the shape, size, internal architecture of the lesion via various sequences. MRI and DL in breast illnesses concentrates on the detection, segmentation, characterization, and categorization of breast lesions, similar to other imaging modalities. A small number of studies have demonstrated that combining volumetric and dynamic DCE-MRI data can enhance CNN-based lesion categorization [162]. CAD-based algorithms can lower false positive rates, increase the sensitivity of breast MRI, and lessen the psychological toll on patients [163]. Additionally, a number of CAD systems have recently been created in the disciplines of contrast-enhanced mammography, elastography, and digital breast tomosynthesis, providing improved accuracy and efficiency [164–166].

2.3.4.7 Applications in Foetal and Gynaecological Imaging

Given the clinical significance of ultrasound in prenatal screening, AI research in prenatal ultrasound has significant practical implications and includes advantages of increasing work efficiency, providing quantitative assessments, standardised measurements, automatic image quality control, and enhancing diagnostic accuracy. Ultrasonography is the most safe, non-invasive investigation of choice in prenatal diagnosis and monitoring of foetal growth. Measurement of foetal structures, assessment of organ functions, and diagnosis of different pathologies are currently performed using two- and three-dimensional ultrasonography [167]. However, there

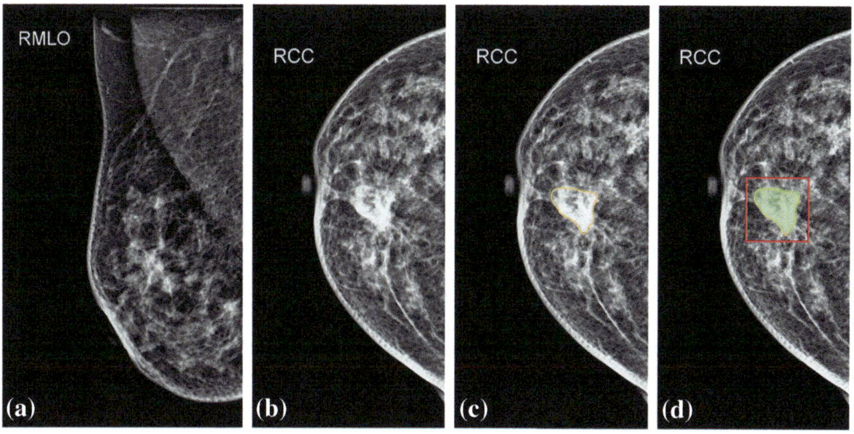

Fig. 2.16 BIRADS category 6 lesion. The yellow outline in image C is the manual delineation by radiologist. The green and red lines in D are the outlined areas observed by AI-based model

are few limitations such as involuntary foetal movements, ultrasound artefacts and advanced growth in second and third trimester that could obscure the structures of interest. Newer technologies can improve the quality of acquired images, help extract and standardize measurements. Automatic measurements and assessments have been introduced, with the help of AI, to improve the diagnostic accuracy and reduce inter-observer and intra-observer variability [168]. Assessment of foetal development during first trimester is quite important. Gestational sac is the first structure that needs to be identified, for confirmation of pregnancy. Automatic AI-based solutions are available to roughly estimate the gestational age by measuring gestational sac diameter or crown-rump length [169]. Three-dimensional-image volume measurements can provide better results than two-dimensional information [170]. In addition, nuchal translucency (NT) measurement in first trimester is critical for identification of chromosomal abnormalities or malformations. Compared to conventional manual methods, measurements taken using automatic approaches were more accurate [171]. These methods can also be used to partition the gestation period and extract the biometric planes of the head, abdomen, and limbs for anatomical evaluation.

Radiologists and sonographers assess the foetal growth and detects anomalies during the second trimester ideally. The foetal brain is the organ of interest where the prenatal ultrasonography examination is the most challenging. Foetal head biometry, including biparietal diameter and head circumference, are two applications of AI-aided ultrasonography in foetal neurosonography that seek for any major abnormality. Furthermore, routine ultrasound imaging of fetal face structures is carried out. AI can aid in congenital deformity identification, facial recognition, and craniofacial development [172]. Pre-processing of ultrasound images can be performed, to improve diagnostic efficiency. Other structures, where measurements have to be standardized, are abdominal circumference and femur length, which can be easily assessed with the help of automatic measurements. Foetal echocardiography is also performed and is essential for prenatal care. Monitoring the cardiac function is challenging due to involuntary foetal movements and abnormal position. Arnaout et al. [173] created a DL method in identifying five most essential views of foetal cardia and segmentation of cardiac structures. Due to frequent updates of AI-based algorithms and data volume amplifications, shortcomings in picture identification and automatic measurements have been addressed. Accurate acquisition of foetal standard planes is crucial for obstetric examination. Feature extraction is performed from ultrasound image dataset to train their intelligent system. Compared to traditional hand-crafted features, deep CNN-based models are more advanced and produce superior results (Fig. 2.17).

Foetal MRI is usually performed to detection of foetal brain disease and placental abnormalities. It is the modality of choice when the use of obstetric ultrasound is limited, for treatment planning and prognostication tool. Foetal MRI provides excellent soft tissue resolution and better visualisation. AI would help in more automated and less time-consuming analysis of the MRI dataset [174, 175]. The first step of AI in foetal MRI is obtaining high quality images, by performing automated correction for foetal motion during the start of the study, which would reduce motion artefacts and better image interpretation [176]. For foetal brain pathologies, there are various models that have been created to automatically segment the foetal brain into three planes, which would reduce the time of image acquisition [177, 178]. Placental segmentation can also be performed using AI, in identifying the severity of placental previa and growth restriction [179, 180].

Gynaecological malignancies can have a considerable impact on morbidity and death, but these effects can be minimised with early detection. Gynaecological malignancies include endometrial, ovarian and cervical cancers. Even though the majority of AI research focuses on breast cancer, ovarian cancer patients are thought to have

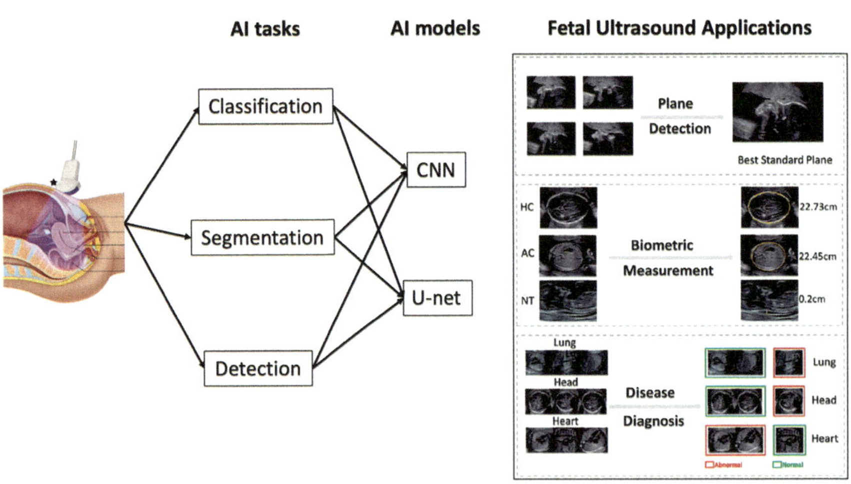

Fig. 2.17 Overview of AI applications in foetal ultrasound. Current AI applications focuses on standard plane detection, biometric measurements and detection of pathologies

the greatest fatality rates. The preferred imaging method for early identification of such malignancies is MRI. AI-based models can detect early disease process, presence of micro-metastases or pelvic lymph nodes metastases, and staging of the disease process. Also, risk prediction and treatment outcomes can also be detected. Comparing CNN-based research to radiomics and conventional ML approaches, they demonstrate great performance [181].

2.3.4.8 Applications in Interventional Radiology

There have been advancements of AI in the field of interventional radiology, even though the main AI applications are prevalent in diagnostic radiology. In addition to computer vision and opinion, implicit AI operations in interventional radiology also model patient selection, personalise treatment alternatives, and increase the delicacy of minimally invasive treatments in image guided therapies. The one field in which complete procedure's record is stored in a standardised manner and is accessible in retrospect, is interventional radiology. With the current developments, new deep learning methods can shrink the datasets, that are needed to develop therapeutically useful algorithms [182].

Patient selection is crucial in the pre-procedural context to help customise treatment decisions and plan for improved clinical outcomes. Before recommending any course of treatment, interventional radiologists do multiparametric risk stratification, using the patient's clinical data. Using the facts at hand, AI can aid in outcome prediction. Incorporating clinical data, radiomics, and genetic data can increase precision of decision-making, assist in prioritising patients for interventional radiology, and facilitate therapy planning by weighing risks and forecasting results [183]. There are various models that have been introduced in predicting the response to TACE for HCC. It is shown that AI can outperform traditional staging systems by utilizing imaging and clinical features [184, 185]. Combined radiomics and clinical model of HCC can provide improved survival estimation. Utilising cutting-edge navigational methods, such as augmented reality or virtual reality, which superimpose virtual pre-procedural 3D anatomic data onto 2D images in real time, enables operators to visualise complex anatomy and enhance procedural planning and skills [186].By accelerating computationally laborious or manual processes, AI can enhance interventional procedures. Protocol automation in CT is performed based on the information gather from health records and is tailored to minimize radiation exposure [36]. Deep learning approaches have been used in identification of catheters, guidewires, sheaths or stents during angiographic procedures [187]. Few AI techniques provide deeper algorithms, including contrast-free virtual road mapping of the vessels. Another crucial element of AI is the choice of endovascular devices. AI-based models can select the optimal stent for vascular stenosis, e.g., coronary vessels, and can guide personalized 3D printing of these stents [188, 189]. Precision guidance for biopsies and local image-guided therapies can be achieved by combining intraoperative 2D fluoroscopic or ultrasound images with pre-procedural 3D images [190]. After therapy, AI is capable of evaluating the treatment's effectiveness, forecasting the future, and making management decisions. AI can uncover any modest changes that indicate the progression or recurrence of the disease in longitudinal investigations conducted during treatment follow-ups. Automated volumetric measurements, or response evaluation criteria in solid tumours (RECIST), can be performed by DL [191]. AI can outperform RECIST criteria and can predict overall survival. Additionally, AI can enhance interobserver reliability, assess therapy responsiveness in acute ischemic stroke, and predict patient outcomes [192]. In addition, AI can reduce the time in interpreting post-treatment images.

2.3.4.9 Other AI Applications in Radiology

AI plays an important role in the field of nephrology and uroradiology. Kidney-related diseases cause great economical problem worldwide. Early diagnosis of acute kidney injury is vital and the current AI researches are focused mainly on risk prediction, detection and automatic alerting. This automated electronic alerts reduced mortality and length of hospital stay in patients with acute kidney injury. The early detection of primary and secondary chronic kidney diseases can potentially be enhanced by AI. End-stage renal disease patients receive specific therapies such as haemodialysis, peritoneal dialysis, or kidney transplantation. AI can help in predicting future haemoglobin concentration, iron dosage, heart rate, blood pressure and appropriate volume load in patients undergoing haemodialysis. Arteriovenous fistula is the first line of access for haemodialysis patients. AI can help predict the life of arteriovenous fistula using ML algorithms. ML algorithms can also be used in predicting the mortality risk of patients undergoing peritoneal dialysis. The application of AI in kidney transplantation include prediction of graft rejection and optimization of post-transplantation immunosuppressive therapy [193]. Genitourinary malignancies include prostate, bladder and renal cancers. The care process for patients with genitourinary malignancies can be improved in a number of ways, thanks to AI. AI can aid in early cancer detection and enable early treatment. AI can also forecast the course of cancers, which allows doctors to make adjustments to the type and extent of their treatment plans. AI can assist in the diagnosis and categorization of prostate cancer by detecting intraprostatic lesions. The laborious process of segmenting prostate and prostatic lesions, in order to prepare for targeted biopsy, can also be automated using AI. It has been demonstrated that AI can aid in more accurate cystic renal mass characterization. AI can also increase imaging diagnostic rates and screen the kidneys for solid masses. Additionally, AI has been used to enhance the effectiveness of cystoscopic imaging of bladder tumours [194]. In head and neck radiology, segmentation process is quite important in identifying borders between the region of interest and its neighbouring areas. Segmentation is performed to identify either normal anatomy or pathological areas. ML and DL segmentation have been applied to areas in head and neck radiation treatment planning. Applying segmentation of abnormal areas in the head and neck regions, allows faster quantification of lesion characteristics, such as size, volume, texture, etc. DL applications have also shown great results in segmentation of tumour volumes, detection of head and neck nodal metastases and identification of extra-nodal extension of tumour. Radiomics is performed for characterization of tumours, including molecular markers. Research in radiomics and detection of subtle features within regions of interest on diagnostic imaging has been enhanced by ML algorithms. Radiation oncology in head and neck malignancies can utilize ML and DL to improve decision-support algorithms, auto-segmentation, treatment planning, prediction of outcome and quality assurance. There are studies where computer-aided diagnosis of benign versus malignant thyroid nodules using ML algorithms showed more accurate results than human interpretation [195]. The most commonly used radiotracer in oncological PET imaging, ^{18}F-FDG, is recommended for diagnosis, therapy response evaluation, and recurring cancer detection. The quantitative evaluation of cancer glucose metabolism is also possible with PET imaging. Lung cancer is the most frequent malignancy examined using AI models, followed by lymphoma, head and neck, pancreas, breast, cervical cancers, and sarcoma. The characterization of lung nodules, tumour identification, distinction of lung cancer subtypes, staging of lung cancer, and response evaluation are just a few of the imaging-related applications where artificial intelligence (AI) approaches have been created. They also show promising value in predicting tumour behaviour and patients' survival. AI can also forecast early metastatic illness. ^{18}F-FDG PET imaging's quality can be enhanced using AI models [196].

2.3.5 *Challenges and Future of AI in Radiology*

There are several challenges to AI implementation in diagnostic and interventional radiology. The exact mechanism of how a neural network arrives at its classification output or, why the parameters end up at their trained values, is poorly understood. In general, the researchers understand that they work, but do not know how or why. This lack of understanding prevents the FDA from approving certain commercial AI-based medical devices and algorithms. Training and validation of medical image algorithm is also a challenge. Even though extensive image datasets are available, procurement of the data is expensive and time-consuming. Lack of training data and transparency can cause problems of algorithmic biases. Despite the exaggerated enthusiasm we have for AI, we must acknowledge that it will play a more significant role in radiology. Despite the early phases of AI development, radiologists have the chance to adapt to the workflow. Radiologists and AI can collaborate to improve image interpretation [197]. Multitask learning combines estimates of models for many tasks. In order to develop a model that performs better than the original task, information from training signals for similar tasks is used [198]. For a variety of image segmentation tasks in a variety of modalities, CNN models can be created. One algorithm's utility in clinical practise increases with the number of tasks it can handle. To facilitate the workflow of radiologists, ML-based or DL-based algorithms should also be fully incorporated into PACS workstations. Before using ML-based applications, radiologists must first build the requisite infrastructure. They should be able to interpret images from PACS and retrieve information from the electronic health records, instead of leaving PACS and logging into a separate system.

2.3.6 Conclusion

The use of artificial intelligence in radiology has a bright future since it will enable better patient care and treatment recommendations. This future is both diverse and exciting. Radiology will be significantly impacted by machine learning and deep learning technologies in the automation of repetitive operations and the provision of assistance in the decision-making process for complex image interpretations. Radiologists are not anticipated to be replaced by artificial intelligence in the near future, but they will be able to offer extremely accurate diagnostic services. These methods can speed up radiology workflow, increase radiologist productivity, enhance patient care, improve detection and diagnosis, and better interpret results. The main challenge is to incorporate artificial intelligence into our professional lives and to embrace the predestined change of radiological practice. Artificial intelligence can escalate radiology to the next level, grant radiologists to upgrade their purpose and value, and redefine imaging care for years to come.

References

1. P. Suetens, *Fundamentals of medical imaging* (2017). https://doi.org/10.1017/9781316671849.
2. I.N. Bankman, *Handbook of medical imaging processing and analysis* (Academic Press, Amsterdam, 2009)
3. K.H. Zou, S.K. Warfield, A. Bharatha, C.M.C. Tempany, M.R. Kaus, S.J. Haker et al., Statistical validation of image segmentation quality based on a spatial overlap index1. Acad. Radiol. **11**(2), 178–189 (2004)
4. O. Zendel, M. Murschitz, M. Humenberger, W. Herzner, How good is my test data? introducing safety analysis for computer vision. Int. J. Comput. Vision **125**(1–3), 95–109 (2017)
5. J. Hajnal, Preparation and display of image data. Med. Image Registration. 71–86 (2001). https://doi.org/10.1201/9781420042474.ch4
6. E.K. Fishman, D.R. Ney, D.G. Heath, F.M. Corl, K.M. Horton, P.T. Johnson, Volume rendering versus maximum intensity projection in CT angiography: What works best, when, and why. Radiographics **26**(3), 905–922 (2006)
7. W. Shi, Y. Miao, C. Zhang, Y. Ding, L. Shi, M. Li et al., *Emerging challenges of AI for Biomedical Image Analysis: A CBCT image noise reduction method based on cgan* (2022)
8. J. Weese, C. Lorenz, Four challenges in medical image analysis from an industrial perspective. Med. Image Anal. **33**, 44–49 (2016)
9. P. Suetens, P. Fua, A.J. Hanson, Computational strategies for object recognition. ACM Comput. Surv. **24**(1), 5–62 (1992)
10. A. Sarvazyan, Model-based imaging. Ultrasound Med. Biol. **32**(11), 1713–1720 (2006)
11. H. Greenspan, B. van Ginneken, R.M. Summers, Guest editorial deep learning in medical imaging: Overview and future promise of an exciting new technique. IEEE Trans. Med. Imaging **35**(5), 1153–1159 (2016)

12. Artificial Intelligence. Merriam-Webster.com. 2017. https://www.merriam-webster.com/dictionary/artificial%20intelligence (2 Nov 2017)
13. J.A. Brink, R.L. Arenson, T.M. Grist, J.S. Lewin, D. Enzmann, Bits and bytes: The future of radiology lies in informatics and information technology. Eur. Radiol. **27**(9), 3647–3651 (2017)
14. On computable numbers, With an application to the Entscheidungsproblem. Ideas That Created the Future. **2021**, 51–60 (1936)
15. I. El Naqa, M.A. Haider, M.L. Giger, R.K. Ten Haken, Artificial intelligence: Reshaping the practice of radiological sciences in the 21st century. Br. J. Radiol. **93**(1106), 20190855 (2020)
16. A. Hosny, C. Parmar, J. Quackenbush, L.H. Schwartz, H.J. Aerts, Artificial Intelligence in radiology. Nat. Rev. Cancer **18**(8), 500–510 (2018)
17. E. Montagnon, M. Cerny, A. Cadrin-Chênevert, V. Hamilton, T. Derennes, A. Ilinca et al., Deep learning workflow in radiology: A primer. Insights Imaging. **11**(1), (2020)
18. R. Takahashi, Y. Kajikawa, Computer-aided diagnosis: A survey with bibliometric analysis. Int. J. Med. Informatics **101**, 58–67 (2017)
19. W.L. Bi, A. Hosny, M.B. Schabath, M.L. Giger, N.J. Birkbak, A. Mehrtash et al., Artificial intelligence in cancer imaging: Clinical challenges and applications. CA Cancer J. Clin. **69**(2), 127–157 (2019)
20. A.L. Samuel, Some studies in machine learning using the game of Checkers. IBM J. Res. Development. **44**(1.2), 206–226 (2000)
21. M. Paliwal, U.A. Kumar, Neural networks and statistical techniques: A review of applications. Expert Syst. Appl. **36**(1), 2–17 (2009)
22. C.M. Bishop, *Pattern recognition and machine learning* (Springer, New York, NY, 2006)
23. P. Saraswat, Supervised machine learning algorithm: A review of Classification Techniques. in *International conference on intelligent emerging methods of artificial intelligence and cloud computing* (2022), 477–482
24. A. Liaw, M. Wiener, Classification and regression by random forest. R News **2**, 18–22 (2002)
25. D.J. Amit, *The basic attractor neural network. Modeling brain function* (Cambridge University Press, 1989), pp. 58–96
26. T. Moldwin, I. Segev, Perceptron learning and classification in a modeled cortical pyramidal cell. Front. Comput. Neuroscience. **14**, (2020)
27. J. Schmidhuber, Deep learning in neural networks: An overview. Neural Netw. **61**, 85–117 (2015)
28. G. Brewka, Artificial intelligence—a modern approach by Stuart Russell and Peter Norvig, Prentice Hall. series in Artificial Intelligence, Englewood Cliffs, NJ. Knowl. Eng. Rev. **11**(1), 78–79 (1996)
29. Y. LeCun, Y. Bengio, G. Hinton, Deep learning. Nature **521**(7553), 436–444 (2015)
30. D. Shen, G. Wu, H.-I. Suk, Deep learning in medical image analysis. Annu. Rev. Biomed. Eng. **19**(1), 221–248 (2017)
31. N. Buduma, N. Buduma, J. Papa, N. Locascio, *Fundamentals of deep learning: Designing next-generation machine intelligence algorithms* (O'Reilly®, Beijing, 2022)
32. K. Eremenko, H. de Ponteves, *Deep learning AZ: hands-on artificial neural networks in Python.* Udemy Inc. https://www.udemy.com/deeplearning. Most popular course 2018
33. A. Vasuki, S. Govindaraju, Deep neural networks for image classification. *Deep Learning for Image Processing Applications*, vol. 31 (IOS Press, 2017), p. 27
34. Hype cycle for emerging technologies 2017. 15–08–2017. www.gartner.com. Accessed 1 May 2018
35. B.J. Erickson, P. Korfiatis, Z. Akkus, T.L. Kline, Machine learning for medical imaging. Radiographics **37**(2), 505–515 (2017)
36. P. Lakhani, A.B. Prater, R.K. Hutson, K.P. Andriole, K.J. Dreyer, J. Morey et al., Machine learning in radiology: Applications beyond image interpretation. J. Am. Coll. Radiol. **15**(2), 350–359 (2018)
37. D. Fohr, O. Mella, I. Illina, New paradigm in speech recognition: deep neural networks. In *IEEE international conference on information systems and economic intelligence* (Marrakech, Morocco, 2017)

38. T. Cai, A.A. Giannopoulos, S. Yu, T. Kelil, B. Ripley, K.K. Kumamaru et al., Natural language processing technologies in radiology research and clinical applications. Radiographics **36**(1), 176–191 (2016)
39. European Society of Radiology (ESR). ESR position paper on imaging biobanks. Insights Imaging. **6**, 403–410 (2015)
40. I. Goodfellow, J. Pouget-Abadie, M. Mirza et al., *Generative adversarial networks* (2014). arXiv:1406.2661
41. A. Alberich-Bayarri, Image interpretation. in *Medical radiology* (Springer, Berlin, 2017)
42. K.S. Choi, L. Sunwoo, Artificial Intelligence in neuroimaging: Clinical applications. Investig. Magn. Reson. Imaging. **26**(1), 1 (2022)
43. D. Lee, J. Lee, J. Ko, J. Yoon, K. Ryu, Y. Nam, Deep learning in MR image processing. Investig. Magn. Reson. Imaging. **23**(2), 81 (2019)
44. X.W. Gao, R. Hui, Z. Tian, Classification of CT brain images based on deep learning networks. Comput. Methods Programs Biomed. **138**, 49–56 (2017)
45. M. Grewal, M.M. Srivastava, P. Kumar, S. Varadarajan, RADnet: Radiologist level accuracy using deep learning for hemorrhage detection in CT scans. in *Proceedings—international symposium on biomedical imaging* (2018), pp. 281–284.
46. P.D. Chang, E. Kuoy, J. Grinband, B.D. Weinberg, M. Thompson, R. Homo et al., Hybrid 3d/2d Convolutional Neural Network for hemorrhage evaluation on head CT. Am. J. Neuroradiol. **39**(9), 1609–1616 (2018)
47. K. Jnawali, M.R. Arbabshirani, N. Rao, A.A. Patel., Deep 3D convolution neural network for CT brain hemorrhage classification. in *Medical Imaging 2018: Computer-Aided Diagnosis*, eds. by K. Mori, N. Petrick
48. M.R. Arbabshirani, B.K. Fornwalt, G.J. Mongelluzzo, J.D. Suever, B.D. Geise, A.A. Patel et al., Advanced machine learning in action: Identification of intracranial hemorrhage on computed tomography scans of the head with clinical workflow integration. Npj Digit. Med. **1**(1), (2018)
49. B. Rao, V. Zohrabian, P. Cedeno, A. Saha, J. Pahade, M.A. Davis, Utility of artificial intelligence tool as a prospective radiology peer reviewer—detection of unreported intracranial hemorrhage. Acad. Radiol. **28**(1), 85–93 (2021)
50. J. Ker, S.P. Singh, Y. Bai, J. Rao, T. Lim, L. Wang, Image thresholding improves 3-dimensional convolutional neural network diagnosis of different acute brain hemorrhages on computed tomography scans. Sensors. **19**(9), 2167 (2019)
51. A.E. Flanders, L.M. Prevedello, G. Shih, S.S. Halabi, J. Kalpathy-Cramer, R. Ball et al., Construction of a machine learning dataset through collaboration: The RSNA 2019 Brain CT Hemorrhage Challenge. Radiol.: Artif. Intell.. **2**(4), (2020)
52. D.T. Ginat, Analysis of head CT scans flagged by deep learning software for acute intracranial hemorrhage. Neuroradiology **62**(3), 335–340 (2019)
53. N.M. Murray, M. Unberath, G.D. Hager, F.K. Hui, Artificial intelligence to diagnose ischemic stroke and identify large vessel occlusions: A systematic review. J. NeuroInterventional Surgery. **12**(2), 156–164 (2019)
54. P.A. Barber, A.M. Demchuk, J. Zhang, A.M. Buchan, Validity and reliability of a quantitative computed tomography score in predicting outcome of hyperacute stroke before thrombolytic therapy. The Lancet. **355**(9216), 1670–1674 (2000)
55. J. Goebel, E. Stenzel, N. Guberina, I. Wanke, M. Koehrmann, C. Kleinschnitz et al., Automated aspect rating: Comparison between the frontier aspect score software and the brainomix software. Neuroradiology **60**(12), 1267–1272 (2018)
56. N. Guberina, U. Dietrich, A. Radbruch, J. Goebel, C. Deuschl, A. Ringelstein et al., Detection of early infarction signs with machine learning-based diagnosis by means of the Alberta stroke program early CT score (aspects) in the clinical routine. Neuroradiology **60**(9), 889–901 (2018)
57. C. Maegerlein, J. Fischer, S. Mönch, M. Berndt, S. Wunderlich, C.L. Seifert et al., Automated calculation of the alberta stroke program early CT score: Feasibility and reliability. Radiology **291**(1), 141–148 (2019)

58. J. You, A.C. Tsang, P.L. Yu, E.L. Tsui, P.P. Woo, C.S. Lui et al., Automated hierarchy evaluation system of large vessel occlusion in acute ischemia stroke. Front. Neuroinformatics. **14**, (2020)
59. A. Nielsen, M.B. Hansen, A. Tietze, K. Mouridsen, Prediction of tissue outcome and assessment of treatment effect in acute ischemic stroke using deep learning. Stroke **49**(6), 1394–1401 (2018)
60. S. Schalekamp, B. van Ginneken, E. Koedam, M.M. Snoeren, A.M. Tiehuis, R. Wittenberg et al., Computer-aided detection improves detection of pulmonary nodules in chest radiographs beyond the support by bone-suppressed images. Radiology **272**(1), 252–261 (2014)
61. J. De Fauw, J.R. Ledsam, B. Romera-Paredes, S. Nikolov, N. Tomasev, S. Blackwell et al., Clinically applicable deep learning for diagnosis and referral in retinal disease. Nat. Med. **24**(9), 1342–1350 (2018)
62. Z. Shi, B. Hu, U.J. Schoepf, R.H. Savage, D.M. Dargis, C.W. Pan et al., Artificial Intelligence in the management of intracranial aneurysms: Current status and future perspectives. Am. J. Neuroradiol. **41**(3), 373–379 (2020)
63. A. Krizhevsky, I. Sutskever, G.E. Hinton, ImageNet classification with deep convolutional Neural Networks. Commun. ACM **60**(6), 84–90 (2017)
64. D. Caudo, Artificial intelligence, machine learning and deep learning in neuroradiology: Current applications. Imaging Med. **14**(10), (2022)
65. I. Blümcke, R. Coras, R.M. Busch, M. Morita-Sherman, D. Lal, R. Prayson et al., Toward a better definition of focal cortical dysplasia: An iterative histopathological and genetic agreement trial. Epilepsia **62**(6), 1416–1428 (2021)
66. B. Caldairou, N.A. Foit, C. Mutti, F. Fadaie, R. Gill, H.M. Lee et al., MRI-based machine learning prediction framework to lateralize hippocampal sclerosis in patients with Temporal Lobe epilepsy. Neurology. **97**(16), (2021)
67. E. Gleichgerrcht, B.C. Munsell, S. Alhusaini, M.K.M. Alvim, N. Bargalló, B. Bender et al., Artificial intelligence for classification of temporal lobe epilepsy with roi-level MRI DATA: A worldwide enigma-epilepsy study. NeuroImage: Clinical. **31**, 102765 (2021)
68. R. Bonacchi, M. Filippi, M.A. Rocca, Role of artificial intelligence in ms clinical practice. NeuroImage: Clinical. **35**, 103065 (2022)
69. R.J. Gillies, P.E. Kinahan, H. Hricak, Radiomics: Images are more than pictures, they are data. Radiology **278**(2), 563–577 (2016)
70. A. Gooya, K.M. Pohl, M. Bilello, L. Cirillo, G. Biros, E.R. Melhem et al., GLISTR: Glioma image segmentation and registration. IEEE Trans. Med. Imaging **31**(10), 1941–1954 (2012)
71. R. Stupp, S. Taillibert, A. Kanner, W. Read, D.M. Steinberg, B. Lhermitte et al., Effect of tumor-treating fields plus maintenance temozolomide vs maintenance temozolomide alone on survival in patients with glioblastoma. JAMA **318**(23), 2306 (2017)
72. M.D. Kuo, N. Jamshidi, Behind the numbers: Decoding molecular phenotypes with radiogenomics—guiding principles and technical considerations. Radiology **270**(2), 320–325 (2014)
73. P.Y. Wen, D.R. Macdonald, D.A. Reardon, T.F. Cloughesy, A.G. Sorensen, E. Galanis et al., Updated response assessment criteria for high-grade gliomas: Response assessment in neuro-oncology working group. J. Clin. Oncol. **28**(11), 1963–1972 (2010)
74. S. Wang, M. Martinez-Lage, Y. Sakai, S. Chawla, S.G. Kim, M. Alonso-Basanta et al., Differentiating tumor progression from pseudoprogression in patients with glioblastomas using diffusion tensor imaging and dynamic susceptibility contrast MRI. Am. J. Neuroradiol. **37**(1), 28–36 (2015)
75. H. Akbari, L. Macyszyn, X. Da, M. Bilello, R.L. Wolf, M. Martinez-Lage et al., Imaging surrogates of infiltration obtained via multiparametric imaging pattern analysis predict subsequent location of recurrence of glioblastoma. Neurosurgery **78**(4), 572–580 (2016)
76. R. Ortiz-Ramón, A. Larroza, S. Ruiz-España, E. Arana, D. Moratal, Classifying brain metastases by their primary site of origin using a radiomics approach based on texture analysis: A feasibility study. Eur. Radiol. **28**(11), 4514–4523 (2018)

77. F.S. Collins, H. Varmus, A new initiative on precision medicine. N. Engl. J. Med. **372**(9), 793–795 (2015)
78. Diagnostic Image Analysis Group, *AI for radiology: an implementation guide* (2020). Accessed 8 Jun 2021
79. M.J. Carreira, D. Cabello, M.G. Penedo, A. Mosquera, Computer-aided diagnoses: Automatic detection of lung nodules. Med. Phys. **25**(10), 1998–2006 (1998)
80. B. van Ginneken, N. Karssemeijer, C.M. Schaefer-Prokop, S. Schalekamp, Chest radiography: New technological developments and their applications. Semin. Respir. Crit. Care Medicine. **35**(01), 003–016 (2014)
81. J.G. Nam, S. Park, E.J. Hwang, J.H. Lee, K.-N. Jin, K.Y. Lim et al., Development and validation of deep learning–based automatic detection algorithm for malignant pulmonary nodules on chest radiographs. Radiology **290**(1), 218–228 (2019)
82. X.S. Wang, Y.F. Peng, L. Lu et al., ChestX-ray: hospital-scale chest X-ray database and benchmarks on weakly-supervised classification and localization of common thorax diseases. Proc CVPR IEEE **2017**, 3462–3471 (2017)
83. Murphy K, Habib SS, Zaidi SM, Khowaja S, Khan A, Melendez J, et al. Computer aided detection of tuberculosis on chest radiographs: An evaluation of the CAD4TB V6 system. Sci. Rep. **10**(1), (2020)
84. E.J. Hwang, S. Park, K-N. Jin, J.I. Kim, S.Y. Choi, J.H. Lee et al., Development and validation of a deep learning–based automated detection algorithm for major thoracic diseases on chest radiographs. JAMA Network Open. **2**(3), (2019)
85. S. Park, S.M. Lee, N. Kim, J. Choe, Y. Cho, K.-H. Do et al., Application of deep learning–based computer-aided detection system: Detecting pneumothorax on chest radiograph after biopsy. Eur. Radiol. **29**(10), 5341–5348 (2019)
86. C. Sheng, L. Li, W. Pei, Automatic detection of supporting device positioning in intensive care unit radiography. Int. J. Med. Robot. Comput. Assist. Surgery. **5**(3), 332–340 (2009)
87. E.-F. Kao, T.-S. Jaw, C.-W. Li, M.-C. Chou, G.-C. Liu, Automated detection of endotracheal tubes in paediatric chest radiographs. Comput. Methods Programs Biomed. **118**(1), 1–10 (2015)
88. D.R. Baldwin, J. Gustafson, L. Pickup, C. Arteta, P. Novotny, J. Declerck et al., External validation of a convolutional neural network artificial intelligence tool to predict malignancy in pulmonary nodules. Thorax **75**(4), 306–312 (2020)
89. E.M. DeBoer, W. Swiercz, S.L. Heltshe, M.M. Anthony, P. Szefler, R. Klein et al., Automated CT scan scores of bronchiectasis and air trapping in cystic fibrosis. Chest **145**(3), 593–603 (2014)
90. M. El Haj, P. Allain, C. Annweiler, C. Boutoleau-Bretonnière, G. Chapelet, K. Gallouj et al., Burnout of healthcare workers in acute care geriatric facilities during the COVID-19 crisis: An online-based study. J. Alzheimer's Disease. **78**(2), 847–852 (2020)
91. Y. Kim, J.Y. Park, E.J. Hwang, S.M. Lee, C.M. Park, Applications of artificial intelligence in the thorax: A narrative review focusing on Thoracic Radiology. J. Thorac. Dis. **13**(12), 6943–6962 (2021)
92. D. Dey, P.J. Slomka, P. Leeson, D. Comaniciu, S. Shrestha, P.P. Sengupta et al., Artificial intelligence in cardiovascular imaging. J. Am. Coll. Cardiol. **73**(11), 1317–1335 (2019)
93. M.T. Nolan, P. Thavendiranathan, Automated quantification in echocardiography. JACC: Cardiovasc. Imaging. **12**(6), 1073–1092 (2019)
94. M.D. Samad, A. Ulloa, G.J. Wehner, L. Jing, D. Hartzel, C.W. Good et al., Predicting survival from large echocardiography and electronic health record datasets. JACC: Cardiovasc. Imaging. **12**(4), 681–689 (2019)
95. M. van Assen, G. Muscogiuri, D. Caruso, S.J. Lee, A. Laghi, C.N. De Cecco, Artificial Intelligence in cardiac radiology. Radiol. Med. (Torino) **125**(11), 1186–1199 (2020)
96. K. Kusunose, A. Haga, T. Abe, M. Sata, Utilization of artificial intelligence in Echocardiography. Circ. J. **83**(8), 1623–1629 (2019)
97. R. Underwood, *Faculty opinions recommendation of ACCF/AHA 2007 clinical expert consensus document on coronary artery calcium scoring by computed tomography in global*

cardiovascular risk assessment and in evaluation of patients with chest pain: A report of the American College of Cardiology Foundation Clinical Expert Consensus Task Force (ACCF/Aha Writing Committee to update the 2000 expert consensus document on Electron Beam computed tomography) developed in collaboration with the Society of Atherosclerosis Imaging and Prevention and the Society of Cardiovascular Computed Tomography. Faculty Opinions—Post-Publication Peer Review of the Biomedical Literature (2007)

98. I. Išgum, A. Rutten, M. Prokop, B. van Ginneken, Detection of coronary calcifications from computed tomography scans for automated risk assessment of coronary artery disease. Med. Phys. **34**(4), 1450–1461 (2007)
99. R.A. Takx, P.A. de Jong, T. Leiner, M. Oudkerk, H.J. de Koning, C.P. Mol et al., Automated coronary artery calcification scoring in non-gated chest CT: Agreement and Reliability. PLoS ONE. **9**(3), (2014)
100. R.C. Cury, S. Abbara, S. Achenbach, A. Agatston, D.S. Berman, M.J. Budoff et al., CAD-RADSTM coronary artery disease—reporting and data system. an expert consensus document of the Society of Cardiovascular Computed Tomography (SCCT), the American College of Radiology (ACR) and the North American Society for Cardiovascular Imaging (NASCI). endorsed by the American College of Cardiology. J. Cardiovasc. Comput. Tomogr., **10**(4), 269–281 (2016)
101. G. Muscogiuri, M. Chiesa, M. Trotta, M. Gatti, V. Palmisano, S. Dell'Aversana et al., Performance of a deep learning algorithm for the evaluation of CAD-RADS classification with CCTA. Atherosclerosis **294**, 25–32 (2020)
102. A.D. Choi, H. Marques, V. Kumar, W.F. Griffin, H. Rahban, R.P. Karlsberg et al., CT evaluation by artificial intelligence for atherosclerosis, stenosis and vascular morphology (clarify): A multi-center, International Study. J. Cardiovasc. Comput. Tomogr. **15**(6), 470–476 (2021)
103. F. Commandeur, M. Goeller, A. Razipour, S. Cadet, M.M. Hell, J. Kwiecinski et al., Fully automated CT quantification of epicardial adipose tissue by deep learning: A multicenter study. Radiol.: Artif. Intelligence. **1**(6), (2019)
104. T. Masuda, T. Nakaura, Y. Funama, T. Okimoto, T. Sato, T. Higaki et al., Machine-learning integration of CT histogram analysis to evaluate the composition of atherosclerotic plaques: Validation with IB-Ivus. J. Cardiovasc. Comput. Tomogr. **13**(2), 163–169 (2019)
105. T.J. Dawes, A. de Marvao, W. Shi, T. Fletcher, G.M. Watson, J. Wharton et al., Machine learning of three-dimensional right ventricular motion enables outcome prediction in Pulmonary hypertension: A cardiac MR imaging study. Radiology **283**(2), 381–390 (2017)
106. T. He, J.N. Fong, L.W. Moore, C.F. Ezeana, D. Victor, M. Divatia et al., An imageomics and multi-network based deep learning model for risk assessment of liver transplantation for hepatocellular cancer. Comput. Med. Imaging Graph. **89**, 101894 (2021)
107. S. Benjamens, P. Dhunnoo, B. Meskó, The state of Artificial Intelligence-based FDA-approved medical devices and algorithms: An online database. Npj Digit. Medicine. **3**(1), (2020)
108. T. Teramoto, T. Shinohara, A. Takiyama, Computer-aided classification of hepatocellular ballooning in liver biopsies from patients with Nash using persistent homology. Comput. Methods Programs Biomed. **195**, 105614 (2020)
109. F. Pérez-Sanz, M. Riquelme-Pérez, E. Martínez-Barba, J. de la Peña-Moral, A. Salazar Nicolás, M. Carpes-Ruiz et al., Efficiency of machine learning algorithms for the determination of macrovesicular steatosis in frozen sections stained with Sudan to evaluate the quality of the graft in liver transplantation. Sensors. **21**(6), 1993 (2021)
110. H. Qu, C.D. Minacapelli, C. Tait, K. Gupta, A. Bhurwal, C. Catalano et al., Training of computational algorithms to predict NAFLD activity score and fibrosis stage from liver histopathology slides. Comput. Methods Programs Biomed. **207**, 106153 (2021)
111. S. Vanderbeck, J. Bockhorst, R. Komorowski, D.E. Kleiner, S. Gawrieh, Automatic classification of white regions in liver biopsies by supervised machine learning. Hum. Pathol. **45**(4), 785–792 (2014)
112. R. Forlano, B.H. Mullish, N. Giannakeas, J.B. Maurice, N. Angkathunyakul, J. Lloyd et al., High-throughput, machine learning–based quantification of steatosis, inflammation, ballooning, and fibrosis in biopsies from patients with nonalcoholic fatty liver disease. Clin. Gastroenterol. Hepatology. **18**(9), (2020)

113. S. Gawrieh, D. Sethunath, O.W. Cummings, D.E. Kleiner, R. Vuppalanchi, N. Chalasani et al., Automated quantification and architectural pattern detection of hepatic fibrosis in NAFLD. Ann. Diagn. Pathol. **47**, 151518 (2020)
114. M. Biswas, V. Kuppili, D.R. Edla, H.S. Suri, L. Saba, R.T. Marinhoe et al., Symtosis: A liver ultrasound tissue characterization and risk stratification in optimized deep learning paradigm. Comput. Methods Programs Biomed. **155**, 165–177 (2018)
115. R. Wang, Y. He, C. Yao, S. Wang, Y. Xue, Z. Zhang et al., Classification and segmentation of hyperspectral data of hepatocellular carcinoma samples using 1-d Convolutional Neural Network. Cytometry A **97**(1), 31–38 (2019)
116. C. Sun, A. Xu, D. Liu, Z. Xiong, F. Zhao, W. Ding, Deep learning-based classification of liver cancer histopathology images using only global labels. IEEE J. Biomed. Health Inform. **24**(6), 1643–1651 (2020)
117. M. Roy, J. Kong, S. Kashyap, V.P. Pastore, F. Wang, K.C. Wong et al., Convolutional autoencoder based model histocae for segmentation of viable tumor regions in liver whole-slide images. Scientific Rep. **11**(1), (2021)
118. R. Yamashita, J. Long, A. Saleem, D.L. Rubin, J. Shen, Deep learning predicts postsurgical recurrence of hepatocellular carcinoma from digital histopathologic images. Scientific Reports. **11**(1), (2021)
119. J. Peng, J. Zhang, Q. Zhang, Y. Xu, J. Zhou, L. Liu, A radiomics nomogram for preoperative prediction of microvascular invasion risk in hepatitis B virus-related hepatocellular carcinoma. Diagn. Interv. Radiology., 121–127 (2018)
120. A. Abajian, N. Murali, L.J. Savic, F.M. Laage-Gaupp, N. Nezami, J.S. Duncan et al., Predicting treatment response to intra-arterial therapies for hepatocellular carcinoma with the use of supervised machine learning—an artificial intelligence concept. J. Vasc. Interv. Radiology. **29**(6), (2018)
121. K.I. Ringe, V.D. Vo Chieu, F. Wacker, H. Lenzen, M.P. Manns, C. Hundt et al., Fully automated detection of primary sclerosing cholangitis (psc)-compatible bile duct changes based on 3D magnetic resonance cholangiopancreatography using machine learning. Eur. Radiol. **31**(4), 2482–2489 (2020)
122. K.G. Foley, R.K. Hills, B. Berthon, C. Marshall, C. Parkinson, W.G. Lewis et al., Development and validation of a prognostic model incorporating texture analysis derived from standardised segmentation of PET in patients with Oesophageal Cancer. Eur. Radiol. **28**(1), 428–436 (2017)
123. Y. Li, M. Beck, T. Päßler, C. Lili, W. Hua, H.D. Mai et al., A FDG-PET radiomics signature detects esophageal squamous cell carcinoma patients who do not benefit from chemoradiation. Sci. Rep. **10**(1), (2020)
124. X. Jin, X. Zheng, D. Chen, J. Jin, G. Zhu, X. Deng et al., Prediction of response after chemoradiation for esophageal cancer using a combination of dosimetry and CT Radiomics. Eur. Radiol. **29**(11), 6080–6088 (2019)
125. J. Li, D. Dong, M. Fang, R. Wang, J. Tian, H. Li et al., Dual-energy CT–based deep learning radiomics can improve lymph node metastasis risk prediction for gastric cancer. Eur. Radiol. **30**(4), 2324–2333 (2020)
126. K.-W. Wang, M. Dong, Potential applications of artificial intelligence in colorectal polyps and cancer: Recent advances and prospects. World J. Gastroenterol. **26**(34), 5090–5100 (2020)
127. A.M. Godkhindi, R.M. Gowda, Automated detection of polyps in CT colonography images using deep learning algorithms in colon cancer diagnosis. in *2017 international conference on energy, communication, data analytics and soft computing (ICECDS)* (2017)
128. R. Ferrari, C. Mancini-Terracciano, C. Voena, M. Rengo, M. Zerunian, A. Ciardiello et al., MR-based artificial intelligence model to assess response to therapy in locally advanced rectal cancer. Eur. J. Radiol. **118**, 1–9 (2019)
129. M.F. Byrne, N. Chapados, F. Soudan, C. Oertel, M. Linares Pérez, R. Kelly et al., Real-time differentiation of adenomatous and hyperplastic diminutive colorectal polyps during analysis of unaltered videos of standard colonoscopy using a deep learning model. Gut **68**(1), 94–100 (2017)

130. G. Pan, G. Yan, X. Song, X. Qiu, BP neural network classification for bleeding detection in wireless capsule endoscopy. J. Med. Eng. Technol. **33**(7), 575–581 (2009)
131. G. Ciuti, K. Skonieczna-Żydecka, W. Marlicz, V. Iacovacci, H. Liu, D. Stoyanov et al., Frontiers of robotic colonoscopy: A comprehensive review of robotic colonoscopes and technologies. J. Clin. Med. **9**(6), 1648 (2020)
132. H. Nakase, T. Hirano, K. Wagatsuma, T. Ichimiya, T. Yamakawa, Y. Yokoyama et al., Artificial intelligence–assisted endoscopy changes the definition of mucosal healing in ulcerative colitis. Dig. Endosc., (2020)
133. P. Harkey, R. Duszak, S. Gyftopoulos, A.B. Rosenkrantz, Who refers musculoskeletal extremity imaging examinations to Radiologists? Am. J. Roentgenol. **210**(4), 834–841 (2018)
134. S.S. Halabi, L.M. Prevedello, J. Kalpathy-Cramer, A.B. Mamonov, A. Bilbily, M. Cicero et al., The RSNA pediatric bone age machine learning challenge. Radiology **290**(2), 498–503 (2019)
135. H.H. Thodberg, S. Kreiborg, A. Juul, K.D. Pedersen, The BONEXPERT method for automated determination of skeletal maturity. IEEE Trans. Med. Imaging **28**(1), 52–66 (2009)
136. D. Kathirvelu, P. Vinupritha, V. Kalpana, A computer aided diagnosis system for measurement of mandibular cortical thickness on dental panoramic radiographs in prediction of women with low bone mineral density. J. Med. Systems. **43**(6), (2019)
137. M. Gundry, K. Knapp, R. Meertens, J.R. Meakin, Computer-aided detection in musculoskeletal projection radiography: A systematic review. Radiography **24**(2), 165–174 (2018)
138. Y. Pan, D. Shi, H. Wang, T. Chen, D. Cui, X. Cheng et al., Automatic opportunistic osteoporosis screening using low-dose chest computed tomography scans obtained for lung cancer screening. Eur. Radiol. **30**(7), 4107–4116 (2020)
139. C.M. Deniz, S. Xiang, R.S. Hallyburton, A. Welbeck, J.S. Babb, S. Honig et al., Segmentation of the proximal femur from MR images using deep convolutional neural networks. Sci. Rep. **8**(1), (2018)
140. S.W. Chung, S.S. Han, J.W. Lee, K.-S. Oh, N.R. Kim, J.P. Yoon et al., Automated detection and classification of the proximal humerus fracture by using deep learning algorithm. Acta Orthop. **89**(4), 468–473 (2018)
141. J. Olczak, N. Fahlberg, A. Maki, A.S. Razavian, A. Jilert, A. Stark et al., Artificial Intelligence for analyzing orthopedic trauma radiographs. Acta Orthop. **88**(6), 581–586 (2017)
142. H. Roth, Y. Wang, J. Yao, L. Lu, J.E. Burns, *Summers R. [PDF] deep convolutional networks for automated detection of posterior-element fractures on spine CT: Semantic scholar1970*
143. J.E. Burns, J. Yao, R.M. Summers, Vertebral body compression fractures and bone density: Automated detection and classification on CT Images. Radiology **284**(3), 788–797 (2017)
144. G.S. Lodwick, C.L. Haun, W.E. Smith, R.F. Keller, E.D. Robertson, Computer diagnosis of primary bone tumors. Radiology **80**(2), 273–275 (1963)
145. J.M. Winfield, A. Miah, D. Strauss, K. Thway, V.A. Morgan, D.J. Collins et al., Supervised machine-learning enables segmentation and evaluation of heterogeneous post-treatment changes in multi-parametric MRI of soft-tissue sarcoma. Front. Oncol. **9**, (2019)
146. K.A. Thomas, L. Kidziński, E. Halilaj, S.L. Fleming, G.R. Venkataraman, E.H. Oei et al., Automated classification of radiographic knee osteoarthritis severity using deep neural networks. Radiol.: Artif. Intelligence. **2**(2), (2020)
147. V. Pedoia, B. Norman, S.N. Mehany, M.D. Bucknor, T.M. Link, S. Majumdar, 3D convolutional neural networks for detection and severity staging of meniscus and PFJ cartilage morphological degenerative changes in osteoarthritis and anterior cruciate ligament subjects. J. Magn. Reson. Imaging **49**(2), 400–410 (2018)
148. J. Li, K. Qian, J. Liu, Z. Huang, Y. Zhang, G. Zhao et al., Identification and diagnosis of meniscus tear by magnetic resonance Imaging using a deep learning model. J. Orthop. Translation. **34**, 91–101 (2022)
149. Z. Akkus, J. Cai, A. Boonrod, A. Zeinoddini, A.D. Weston, K.A. Philbrick, B.J. Erickson, A survey of deep-learning applications in ultrasound: Artificial intelligence-powered ultrasound for improving clinical workflow. J. Am. Coll. Radiol. **16**(9), 1318–1328 (2019)

150. Y. Hu, Y. Guo, Y. Wang, J. Yu, J. Li, S. Zhou et al., Automatic tumor segmentation in breast ultrasound images using a dilated fully convolutional network combined with an active contour model. Med. Phys. **46**(1), 215–228 (2018)
151. S.-M. Hsu, W.-H. Kuo, F.-C. Kuo, Y.-Y. Liao, Breast tumor classification using different features of quantitative ultrasound parametric images. Int. J. Comput. Assist. Radiol. Surg. **14**(4), 623–633 (2019)
152. H.J. Park, S.M. Kim, B. La Yun, M. Jang, B. Kim, J.Y. Jang et al., A computer-aided diagnosis system using artificial intelligence for the diagnosis and characterization of breast masses on ultrasound. Medicine. **98**(3), (2019)
153. S P, N KV, S S. Breast cancer detection using crow search optimization based intuitionistic fuzzy clustering with neighborhood attraction. Asian Pac. J. Cancer Prev. **20**(1), 157–165 (2019)
154. M.A. Al-antari, M.A. Al-masni, M.-T. Choi, S.-M. Han, T.-S. Kim, A fully integrated computer-aided diagnosis system for digital X-ray mammograms via deep learning detection, segmentation, and classification. Int. J. Med. Informatics **117**, 44–54 (2018)
155. M.F. Inciardi, A.V. Edwards, J. Papaioannou, Interpretation time using a concurrent-read computer-aided detection system for automated breast ultrasound in breast cancer screening of women with dense breast tissue. Am. J. Roentgenol. **211**(2), 452–461 (2018)
156. H. Cai, Q. Huang, W. Rong, Y. Song, J. Li, J. Wang et al., Breast microcalcification diagnosis using deep convolutional neural network from digital mammograms. Comput. Math. Methods Med. **2019**, 2717454 (2019)
157. N.H. Anderson, P.W. Hamilton, P.H. Bartels, D. Thompson, R. Montironi, J.M. Sloan, Computerized scene segmentation for the discrimination of architectural features in ductal proliferative lesions of the breast. J. Pathol. **181**(4), 374–380 (1997)
158. E.E. Fowler, A.M. Smallwood, N.Z. Khan, K. Kilpatrick, T.A. Sellers, J. Heine, Technical challenges in generalizing calibration techniques for breast density measurements. Med. Phys. **46**(2), 679–688 (2019)
159. A.A. Mohamed, Y. Luo, H. Peng, R.C. Jankowitz, S. Wu, Understanding clinical mammographic breast density assessment: A deep learning perspective. J. Digit. Imaging **31**(4), 387–392 (2017)
160. M. Sepandi, M. Akrami, H. Tabatabaee, A. Rajaeefard, S. Tahmasebi, K.A. Angali et al., Breast cancer risk factors in women participating in A breast screening program: A study on 11,850 Iranian females. Asian Pac. J. Cancer Prev. **15**(19), 8499–8502 (2014)
161. D. Sheth, M.L. Giger, Artificial intelligence in the interpretation of breast cancer on MRI. J. Magn. Reson. Imaging **51**(5), 1310–1324 (2019)
162. N. Antropova, H. Abe, M.L. Giger, Use of clinical MRI maximum intensity projections for improved breast lesion classification with deep convolutional neural networks. J. Med. Imaging. **5**(01), 1 (2018)
163. A. Meyer-Bäse, L. Morra, U. Meyer-Bäse, K. Pinker, Current status and future perspectives of artificial intelligence in Magnetic Resonance Breast Imaging. Contrast Media Mol. Imaging **2020**, 1–18 (2020)
164. I. Sechopoulos, J. Teuwen, R. Mann, Artificial intelligence for breast cancer detection in mammography and digital breast tomosynthesis: State of the art. Semin. Cancer Biol. **72**, 214–225 (2021)
165. Q. Zhang, S. Song, Y. Xiao, S. Chen, J. Shi, H. Zheng, Dual-mode artificially-intelligent diagnosis of breast tumours in shear-wave elastography and B-mode ultrasound using deep polynomial networks. Med. Eng. Phys. **64**, 1–6 (2019)
166. M. Adachi, T. Fujioka, M. Mori, K. Kubota, Y. Kikuchi, W. Xiaotong et al., Detection and diagnosis of breast cancer using artificial intelligence based assessment of maximum intensity projection dynamic contrast-enhanced magnetic resonance images. Diagnostics. **10**(5), 330 (2020)
167. C.L. Ondeck, D. Pretorius, J. McCaulley, M. Kinori, T. Maloney, A. Hull et al., Ultrasonographic prenatal imaging of fetal ocular and orbital abnormalities. Surv. Ophthalmol. **63**(6), 745–753 (2018)

168. M. Pramanik, M. Gupta, K.B. Krishnan, Enhancing reproducibility of ultrasonic measurements by new users. in *Medical imaging 2013: Image perception, observer performance, and technology assessment* (2013)
169. L. Zhang, S. Chen, C.T. Chin, T. Wang, S. Li, Intelligent Scanning: Automated standard plane selection and biometric measurement of early gestational sac in routine ultrasound examination. Med. Phys. **39**(8), 5015–5027 (2012)
170. N.A. Smeets, N.A. Dvinskikh, B. Winkens, S.G. Oei, A new semi-automated method for fetal volume measurements with three-dimensional ultrasound: Preliminary results. Prenat. Diagn. **32**(8), 770–776 (2012)
171. Y. Deng, Y. Wang, P. Chen, J. Yu, A hierarchical model for automatic nuchal translucency detection from ultrasound images. Comput. Biol. Med. **42**(6), 706–713 (2012)
172. Z. Yu, E.-L. Tan, D. Ni, J. Qin, S. Chen, S. Li et al., A deep convolutional neural network-based framework for automatic fetal facial standard plane recognition. IEEE J. Biomed. Health Inform. **22**(3), 874–885 (2018)
173. R. Arnaout, L. Curran, Y. Zhao, J.C. Levine, E. Chinn, A.J. Moon-Grady, An ensemble of neural networks provides expert-level prenatal detection of complex congenital heart disease. Nat. Med. **27**(5), 882–891 (2021)
174. L.I. Lee, S. Kanthasamy, R.S. Ayyalaraju, R. Ganatra, The current state of artificial intelligence in medical imaging and nuclear medicine. BJR|Open. **1**(1), 20190037 (2019)
175. B. Letzen, C.J. Wang, J. Chapiro, The role of artificial intelligence in interventional oncology: A primer. J. Vasc. Interv. Radiology. **30**(1), (2019)
176. J. Xu, M. Zhang, E.A. Turk, L. Zhang, P.E. Grant, K. Ying et al., *Fetal pose estimation in volumetric MRI using a 3D convolution neural network. Lecture Notes in Computer Science* (2019), pp. 403–410
177. H. Li, G. Yan, W. Luo, T. Liu, Y. Wang, R. Liu et al., Mapping Fetal Brain Development based on automated segmentation and 4D brain atlasing. Brain Struct. Funct. **226**(6), 1961–1972 (2021)
178. N. Khalili, E. Turk, M.J.N.L. Benders, P. Moeskops, N.H.P. Claessens, R. de Heus et al., Automatic extraction of the intracranial volume in fetal and neonatal MR scans using convolutional neural networks. NeuroImage: Clin. **24**, 102061 (2019)
179. S. Dahdouh, N. Andescavage, S. Yewale, A. Yarish, D. Lanham, D. Bulas et al., In vivo placental MRI shape and textural features predict fetal growth restriction and postnatal outcome. J. Magn. Reson. Imaging **47**(2), 449–458 (2017)
180. J.R. Leyendecker, M. DuBose, K. Hosseinzadeh, R. Stone, J. Gianini, D.D. Childs et al., MRI of pregnancy-related issues: Abnormal placentation. Am. J. Roentgenol. **198**(2), 311–320 (2012)
181. P. Shrestha, B. Poudyal, S. Yadollahi, E.D. Wright, V.A. Gregory, D.J. Warner et al., A systematic review on the use of artificial intelligence in gynecologic imaging—background, state of the art, and Future Directions. Gynecol. Oncology. **166**(3), 596–605 (2022)
182. Y. Wang, Q. Yao, J.T. Kwok, L.M. Ni, Generalizing from a few examples. ACM Comput. Surv. **53**(3), 1–34 (2020)
183. S.N. Keshava, S. Kalva, Artificial intelligence in interventional radiology. J. Clin. Interv. Radiol. ISVIR. **03**(02), 071–071 (2019)
184. A. Morshid, K.M. Elsayes, A.M. Khalaf, M.M. Elmohr, J. Yu, A.O. Kaseb et al., A machine learning model to predict hepatocellular carcinoma response to transcatheter arterial chemoembolization. Radiol.: Artif. Intelligence. **1**(5), (2019)
185. J. Peng, S. Kang, Z. Ning, H. Deng, J. Shen, Y. Xu et al., Residual convolutional neural network for predicting response of transarterial chemoembolization in hepatocellular carcinoma from CT imaging. Eur. Radiol. **30**(1), 413–424 (2019)
186. R.N. Uppot, B. Laguna, C.J. McCarthy, G. De Novi, A. Phelps, E. Siegel et al., Implementing virtual and augmented reality tools for Radiology Education and training, communication, and clinical care. Radiology **291**(3), 570–580 (2019)
187. P. Ambrosini, D. Ruijters, W.J. Niessen, A. Moelker, T. van Walsum, *Fully automatic and real-time catheter segmentation in X-ray fluoroscopy. Lecture Notes in Computer Science* (2017), pp. 577–585

188. S. Yang, J. Kweon, J-H. Roh, J-H. Lee, H. Kang, L-J. Park et al., Deep learning segmentation of major vessels in X-ray coronary angiography. Sci. Rep. **9**(1), (2019)
189. Y. Lee, K. Veerubhotla, M.H. Jeong, C.H. Lee, Deep learning in personalization of cardiovascular stents. J. Cardiovasc. Pharmacol. Ther. **25**(2), 110–120 (2019)
190. C. Floridi, A. Radaelli, F. Pesapane, E.M. Fumarola, M. Lecchi, A. Agostini et al., Clinical impact of cone beam computed tomography on iterative treatment planning during ultrasound-guided percutaneous ablation of liver malignancies. Med. Oncol. **34**(6), (2017)
191. Y. Tang, A.P. Harrison, M. Bagheri, J. Xiao, Summers RM. Semi-automatic recist labeling on CT scans with cascaded convolutional neural networks. In *Medical Image Computing and Computer Assisted Intervention—MICCAI 2018* (2018), pp. 405–413
192. R. Su, S.A. Cornelissen, M. van der Sluijs, A.C. van Es, W.H. van Zwam, D.W. Dippel et al., AutoTICI: Automatic brain tissue reperfusion scoring on 2D DSA images of acute ischemic stroke patients. IEEE Trans. Med. Imaging **40**(9), 2380–2391 (2021)
193. L. Yao, H. Zhang, M. Zhang, X. Chen, J. Zhang, J. Huang et al., Application of artificial intelligence in renal disease. Clinical eHealth. **4**, 54–61 (2021)
194. E.C. Yilmaz, M.J. Belue, B. Turkbey, C. Reinhold, P.L. Choyke, A brief review of artificial intelligence in genitourinary oncological imaging. Can. Assoc. Radiol. J., 084653712211357 (2022)
195. K. Werth, L. Ledbetter, Artificial Intelligence in head and neck imaging. Neuroimaging Clin. N. Am. **30**(3), 359–368 (2020)
196. M.S. Sadaghiani, S.P. Rowe, S. Sheikhbahaei, Applications of artificial intelligence in Oncologic 18F-FDG PET/CT Imaging: A systematic review. Ann. Transl. Med. **9**(9), 823–823 (2021)
197. A. Tang, R. Tam, A. Cadrin-Chênevert, W. Guest, J. Chong, J. Barfett et al., Canadian association of radiologists white paper on artificial intelligence in radiology. Can. Assoc. Radiol. J. **69**(2), 120–135 (2018)
198. S. Ruder, *An overview of multi-task learning in deep neural networks* (2017)

Chapter 3
Challenges in Accurately Using Artificial Intelligence and Machine Learning in Biomedical Imaging

Mayuri Sharma, Bidisha Goswami, Nabajyoti Goswami, Saurov Mahanta, and Yumnam Jayanta Singh

Abstract Biomedical imaging plays an important role in advancing the quality of health care. It aids in diagnosing and treating patients without the need for surgical intervention. Various types of biomedical imaging techniques help in the diagnosis of various diseases. In the present day, artificial intelligence has been implemented to improve the healthcare sector. Many researchers have various opinions on the applications of Artificial Intelligence and Machine Learning (AI/ML) in healthcare. Artificial Intelligence (AI) in biomedical imaging is currently one of the most popular research topics in healthcare, and it's generating a lot of interest with many scientific articles being published on this subject. While it offers numerous benefits, there are also accompanying challenges that require attention. The primary objective of this chapter is to emphasize the limitations in applying AI/ML in biomedical imaging. To understand the numerous challenges linked to the application of AI/ML in biomedical imaging. This will assist researchers in maintaining awareness of the challenges encountered when utilizing AI/ML in the context of biomedical imaging.

3.1 Introduction

Biomedical imaging encompasses a range of imaging methods and advanced technologies employed to observe and record complicated visuals of the internal structures, functions and physiological activities within the bodies of both humans and animals. These methodologies offer pivotal insights for medical diagnostics, scientific investigations and the development of treatment strategies. In modern healthcare, biomedical imaging plays substantially important roles because it empowers doctors to non-invasively explore the body, which is essential for both diagnosis and treatment. This information equips them to identify issues and address a wide range of medical conditions. Biomedical imaging encompasses a diverse array of imaging

M. Sharma · B. Goswami · N. Goswami · S. Mahanta (✉) · Y. J. Singh
National Institute of Electronics and Information Technology (NIELIT), Guwahati, Assam, India
e-mail: saurov.mahanta@gmail.com

A. Gogoi and N. Mazumder (eds.), *Biomedical Imaging*, Biological and Medical Physics, Biomedical Engineering, https://doi.org/10.1007/978-981-97-5345-1_3

modalities, each offering unique capabilities and applications in the field of medicine [1].

1. **X-ray Imaging**: X-rays employ high-energy electromagnetic waves to create detailed images of bones, soft tissues, and organs. They are invaluable for detecting fractures, tumors, and various pathological conditions.
2. **Computed Tomography (CT)**: CT scans combine X-ray technology with computer processing to create cross-sectional images of the body. It is excellent for visualizing detailed anatomy, detecting tumors, and assessing injuries.
3. **Magnetic Resonance Imaging (MRI)**: MRI utilizes powerful magnets and radio waves to generate highly detailed images of soft tissues within the body. It is particularly useful for examining the brain, muscles, and internal organs, providing crucial diagnostic insights.
4. **Ultrasound**: Ultrasound imaging relies on high-frequency sound waves to produce real-time images. It is commonly used for visualizing organs, monitoring fetal development during pregnancy, and assessing blood flow. Its non-invasive nature makes it ideal for various medical applications.
5. **Positron Emission Tomography (PET)**: PET scans involve the injection of a small amount of radioactive material, known as a tracer, which emits positrons. These positrons interact with surrounding tissues and produce signals that can be used to create images of metabolic processes in the body, aiding in the diagnosis and monitoring of diseases like cancer.
6. **Single Photon Emission Computed Tomography (SPECT)**: Similar to PET, SPECT imaging uses radioactive tracers to visualize blood flow and organ function. It is often employed in cardiology, neurology, and oncology to assess various conditions.
7. **Fluoroscopy**: Fluoroscopy is a real-time X-ray technique used for dynamic imaging during procedures such as angiography (imaging of blood vessels) and gastrointestinal studies. It provides continuous visual feedback, assisting in the guidance of medical interventions.
8. **Nuclear Medicine**: Nuclear medicine combines radioactive materials with specialized detectors to create images of organ function and disease. It is particularly useful in diagnosing conditions like thyroid disorders and bone abnormalities.
9. **Optical Imaging**: Optical imaging techniques, such as endoscopy and optical coherence tomography (OCT), employ light to visualize internal structures within the body. Endoscopy is commonly used for examining the gastrointestinal tract, while OCT aids in eye examinations and assessing skin lesions.
10. **Functional MRI (fMRI)**: fMRI not only captures the anatomical structure but also the functional activity of the brain. It does so by measuring changes in blood flow related to neural activity, making it an invaluable tool for neuroscience research and understanding brain disorders.

Each of these imaging technologies offers unique capabilities and advantages, allowing healthcare professionals to tailor their use to specific diagnostic and research needs including various types of cancers, cardiovascular diseases, coronary artery diseases, heart failure, stroke, and brain tumors [1, 2].

Throughout the past decade, we have witnessed a noticeable and increasing prevalence of the term Artificial Intelligence (AI), not only within the realm of scientific literature but also in various other domains of applied technologies including healthcare, image processing, medical physics and computer vision. AI encompasses computer algorithms capable of replicating human-like intelligence attributes, such as problem-solving and learning. The recent achievements in AI are attributable to significant advancements in computational capabilities and the abundance of available data. The proliferation of healthcare data, including a significant volume of medical images, has made it imperative to adopt innovative approaches using big data technologies, Machine Learning (ML), a subset of AI [2]. These technologies have the ability to process and learn from large volumes of medical data, such as radiological images, patient records, and clinical notes. Furthermore, presently, AI algorithms are employed to examine millions of medical images in real-time, facilitating the early detection of diseases and medical conditions. In particular, AI is applied to enhance the precision of mammograms, which are essential for identifying breast cancer, and for conducting advanced evaluations of MRIs to aid in the identification and treatment of prostate cancers. These algorithms excel in image analysis, delivering quicker and more precise results compared to human radiologists, thereby diminishing the need for exclusive dependence on manual image interpretation and mitigating the potential for overlooked diagnoses.

AI-based methods have been widely acknowledged for their success in reconstructing medical images. However, their extensive adoption in clinical settings remains unrealized. One significant obstacle is the challenge of interpreting the models. Because various AI algorithms inherently operate as "black boxes," they lack a direct physical or theoretical mechanism to elucidate how inputs are transformed into outputs [3]. Consequently, AI-based models encounter resistance among clinicians. Lately, there has been a growing focus on enhancing model interpretability by either developing transparent neural networks or employing various visualization techniques, particularly in the realm of natural image analysis based on Deep Learning (DL)—another subset of AI. Likewise, it is vitally important to invest more effort in constructing AI-based image reconstruction models that are not only interpretable but also capable of high performance. In order to process the medical images for better prediction and accuracy of diagnosis, ML and DL techniques are used as shown in Fig. 3.1. As input, medical images from various modalities are taken into consideration, and then algorithms are applied to these images. Further, the input image is segmented based on various factors, these segments were used to extract the essential and maximum features using feature extraction techniques. Following the extraction of essential features, they undergo additional refinement to acquire the precise features utilized in disease identification [4]. As the integration of AI/ML technologies into the field of biomedical imaging advances, it unveils a dynamic landscape filled with both promising benefits and intricate limitations that warrant thorough examination. Among these challenges, data privacy concerns stand out, demanding vigilant safeguarding of sensitive information [5]. Additionally, the demand for perfectly curated datasets of the highest quality imposes a significant hurdle, necessitating meticulous data collection and curation efforts. The labyrinth of issues surrounding algorithmic interpretability further complicates matters, as it involves deciphering the complex decision-making processes of AI models. Moreover, the persistent presence of biases within these algorithms adds an extra layer of complexity, requiring continuous scrutiny and mitigation. This complex web of challenges underscores the multifaceted nature of implementing AI/ML within the dynamic field of biomedical imaging. Some researchers argue that the potential impact of AI/ML in medical applications on real-world outcomes may not be as significant because the accuracy of disease diagnosis doesn't progress as rapidly when using real-world data compared to research data [6]. While AI/ML offers several advantages, there are also drawbacks, such as the high implementation costs Additionally, it is challenging to provide comprehensive training to everyone in the pharmaceutical sector, especially in a rapidly growing nation like India [7].

AI has further limitations, including its inability to replicate human intelligence, as well as a lack of creativity and adaptability. These challenges highlight the unique risks associated with AI, including its potential effects on employment and ethical concerns related to decision-making processes [8]. While AI methods have made significant strides, the automation of data collection continues to be deficient, leading to excessively lengthy periods of time spent on data collection and curation. Indeed, many cutting-edge AI algorithms can be trained within a few hours, whereas the construction of a comprehensive, meticulously curated database can span several months. Hence, in the era dominated by AI, clinical teams should start becoming familiar with guiding principles for data management and curation, just as physicians

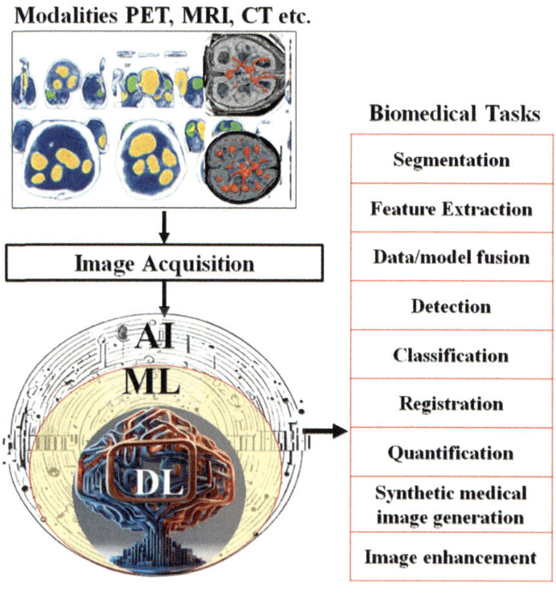

Fig. 3.1 AI in biomedical imaging

are familiar with planning protocols or delineation of guidelines. The FAIR (Findability, Accessibility, Interoperability, and Reusability) data principles are indeed the most popular and broadly applicable, but it is essential for the medical community to prioritize adapting these principles to the specific requirements of the medical field [9–12].

In this chapter, we thoroughly explore the various challenges associated with the applications of AI/ML in biomedical imaging. We carefully examine the risks and accompanying challenges in the field of biomedical imaging when applying AI/ML. Gaining insight into these challenges will assist researchers in improving their outcomes and implementing necessary refinements in their research. The rest of the chapter is organized as follows. Section 3.2 provides an overview of AI/ML in biomedical imaging including a comprehensive survey on the AI/ML approaches which have been concurrently trending and popularly employed into the field. Section 3.3 highlighted the various challenges associated with biomedical imaging addressing several factors affecting the accuracy of various AI/ML methods used for image reconstruction in the medical sector. Finally, we conclude the chapter on Sect. 3.4 by describing the pitfalls and alternatives to overcome limitations as well as to improve accuracy of the algorithms used in the area of biomedical image recognition.

3.2 AI and ML in Biomedical Imaging

The application of AI and ML in biomedical imaging has led to significant advancements in the healthcare industry. Globally, around 20 million lives are lost each year due to diagnostic inaccuracies [13]. These inaccuracies primarily occur when healthcare professionals interpreting medical images occasionally make incorrect assessments. To tackle this challenge, AI/ML are being extensively employed in biomedical imaging. AI/ML plays a critical role by processing digital medical images such as X-rays and MRIs, generating detailed visuals that assist healthcare providers in making diagnostic decisions and predicting disease progression in patients. Furthermore, when examining biomedical images, there is no standardized approach, and different experts may focus on various aspects, leading to uncertainties regarding the accuracy of their assessments.

AI algorithms can be categorized into two groups: those that rely on manually crafted features and those that work directly with raw data. The former falls under the domain of traditional or conventional AI, while the latter is associated with modern DL approaches as depicted in Figs. 3.2 and 3.3. The following subsections are focused on providing an outline of the most popular AI sub-types extensively applied in medical imaging and disease diagnosis.

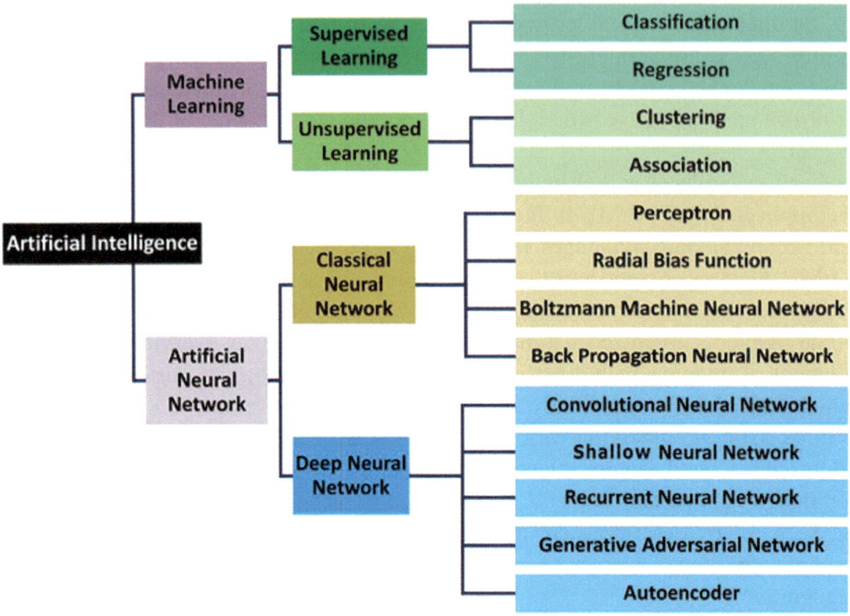

Fig. 3.2 Categorization of AI algorithms

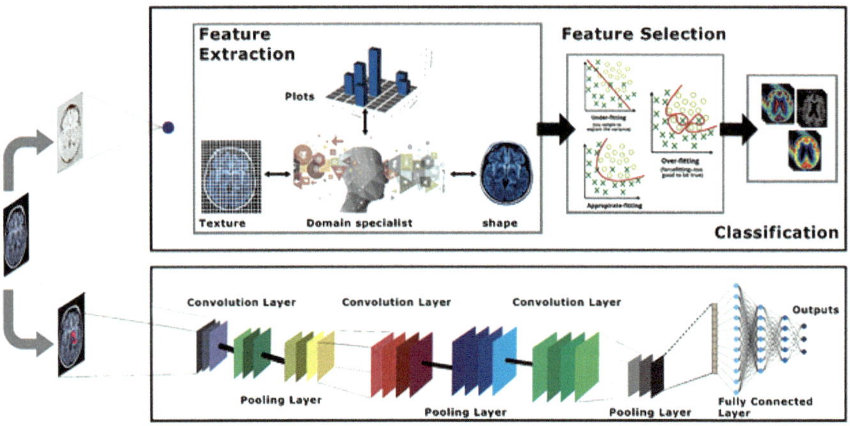

Fig. 3.3 Exploring Machine Learning (ML) and Deep Learning (DL) techniques in medical imaging data analysis. The top row of the image illustrates the conventional AI/ML pipeline, while the bottom row showcases the Convolutional Neural Network (CNN) architecture, a specific instance of Deep Learning (DL)

3.2.1 Traditional Machine Learning Algorithms

Machine Learning (ML) is a concept introduced by Arthur Samuel in 1959 as a subset of AI [14]. It is a method for identifying patterns in provided data, learning these patterns without explicit programming, and solving problems based on the input [15–17]. In the context of medical imaging, ML algorithms take manually extracted features from images, like area, shapes, texture, and pixel histograms, as input [18]. These features can go through a selection process to choose the most relevant ones. Ultimately, the ML algorithms aim to combine these selected features into a single value, such as a tumor signature, which can indicate the likelihood of a disease [19].

ML algorithms are broadly divided into two main categories: supervised learning and unsupervised learning. These two categories are then further divided based on the type of output they produce, resulting in classification and regression for supervised learning and clustering and association for unsupervised learning. In supervised learning, the algorithm is trained using labeled data, where it learns from existing knowledge or from the provided probability distribution of classes, such as predicting diseases from medical images. In contrast, unsupervised learning, as the name implies, doesn't receive labeled data; instead, it discovers patterns within the data and associates them with newly found clusters of data points (Fig. 3.2).

Different machine learning algorithms that depend on manually designed features have been employed in breast cancer detection and analysis. A variety of traditional machine learning methods applied in recent years for breast cancer diagnosis has been highlighted by Yassin et al. [20]. These algorithms encompass Decision Trees (DT), Random Forests (RF), Support Vector Machines (SVM), Naive Bayes (NB), K-Nearest Neighbors (KNN), Linear Discriminant Analysis (LDA), and Logistic Regression (LR), among others [21, 22].

Some of the important points have been emphasized regarding the limitations of handcrafted features in image analysis and the necessity of classifiers in the diagnostic process [23]:

- Dependency on Expert Definitions: Handcrafted features often rely on domain-specific knowledge and expert-defined algorithms. This can be limiting because it may not account for all relevant information in the data. Furthermore, different experts may define features differently, leading to inconsistencies.
- Limited Focus of Feature Extracting Algorithms: Feature extracting algorithms like GLCM, LBP, GRLM, and HOG tend to focus on specific aspects of images (e.g., texture, edges, gradients). This specialization means that a feature extractor designed for one type of data may not perform well on a different type of data. For instance, an algorithm designed for extracting shape-based features may struggle to capture texture information effectively.
- Non-Adaptability to Unknown Data: Systems relying solely on handcrafted features can be rigid and non-adaptable to new or unseen data. They are usually designed and trained for specific data distributions, and when faced with data outside that distribution, their performance can significantly degrade.

- Need for Classifiers: To address the limitations of handcrafted features, classifiers are essential. Classifiers are responsible for making decisions based on the feature space. They take the extracted features and learn how to discriminate between different classes or categories. However, selecting the right classifier can be a complex task because different classifiers have their own strengths and weaknesses, and their performance can vary depending on the data and the problem at hand.

To overcome these challenges, researchers often turn to machine learning techniques that can automatically learn relevant features from the data (feature learning or deep learning). Convolutional Neural Networks (CNNs) are a prime example of deep learning models used in image analysis. CNNs can learn hierarchical representations of images, capturing both low-level features like edges and high-level features like object shapes. This approach can be more adaptable to diverse image types and potentially reduce the need for manual feature engineering. Additionally, ensemble methods that combine the outputs of multiple classifiers or feature learning techniques can enhance classification performance and robustness. The choice of classifier or combination of classifiers should be guided by rigorous experimentation and cross-validation to ensure the best results for a specific problem. In summary, while handcrafted features have their limitations, they can still be valuable in certain scenarios, especially when combined with appropriate classifiers. However, with the advancements in deep learning and feature learning, the field of image analysis is increasingly moving toward more automated and adaptable approaches that can better handle the complexities of diverse image data [23].

3.2.2 *Deep Learning Based AI Algorithms*

Deep Learning (DL) based AI algorithms represent a cutting-edge approach that has emerged to overcome the limitations inherent in traditional ML methods (Fig. 3.2 for algorithmic hierarchy). Recent research endeavors have been increasingly directed towards harnessing the potential of DL-based techniques. Unlike conventional ML algorithms, DL algorithms have a distinct advantage: they can directly process raw data without necessitating manual and explicit feature extraction from the input data to generate predictions or outputs. This pivotal characteristic of DL algorithms enables them to adapt seamlessly to a diverse array of input data types, spanning images, text, audio, and more [24–28]. As a result, DL algorithms consistently yield more robust results in complex tasks such as segmentation and classification when compared to their traditional ML counterparts. Notably, DL algorithms have found remarkable success in the realm of medical imaging, where they have been instrumental in the evaluation and analysis of various medical modalities, including but not limited to Computed Tomography (CT), Magnetic Resonance Imaging (MRI), Ultrasound (US), and Histopathology (HP) [29–31].

The foundation of DL algorithms can be traced back to Artificial Neural Networks (ANN), a subset of AI algorithms. ANNs are designed to emulate the basic functioning of biological neurons within the human brain. At their core, artificial neurons serve as the fundamental building blocks of ANNs where weighted inputs are fed into the neuron, and these inputs are then aggregated to produce an output or prediction. Crucially, the neuron applies a nonlinear activation function—options include Sigmoid, Tanh, Gaussian, or ReLu, among others—to the aggregated inputs, yielding an output response typically within the range of 0–1 [32]. This process mimics the behavior of biological neurons and forms the basis for the transformative capabilities of DL algorithms in processing complex data and performing advanced tasks.

In ANNs, neurons, also known as nodes, establish connections with each other, forming single or multiple hierarchical layers. The decision to accept or reject the incoming signals of neurons hinges on the nonlinear activation function's output. Each neuron or node in the network is associated with a weight that can influence the input it receives, facilitating the transfer of data to the output layer. The input layer conveys inputs or data in the form of a feature vector, complete with weighted values, to the hidden layer. The hidden layer, comprising activation units, takes the feature vector from the first layer, incorporating the weighted values, and performs calculations to produce an output. Meanwhile, the output layer consists of activation units, each corresponding to a label or class in the dataset, carrying the weighted output from the hidden layer to make predictions about the respective class [32]. During the training phase, ANNs employ the back-propagation technique to minimize the error function [33]. This entails reducing errors by updating the weight values across each layer.

Two primary types of ANN have been employed in breast cancer classification: Shallow Neural Networks (SNN) and Deep Neural Networks (DNN) (please refer to Fig. 3.2 for more details on classification [34, 35]. Of these two neural network variations, DNN emerges as the most widely adopted type of ANN, with widespread application in the field of medical image analysis and diagnosis [36]. DNN possesses the capability to autonomously extract pertinent features from raw input data, without requiring expert knowledge or human intervention. Owing to these aforementioned qualities, deep learning techniques play a pivotal role in elevating diagnostic precision, bolstering analytical capabilities, and streamlining decision-making in clinical applications that rely on medical imaging data.

Following the achievement of cutting-edge outcomes in the field of image analysis, various adaptations of deep neural networks (DNNs) have been utilized in the literature to diagnose disease classes or stages from medical imaging data, such as the classification of breast cancer. These DNN variants encompass the Multi-Layer Network (ML-NN) [37, 38], Deep Belief Network (DBN) [39, 40], Principal Component Analysis Net (PCA.Net) [41] Stacked Denoising Autoencoder (SDAE) [42] and various iterations of Convolutional Neural Network (CNN) [43–46]. Each of these deep learning algorithms possesses distinct strengths and limitations when applied to the analysis of medical images. In the literature, it is common to observe that some variants of CNNs are frequently employed for the purpose of analyzing breast cancer images. Within the set of previously mentioned deep learning architectures, the CNN is widely acclaimed as the foremost, highly efficient, and extensively applied framework for the medical image analysis, specifically for the analysis of breast imaging for cancer detection. In the following subsection, we will provide an extensive examination of CNN architecture, specifically in its relevance to the analysis of medical imaging.

3.2.3 Convolutional Neural Networks for Medical Imaging Analysis

Within the field of medical image analysis, CNNs offer a distinct advantage compared to traditional Artificial Neural Networks (ANNs). One of the key distinctions lies in CNN's ability to directly process entire datasets, specifically images. The training of CNN architectures typically involves the utilization of two primary techniques: transfer learning (TL) and de novo training. The de novo technique involves training the CNN architecture from scratch [47]. This approach combines different trained models to create an optimal model. In contrast, transfer learning leverages pre-trained models for analysis. While transfer learning is a convenient method, it tends to be most effective when dealing with limited or small datasets [48].

3.2.3.1 Components of CNN Architecture

A CNN architecture is composed of several key components, each playing a vital role in its functionality:

- Convolutional layers: These layers are the workhorses of CNNs and contain convolutional filters, often referred to as kernels. Convolution operations are applied to the input data, enabling the network to detect high-level features such as edges, colors, shapes, and blobs. These features are essential for accurate image analysis.
- Pooling operations: Pooling operations serve to reduce the size of the image representation, making it computationally more manageable. By aggregating information from neighboring pixels, pooling operations help retain critical information while reducing redundancy.
- Hierarchical layers: CNNs are designed with a hierarchical structure, where layers are stacked on top of each other. This stacking increases the depth of the network and corresponds to an escalating level of feature extraction. As one progresses further into the network, the features take on a more abstract and high-level nature, enabling the network to grasp complex patterns and structures within the data.
- Classification layer: At the end of the CNN architecture, a classification layer is responsible for computing the probability or score associated with learned classes based on the features extracted during the previous layers [49]. This step

is crucial for making predictions in medical imaging analysis tasks, such as disease detection.

The bottom row of Fig. 3.3 provides a schematic representation of a deep CNN (often referred to as DCNN), which encompasses multiple convolutional and pooling layers. Deep CNNs are particularly suited for image classification tasks, and they have been extensively used in medical imaging, especially for disease detection, such as cancer. The input layer of the CNN architecture accepts the medical image under analysis. Convolutional layers, housing convolutional filters or kernels, play a pivotal role in identifying intricate patterns and features within the image data. The arrangement of these layers is hierarchical, mirroring the progression from low-level features (e.g., edges) to high-level features (e.g., complex shapes).

In essence, CNNs have transformed the landscape of medical imaging analysis by excelling in the efficient processing of image data and the extraction of significant features. Regardless of whether one opts for transfer learning or initiates de novo training approaches, CNNs undeniably emerge as indispensable tools in the pursuit of precise disease detection and diagnosis. Their hierarchical architecture and specialized layers enable them to uncover complicated patterns within medical images, making them a vital asset in the medical field's ongoing pursuit of more effective diagnostics and treatments [47–49]. Currently, we have supplied ample references demonstrating the utilization of CNN in breast cancer imaging (mammography). However, to offer further supporting evidence, we are concisely introducing the CNN techniques predominantly employed in medical imaging to detect and diagnose various complications and diseases in the following subsections.

3.2.3.2 CNN in Neuroimaging

AI has been widely employed in brain research, with numerous studies dedicated to classifying Alzheimer's disease (AD), segmenting brain regions for anatomical analysis, and detecting tumors. The successful classification of AD, mild cognitive impairment, and healthy control individuals was accomplished through the application of the Gaussian restricted Boltzmann machine (RBM) to uncover feature representations within volume patches of MRI and PET images [50]. The 3D CNN has demonstrated its superiority over other algorithm classifiers in AD classification, as evidenced in the Computing Research Repository (CoRR) for the years 2015 and 2016 [51, 52]. These 3D CNNs exhibit the capability to automatically delineate the regions of interest within magnetic resonance (MR) images of the human brain through Convolutional Neural Networks (CNNs). Specifically, the segmentation of the striatum, a complex brain structure, was effectively accomplished through the utilization of deep CNNs, and the outcomes were subsequently juxtaposed with those generated by FreeSurfer [53, 54]. Manual segmentation in the brain region is labor-intensive and susceptible to inter-individual variations, whereas automatic segmentation faces significant challenges, particularly in complex anatomical structures. A noteworthy achievement was the development of a network comprising twenty-five deep layers, referred to as the 'voxel-wise residual network' (VoxResNet), which successfully automated the segmentation process [55]. To demonstrate the seamless nonlinear mapping from MR images to CT images, a 3D fully CNN was employed and validated using actual pelvic CT/MRI data, and the integration of two-volume CNNs for input and output enhancement led to remarkable improvements in performance, as evidenced by the evaluation of input and output formats applied to MRI and PET images from the ADNI database [56, 57].

3.2.3.3 CNN in Thoracic Radiology Imaging

A Deconvolutional Neural Network (de-CNN) is constructed to generate the heat map of anomalous regions by introducing the multiple-instance learning framework [58]. Using a distinctive collection of publicly available radiologic datasets containing chest X-rays and their corresponding reports, researchers identified and reported 17 distinct patterns through the application of CNN algorithms [59]. Furthermore, researchers applied CNN algorithms to classify lung texture patterns and employed a segmentation-based label propagation method on a dedicated dataset for interstitial lung disease to identify the presence of interstitial patterns, as reported in the studies [60, 61]. A reported approach involves utilizing deep learning techniques to classify frontal and lateral chest X-ray images while also automating the process of annotating metadata [62]. Additionally, a novel method has been introduced, proposing the utilization of a 3D CNN to reduce false positives in the automatic detection of pulmonary nodules within volumetric CT scans. The 3D CNN possesses the capability to incorporate a greater amount of spatial information and extract more meaningful characteristics by utilizing a hierarchical architecture that has been trained with three-dimensional samples. This algorithm, as proposed, has not only attained impressive competition metric scores but has also undergone thorough evaluation in the LUNA16 Challenge, demonstrating its effectiveness in the analysis of three-dimensional PET images.

3.2.3.4 CNN in Cardiovascular Imaging

Research in the field of cardiac artificial intelligence spans a multitude of specialized domains, each with distinct objectives and challenges. Within these domains, there are a variety of tasks, including the essential activity of left ventricle segmentation, which is a critical aspect of cardiac image analysis [63]. Moreover, researchers tackle slice classification, a process involving the categorization of cardiac image sections for diagnostic purposes [64]. Ensuring the quality of acquired images is another pivotal aspect, involving image quality assessment to eliminate artifacts and noise [65]. Furthermore, automated calcium scoring is a crucial area of investigation, integral to assessing coronary artery disease risk [66]. Additionally, researchers focus on coronary centerline tracking, a complex endeavor aimed at accurately mapping the pathways of coronary arteries [67]. The field incorporates Super-Resolution (SR) techniques to enhance image resolution and detail, facilitating improved diagnosis and treatment planning [68]. 2D and 3D CNN techniques are primarily employed for classification tasks, while deep learning methods such as the U-net segmentation algorithm are utilized for segmentation purposes. To enhance the quality of the high-resolution 3D volume within the 2D image stack, a novel image SR approach has been applied [69]. The image quality significantly improves upon that of the SR method due to the computational efficiency of the CNN model; nevertheless, SR-CNN exhibits its advantages in tasks related to image segmentation and motion tracking [70]. By employing a multi-stream CNN approach involving three different views, recent reports indicate that deep learning can achieve high accuracy in identifying low-dose chest CT scans as potential coronary artery calcification candidates when the Region of Interest (ROI) has a Hounsfield unit (HU) value exceeding 130 [71]. Additionally, the detection of coronary calcium in gated cardiac CT angiography has been accomplished using a combination of 3D CNN and multi-stream 2D CNN techniques [72].

3.2.3.5 CNN in Musculoskeletal Imaging

Deep-learning algorithms are used to analyze images of the musculoskeletal system, with a focus on identifying bone, joint, and soft tissue abnormalities. Specifically, a 3D CNN architecture has been created to perform supervised segmentation of vertebral bodies from 3D MR spine images, achieving a high similarity coefficient of 93.4% [73]. Recognizing the spine, including its location and naming in multiple images, poses challenges due to variations in spine shapes and postures. To address this, a deep-learning method called the transformed deep convolutional network is employed to automatically correct the spine's posture during image processing [74]. While CNN regression has limitations in terms of computation time and capture range for intensity-based 2D/3D registration technology, it is reported that highly accurate real-time 2D/3D registration can be achieved with an enlarged capture range [75]. Additionally, various deep-learning approaches have been developed to automatically determine the age of skeletal bones through X-ray imaging, demonstrating impressive accuracy with an average discrepancy of approximately 9.6 months [76].

As elucidated in the preceding discussion, it is evident that CNNs occupy a central and widely adopted role within the field of medical image processing. These advanced AI techniques have become the cornerstone of analyzing and interpreting medical images, playing a pivotal role in the accurate diagnosis and treatment of various medical conditions. Continuing the exploration of the widespread utilization of these methods, we address the challenges associated with applying AI/ML methods in biomedical imaging in the following section.

3.3 Challenges in Application of AI/ML in Biomedical Imaging

Research in machine learning for medical imaging represents a novel approach for conventional radiology investigators. It involves mastering fresh procedures, recognizing constraints, and addressing emerging challenges when engaging in cutting-edge data science analysis. The availability of readily accessible, well-organized, and accurately labeled high-quality data is essential for conducting effective research in this domain. It is also important that while striving to obtain ample quantities of high-quality data, the radiology community should maintain an awareness of our patients' privacy concerns. In the following subsections, we thoroughly discuss some of the unavoidable challenges or obstacles to address several issues while applying AI/ML in medical imaging [1, 5, 6, 77–80].

3.3.1 Intricacy and Data Heterogeneity of Medical Images

Medical images are typically saved in the Digital Imaging and Communications in Medicine (DICOM) format, which differs fundamentally from the image formats commonly employed in the application of AI/ML techniques for recognition. The DICOM standard was created to facilitate various functions in medical imaging and has played a pivotal role in numerous advancements in the field, especially in the process of digitization within radiology. This standard has the capacity to accommodate diverse resolutions and allocate bit depths, such as in mammographic studies where resolutions can reach up to 3000 $\times$ 4000 pixels [81]. Although most medical imaging techniques yield black-and-white pictures, a few, like Doppler US, PET/CT and secondary capture objects, which may include advanced visualization images, display and save images in colored format.

Obtaining medical images is not quite the same as capturing regular photographic images. In some medical studies like radiography, multiple angles are needed to understand the 3D position of body structures. Similarly, imaging methods like CT and MRI record images across the entire volume to visualize the cross-sectional depiction of the same region within the body. Moreover, in multi-sequence modalities like MRI, there are multiple types of images for the same body part, allowing for the extraction of specific features from the tissue being imaged. The importance of image comparison in medical imaging for monitoring changes in patients' health is rarely explored in non-medical fields. Also, medical imaging findings are often not specific to a single disease entity and can be identified in a variety of diseases. For instance, a spot on a chest X-ray could indicate infection, inflammation, bleeding, scarring from past injury, or cancer, and requires consideration of other factors like symptoms and laboratory results. A diagnosis is typically confirmed through pathology, but in cases like pneumonia, imaging findings are often used without confirmation. Additionally, there are anatomical variations that may mimic diseases in imaging but don't affect the patient's health [78].

3.3.2 Diverse Tasks in Various Clinical Settings

Medical imaging serves a wide range of clinical purposes, from disease detection to monitoring. These scenarios involve specific tasks that can be addressed by machine learning, including disease detection, lesion segmentation, classification, and quantification. For instance, chest CT scans are used to detect and classify pulmonary nodules, while other tasks involve measuring anatomical structures like hippocampal volume [82, 83]. Some cases require numerical calculations, like assessing bone age from hand radiographs [84, 85]. Image-based predictions, such as hematoma expansion likelihood from head CT scans, are also relevant [86]. The diversity of clinical scenarios and associated tasks is vast, making it infeasible for a single individual or organization to handle them all with traditional methods.

3.3.3 Challenges in Curation of Medical Imaging Data

ImageNet revolutionized computer vision research by emphasizing the significance of data curation alongside feature and algorithm development [87]. Its success stemmed from the collective contributions of many individuals, resulting in a vast database of image classes [88, 89]. While the general public was well-suited for annotating standard photographic images with easily recognizable objects, labeling medical images demands radiological expertise, making it dependable on trained domain experts. Additionally, even expert radiologists can exhibit substantial variability in detecting subtle findings, impacting the interpretation process. Inter- and interobserver agreement can be notably low, even among experts, for specific clinical conditions [90, 91]. The specialization required for annotation limits the pool of contributors, making it challenging to crowdsource this vital information.

3.3.4 The Traditional Hypothesis-Driven Research Model Delays Knowledge Dissemination

Recognizing a considerable time lag between medical research knowledge discovery and its clinical implementation has been an enduring concern. Existing evidence suggests that this gap, notably in the public health sector, can extend up to 17 years [92]. One primary factor contributing to the delayed dissemination of knowledge is the sequential nature of traditional research methodologies. Conventional research typically follows a hypothesis-driven process where researchers or research groups within an institution formulate a research question to be proven or disproven. Following this, they seek approval from their institutional review board to execute the research within their organization. Upon securing approval, they conduct the research and subsequently publish their findings in a journal, often accessible to only a limited audience. Upon becoming aware of these findings, researchers from other organizations endeavor to validate them and publish their own experiences. This cycle may iterate multiple times until a reasonable level of confidence in the generalizability of the initial findings is achieved. While this process undeniably holds immense significance and has underpinned scientific progress for decades, it may benefit from a more decentralized approach to initial knowledge discovery, such as image analysis competitions. In this alternative approach, multiple individuals or groups engage in competitive efforts to find the most effective solution to a specific problem, openly sharing their newfound knowledge, techniques, and methodologies. These insights can subsequently undergo further scrutiny using conventional scientific methods.

3.3.5 Algorithmic Design and Performance Assessment in Medical Imaging

When exploring algorithmic design and performance metrics within the domain of medical imaging, it is vital to acknowledge the distinct and rigorous requirements that differentiate it from other areas where AI algorithms are utilized. Creating tools for medical use demands an exceptionally meticulous approach due to the critical nature of healthcare decisions. One of the primary considerations is the selection of performance metrics for a medical AI algorithm. These metrics are the yardstick by which the algorithm's effectiveness is measured. In the case of medical applications, the choice of metrics should be driven by their alignment with clinical objectives. For example, an algorithm designed to assist doctors in making rapid decisions when analyzing coronary CT angiography scans in patients suspected of having a myocardial infarction (heart attack). In this scenario, getting the diagnosis wrong can have dire consequences. Therefore, it is crucial to ensure that the algorithm minimizes false negatives, which are instances where it fails to detect a true positive case. However, medical data often exhibit a significant class imbalance, meaning that positive cases (in this case, patients with myocardial infarction) are much less common than normal cases. Traditional metrics like accuracy (the ratio of correct predictions to total predictions) and negative predictive value (the ability to correctly identify negative cases) may not be suitable in this context. They can be misleading because an algorithm that labels most cases as negative could still achieve high accuracy due to the data imbalance. Instead, metrics like Recall (also known as Sensitivity) and the Area Under the Curve of the Receiver Operating Characteristic (AUC-ROC) become more relevant in such scenarios. Recall measures the algorithm's ability to correctly identify all true positive cases, while AUC-ROC assesses its overall discriminatory power. These metrics are better suited to gauge performance in situations where false negatives can have severe consequences. Furthermore, the scope of the algorithm is critical. A narrowly-focused algorithm, designed solely for recognizing pneumonia on chest radiographs, might excel in this specific task. However, it may fall short when confronted with broader clinical situations. For instance, it may not recognize other potentially life-threatening conditions, such as pneumoperitoneum, which is the presence of air in the abdominal cavity. Overlooking such findings could lead to tragic outcomes. In essence, designing AI algorithms for medical imaging requires a deep understanding of both the clinical context and the data's characteristics. Performance metrics must be carefully chosen to align with clinical needs, and the algorithm's scope should be defined to ensure it doesn't miss critical findings outside its primary area of focus. This detailed approach is indispensable in the critical arena of medical diagnostics and decision-making [78].

3.3.6 Challenges in Algorithm Transparency, Validation, and Testing

Lack of transparency in the underlying methodologies employed to create existing algorithm design mechanisms and the challenges associated with clinical implementation are significant issues. Reproducing the testing of a proprietary algorithm can be exceedingly complex, particularly when extending evaluations to different sites and datasets. This complexity is exacerbated because machine learning applications often lack a standardized process, with different pipelines from data ingestion to output even for algorithms with similar performance tackling the same problem. Unique preprocessing methodologies may be necessary before inference, requiring dedicated servers or virtual environments for each application, hindering scalability. One potential solution to these implementation challenges involves the creation of separate application containers. These containers allow multiple isolated applications or services to run on a single host, accessing the same operating system kernel. However, even within this framework, integrating these applications with local clinical systems can prove daunting without the proper infrastructure in place. Machine learning algorithms are typically developed with the assumption that they will be applied to datasets with consistent characteristics and probability distributions, a property known as algorithm generalizability. However, practical realities in healthcare settings, such as variations in patient populations, image acquisition devices, and imaging protocols across institutions, can greatly differ. Consequently, ensuring the transferability of an algorithm—its ability to maintain high performance across data with varying probability distributions—can be challenging, even if the algorithm performs excellently on data from a single source. Unfortunately, no statistical method exists to test algorithm transferability, leaving the only option as testing the algorithm on new data at disparate locations [88–93]. We emphasize some of the challenges associated with algorithmic design and performances as follows:

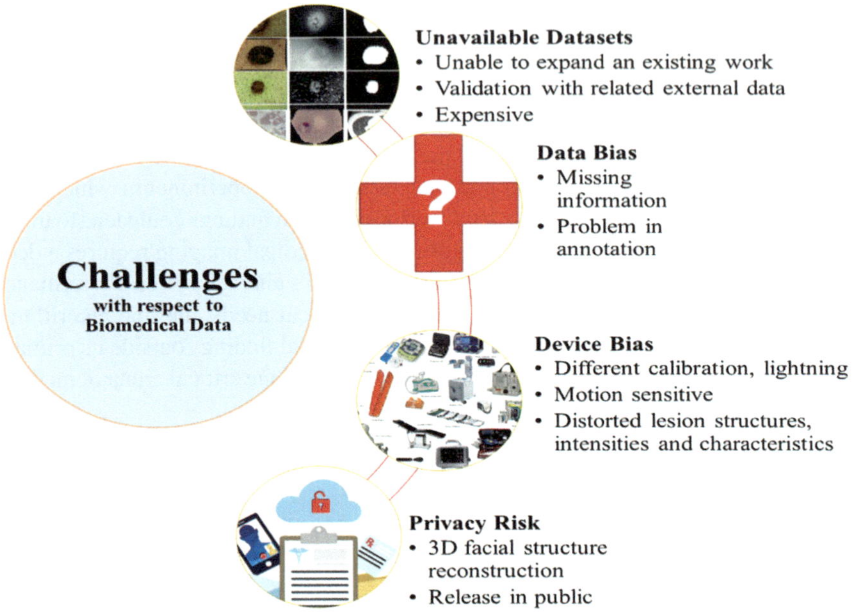

Fig. 3.4 Key challenges observed with respect to biomedical data

3.3.6.1 Challenges with Respect to Dataset

Currently, biomedical imaging contains a large quantity of information, which has the potential to change healthcare, deepen our understanding of diseases, and enable personalized medicine. A number of obstacles are observed during the utilization of biomedical imaging datasets which are addressed in the following subsections. Figure 3.4 provides a visualization of these challenges.

Medical Data Bias

It has been observed that the majority of studies have avoided publishing biomedical datasets regarding demographic information [1, 5]. The following details could be incomplete:

- The location of the patients' collected data.
- Patients' age, gender, and social background etc.

Because of these knowledge gaps, ML researchers may encounter challenges when attempting to apply symptom and disease generalization across diverse age groups. Moreover, they may lack awareness regarding the prevalence of specific diseases within particular age demographics. Consequently, it is crucial to establish precise definitions for the raw medical data. Given that ML researchers often lack expertise in data labeling, they rely on radiologists for image annotation. This process consumes a significant amount of time and effort for both the researchers and medical professionals. Furthermore, disease annotation is typically limited to specialized individuals, restricting the pool of contributors involved in the preparation of medical data [77, 93].

Data Interoperability and Data Acquisition Tools

There is a wide variety of formats, resolutions, and modalities associated with biomedical imaging data like X-rays, MRIs, and CT scans. Preparing the data can be very challenging across these different datasets. Another problem that researchers are unable to solve is the space constraint caused by medical metadata. Occasionally, medical datasets are converted to PNG or JPEG formats to hide the metadata. As a result, the metadata sharing becomes unavailable in the public domain [5–7]. Researchers cannot expand on an existing work using a certain dataset if they do not have access to it. Additionally, this prevents validation of the current experiment with related external data. It can be quite expensive to gain access to the datasets that are integrated to create biomedical big data. This limits the scope of future health economic analyses [6, 79–81].

Medical devices such as MRIs, CTs, and imaging cameras may differ in their calibration. It is possible that these cameras are sensitive to motion and offer a variety of data resolutions depending on lighting conditions. This results in a variety of effects, including distorted lesion structures, altered intensity values, and diminished features [6].

Risk of Patient's Privacy

Medical records may conceal sensitive patient-information such as names, date of birth, medical history etc. AI-based technologies based on facial recognition and 3D reconstruction systems allow this information to be recovered from facial structures, body marks, body tattoos etc. which can be extracted from social media and patients' medical records. Healthcare professionals and researchers frequently exchange biomedical imaging data in order to carry out team research projects and make medical diagnosis. If information is handled carelessly or accessed without authority, data breaches may happen [1, 78, 93]. Bias in the representation of data and the training of algorithms also lead to unfair or unnecessary treatments, which in turn leads to the violation of patient privacy. Generally, patients fail to understand the decision algorithms of ML/DL models. As a result, lack of trust exists between patients and medical systems. The privacy of these patients is more challenging to protect using cyber security, once the data or the analyzed result has been made public [94, 95].

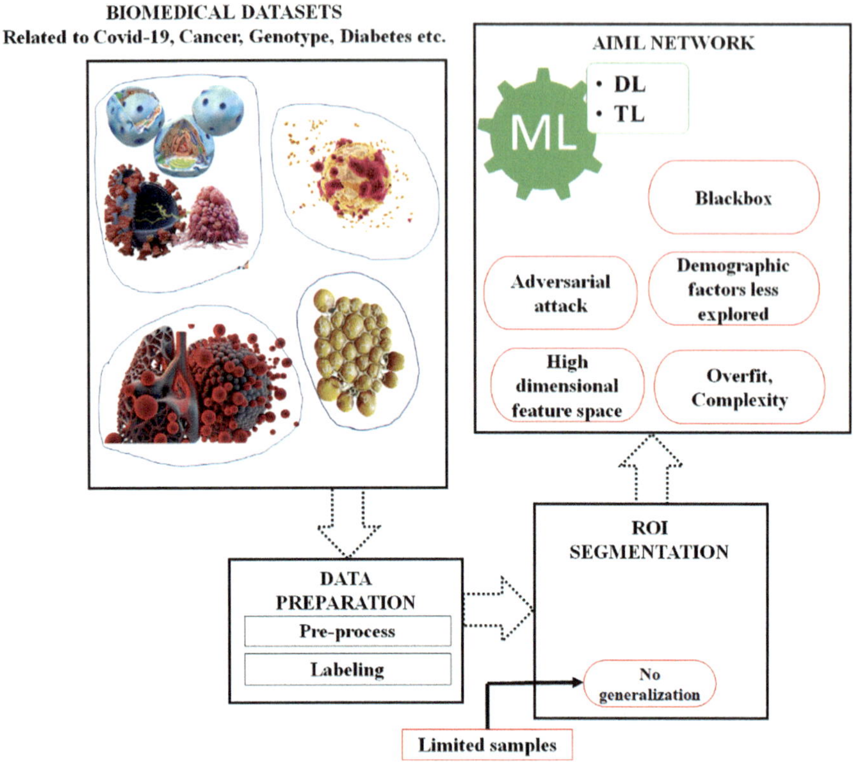

Fig. 3.5 Key challenges (outlined in red) with respect to the ML network

Challenges with Respect to ML/DL Models

In the field of biomedical imaging, ML is capable of performing a variety of tasks, including delineating the boundaries of abnormal patterns, extracting features, locating lesions, and accurately categorizing them. However, these tasks are confronted with a number of challenges (Fig. 3.5) which are discussed in this section [94–100].

a. Feature Extraction: High-resolution biomedical images, such as 3D histopathology scans, volumetric MRI etc. can be used to extract properties like color, texture etc. creating a high-dimensional feature space. This can result in high computation and overfitting of the ML/DL model. A thorough investigation of the connection between imaging characteristics and treatment outcome is still missing in this aspect [94–96]. Tumors with scattered lesions and variable shapes are challenging to label manually. The automated ML/DL makes this task simple. With ML/DL, automated feature extraction is possible without the requirement to manually create a set of features. However, the extremely small lesions from imaging scans might be missed by the automatic region of interest (RoI) segmentation. The models may find it difficult to generalize unseen data when trained on a limited range of characteristics (lesion shapes/sizes) [98–100]. The majority of ML models perform simple bivariate comparisons between clinical outcomes and biomedical image features. Multivariate modeling has also been used in clinical and epidemiological research. There is a substantial omission of demographic factors such as age, gender, and ethnicity in most deep learning (DL)-based medical research. Based on these demographic variables, ML/DL models (XGBoost, Logistic Regression, and Random Forest) performed worse than those trained on imaging data [13, 78].

b. Interoperability: The field of biomedical demands for transparent and explainable approaches. Despite the expertise of the most skilled radiologist, it can be challenging to understand the rationale behind ML-based feature extraction methods. As ML/DL models contain numerous parameters, medical experts have difficulty understanding how they assess data and make decisions. To make the ML/DL system more transparent, certain techniques like the heatmap and confusion matrix have been used in biomedical imaging. However, DL has been criticized as it fails to explain how inputs are converted into outputs. These models are also vulnerable to an adversarial attack when the input is changed by augmentation, noise addition, etc. [97, 98]. Transfer learning is another technique where previously trained parameters are applied from one task to another. For example, the knowledge gathered in medical gene prediction can be utilized to predict if a tumour is malignant. The disadvantage of transfer learning is that certain biomedical datasets, such as 3D medical MRI data, do not contain pre-trained DL models. Thus, transfer learning may have difficulty adapting to different medical domains when the data distribution in the source is different from that in the target. Data and predictions may also be subject to bias. Due to the fact that this learning creates the need for diverse samples in the dataset, the model will pick up more irrelevant features and will therefore have a greater level of complexity. Choosing the correct strategy for fine-tuning and hypermeter setting requires a great deal of computational resources like Graphics Processing Unit (GPU) and Tensor Processing Unit (TPU). The complexity of the model makes it increasingly difficult for medical experts to understand the rationale behind the transfer learning's decisions [99–103].

3.4 Conclusion

In order to improve the healthcare scenario, it is essential to integrate technology into the field. Biomedical imaging is highly effective in the field of healthcare for the diagnosis of diseases. This technology includes a variety of imaging methods, including ultrasound, MRI, and CT scans that let health care providers observe and examine the internal organs and functions of the body. AI/ML is widely used in biomedical imaging for purposes such as diagnosis and research, including tasks like image enhancement, disease identification, and image segmentation. When AI/ML are utilized effectively and with careful consideration, they have the potential to greatly enhance healthcare outcomes. However, it's critical to recognise that while AI/ML provide biomedical imaging with a number of benefits, they also have drawbacks that must be taken into consideration. The main focus of this chapter is to highlight the challenges faced during the implementation of AI/ML in biomedical imaging. A detailed study is provided in the chapter discussing various challenges associated with AI/ML in biomedical imaging, this will benefit researchers to consider these challenges, improve limitations, and enhance the overall healthcare scenario.

References

1. A.K. Sharma, P.K. Tripathi, S. Sharma, Role of artificial intelligence in biomedical imaging. in *Advancements in bio-medical image processing and authentication in telemedicine* (IGI Global, 2023), pp. 17–34
2. C. Tchito Tchapga, T.A. Mih, A. Tchagna Kouanou, T. Fozin Fonzin, P. Kuetche Fogang, B.A. Mezatio, D. Tchiotsop, Biomedical image classification in a big data architecture using machine learning algorithms. J. Healthc. Eng. **2021**, 1–11 (2021)
3. P.W. Koh, P. Liang, Understanding black-box predictions via influence functions. in *International conference on machine learning* (PMLR, 2017), pp. 1885–1894
4. S.M. Nagarajan, V. Muthukumaran, R. Murugesan, R.B. Joseph, M. Munirathanam, Feature selection model for healthcare analysis and classification using classifier ensemble technique. Int. J. Syst. Assur. Eng. Manag., 1–12 (2021)
5. C.J. Kelly, A. Karthikesalingam, M. Suleyman, G. Corrado, D. King, Key challenges for delivering clinical impact with artificial intelligence. BMC Med. **17**, 1–9 (2019)
6. G. Varoquaux, V. Cheplygina, Machine learning for medical imaging: methodological failures and recommendations for the future. NPJ Digit. Med. **5**(1), 48 (2022)
7. R.S.K. Boddu, S. Ahamad, K.P. Kumar, M. Ramalingam, L.K. Pallathadka, F.P. Tupas, Analysis of robotics, artificial intelligence and machine learning in the field of healthcare sector. Mater. Today: Proc. **56**, 2323–2327 (2022)
8. R.R. Patil, A.U. Ruby, B.N. Chaithanya, S.J. TJ, K. Geetha, Review of fundamentals of Artificial Intelligence and application with medical data in healthcare. J. Integr. Sci. Technol., **10**(2), 126–133 (2022)
9. M.D. Wilkinson, M. Dumontier, I.J. Aalbersberg, G. Appleton, M. Axton, A. Baak, N. Blomberg, J.W. Boiten, L.B. da Silva Santos, P.E. Bourne, J. Bouwman, The FAIR guiding principles for scientific data management and stewardship. Scientific data **3**(1), 1–9 (2016)
10. M.D. Kohli, R.M. Summers, J.R. Geis, Medical image data and datasets in the era of machine learning—whitepaper from the 2016 C-MIMI meeting dataset session. J. Digit. Imaging **30**, 392–399 (2017)
11. H. Harvey, B. Glocker, A standardised approach for preparing imaging data for machine learning tasks in radiology. Artif. Intell. Med. Imaging: Oppor., Appl. Risks, 61–72 (2019)
12. M. Kortesniemi, V. Tsapaki, A. Trianni, P. Russo, A. Maas, H.E. Källman, M. Brambilla, J. Damilakis, The European Federation of Organisations for Medical Physics (EFOMP) White Paper: Big data and deep learning in medical imaging and in relation to medical physics profession. Physica Med. **56**, 90–93 (2018)
13. J. Selvaraj, T. Prabha, T.D.A. Kumar, S. Palaniappan, Artificial intelligence in biomedical image processing. In *Machine learning and systems biology in genomics and health* (Springer Nature Singapore, Singapore, 2022), pp. 147–188

14. A.L. Samuel, Some studies in machine learning using the game of checkers. IBM J. Res. Dev., **44**(1.2), 206–226 (2000)
15. J.G. Lee, S. Jun, Y.W. Cho, H. Lee, G.B. Kim, J.B. Seo, N. Kim, Deep learning in medical imaging: general overview. Korean J. Radiol. **18**(4), 570–584 (2017)
16. B.J. Erickson, P. Korfiatis, Z. Akkus, T.L. Kline, Machine learning for medical imaging. Radiographics **37**(2), 505–515 (2017)
17. A.C. Müller, S. Guido, *Introduction to machine learning with Python: a guide for data scientists* (O'Reilly Media, Inc, 2016)
18. X. Tang, The role of artificial intelligence in medical imaging research. BJR|Open, **2**(1), 20190031
19. K. Clark, B. Vendt, K. Smith, J. Freymann, J. Kirby, P. Koppel, S. Moore, S. Phillips, D. Maffitt, M. Pringle, L. Tarbox, The Cancer Imaging Archive (TCIA): maintaining and operating a public information repository. J. Digit. Imaging **26**, 1045–1057 (2013)
20. N.I. Yassin, S. Omran, E.M. El Houby, H. Allam, Machine learning techniques for breast cancer computer aided diagnosis using different image modalities: A systematic review. Comput. Methods Programs Biomed. **156**, 25–45 (2018)
21. D. Bazazeh, R. Shubair, Comparative study of machine learning algorithms for breast cancer detection and diagnosis. in *2016 5th international conference on electronic devices, systems and applications (ICEDSA)* (IEEE, 2016), pp. 1–4
22. A. Nazir, R.A. Khan, Network intrusion detection: Taxonomy and machine learning applications. Mach. Intell. Big Data Anal. Cybersecur. Appl., 3–28 (2021)
23. S. Sharma, R. Mehra, Conventional machine learning and deep learning approach for multi-classification of breast cancer histopathology images—a comparative insight. J. Digit. Imaging **33**, 632–654 (2020)
24. T.H. Chan, K. Jia, S. Gao, J. Lu, Z. Zeng, Y. Ma, PCANet: A simple deep learning baseline for image classification? IEEE Trans. Image Process. **24**(12), 5017–5032 (2015)
25. Y. Chen, Z. Lin, X. Zhao, G. Wang, Y. Gu, Deep learning-based classification of hyperspectral data. IEEE J. Sel. Top. Appl. Earth Obs. Remote. Sens. **7**(6), 2094–2107 (2014)
26. L.C. Chen, G. Papandreou, I. Kokkinos, K. Murphy, A.L. Yuille, Deeplab: Semantic image segmentation with deep convolutional nets, atrous convolution, and fully connected crfs. IEEE Trans. Pattern Anal. Mach. Intell. **40**(4), 834–848 (2017)
27. K. He, X. Zhang, S. Ren, J. Sun, Deep residual learning for image recognition. In *Proceedings of the IEEE conference on computer vision and pattern recognition* (2016), pp. 770–778
28. S. Ren, K. He, R. Girshick, J. Sun, Faster r-cnn: Towards real-time object detection with region proposal networks. Adv. Neural Inf. Process. Syst., **28**
29. Y. Qin, Y. Deng, H. Jiang, N. Hu, B. Song, Artificial intelligence in the imaging of gastric cancer: current applications and future direction. Front. Oncol., 11, 631686 (2021). J.Z. Cheng, D. Ni, Y.H. Chou, J. Qin, C.M. Tiu, Y.C. Chang, C.S. Huang, D. Shen, C.M. Chen, Computer-aided diagnosis with deep learning architecture: applications to breast lesions in US images and pulmonary nodules in CT scans. Sci. Rep., **6**(1), p.24454 (2016)
30. G. Litjens, C.I. Sánchez, N. Timofeeva, M. Hermsen, I. Nagtegaal, I. Kovacs, C. Hulsbergen-Van De Kaa, P. Bult, B. Van Ginneken, J. Van Der Laak, Deep learning as a tool for increased accuracy and efficiency of histopathological diagnosis. Sci. Rep. **6**(1), 26286 (2016)
31. Y. Todoroki, X.H. Han, Y. Iwamoto, L. Lin, H. Hu, Y.W. Chen, Detection of liver tumor candidates from CT images using deep convolutional neural networks. in *Innovation in Medicine and Healthcare 2017: Proceedings of the 5th KES International Conference on Innovation in Medicine and Healthcare (KES-InMed 2017) 5* (Springer International Publishing, 2017), pp. 140–145
32. J. Heaton, Ian Goodfellow, Yoshua Bengio, and Aaron Courville: Deep learning: The MIT Press, 2016, 800 pp, ISBN: 0262035618. Genet. Program. Evolvable Mach., **19**(1–2), 305–307
33. R. Hecht-Nielsen, Theory of the backpropagation neural network. in *Neural networks for perception* (Academic Press, 1992), pp. 65–93
34. Y. Bengio, Learning deep architectures for AI, Foundations and trends R in Machine Learning, vol. 2 (2009)

35. D. Svozil, V. Kvasnicka, J. Pospichal, Introduction to multi-layer feed-forward neural networks. Chemom. Intell. Lab. Syst. **39**(1), 43–62 (1997)
36. M. Bakator, D. Radosav, Deep learning and medical diagnosis: A review of literature. Multimodal Technol. Interact. **2**(3), 47 (2018)
37. E.J. Humphrey, J.P. Bello, Y. LeCun, Feature learning and deep architectures: New directions for music informatics. J. Intell. Inf. Syst. **41**, 461–481 (2013)
38. D. Arefan, A. Talebpour, N. Ahmadinejhad, A.K. Asl, Automatic breast density classification using neural network. J. Instrum. **10**(12), T12002 (2015)
39. A. Fischer, C. Igel, An introduction to restricted Boltzmann machines. in *Progress in pattern recognition, image analysis, computer vision, and applications: 17th Iberoamerican congress, CIARP 2012, Buenos Aires, Argentina, September 3-6, 2012. Proceedings 17* (Springer Berlin Heidelberg, 2016), pp. 14–36
40. Q. Zhang, Y. Xiao, W. Dai, J. Suo, C. Wang, J. Shi, H. Zheng, Deep learning based classification of breast tumors with shear-wave elastography. Ultrasonics **72**, 150–157 (2016)
41. J. Wu, J. Shi, Y. Li, J. Suo, Q. Zhang, Histopathological image classification using random binary hashing based PCANet and bilinear classifier. in *2016 24th European signal processing conference (EUSIPCO)* (IEEE, 2016), pp. 2050–2054
42. J.Z. Cheng, D. Ni, Y.H. Chou, J. Qin, C.M. Tiu, Y.C. Chang, C.S. Huang, D. Shen, C.M. Chen, Computer-aided diagnosis with deep learning architecture: applications to breast lesions in US images and pulmonary nodules in CT scans. Sci. Rep. **6**(1), 24454 (2016)
43. K. Fukushima, Neocognitron: A self-organizing neural network model for a mechanism of pattern recognition unaffected by shift in position. Biol. Cybern. **36**(4), 193–202 (1980)
44. A. Krizhevsky, I. Sutskever, G.E. Hinton, Imagenet classification with deep convolutional neural networks. Adv. Neural Inf. Process. Syst., **25**, (2012)
45. T. Wan, J. Cao, J. Chen, Z. Qin, Automated grading of breast cancer histopathology using cascaded ensemble with combination of multi-level image features. Neurocomputing **229**, 34–44 (2017)
46. Z. Han, B. Wei, Y. Zheng, Y. Yin, K. Li, S. Li, Breast cancer multi-classification from histopathological images with structured deep learning model. Sci. Rep. **7**(1), 4172 (2017)
47. R.A. Khan, A. Crenn, A. Meyer, S. Bouakaz, A novel database of children's spontaneous facial expressions (LIRIS-CSE). Image Vis. Comput. **83**, 61–69 (2019)
48. Y. Liu, X. Yao, Ensemble learning via negative correlation. Neural Netw. **12**(10), 1399–1404 (1999)
49. Y. LeCun, Y. Bengio, G. Hinton, Deep learning. Nature **521**(7553), 436–444 (2015)
50. H.I. Suk, S.W. Lee, D. Shen, A.D.N. Initiative, Hierarchical feature representation and multimodal fusion with deep learning for AD/MCI diagnosis. Neuroimage **101**, 569–582 (2014)
51. A. Payan, G. Montana, *Predicting Alzheimer's disease: a neuroimaging study with 3D convolutional neural networks.* arXiv preprint arXiv:1502.02506
52. E. Hosseini-Asl, G. Gimel'farb, A. El-Baz, *Alzheimer's disease diagnostics by a deeply supervised adaptable 3D convolutional network* (2016). arXiv preprint arXiv:1607.00556
53. A. de Brebisson, G. Montana, Deep neural networks for anatomical brain segmentation. In *Proceedings of the IEEE conference on computer vision and pattern recognition workshops* (2015), pp. 20–28
54. H. Choi, K.H. Jin, Fast and robust segmentation of the striatum using deep convolutional neural networks. J. Neurosci. Methods **274**, 146–153 (2016)
55. H. Chen, Q. Dou, L. Yu, J. Qin, P.A. Heng, VoxResNet: Deep voxelwise residual networks for brain segmentation from 3D MR images. Neuroimage **170**, 446–455 (2018)
56. D. Nie, X. Cao, Y. Gao, L. Wang, D. Shen, Estimating CT image from MRI data using 3D fully convolutional networks. in *Deep Learning and Data Labeling for Medical Applications: First International Workshop, LABELS 2016, and Second International Workshop, DLMIA 2016, Held in Conjunction with MICCAI 2016, Athens, Greece, October 21, 2016, Proceedings 1* (Springer International Publishing, 2016), pp. 170–178

57. R. Li, W. Zhang, H.I. Suk, L. Wang, J. Li, D. Shen, S. Ji, Deep learning based imaging data completion for improved brain disease diagnosis. In *Medical Image Computing and Computer-Assisted Intervention–MICCAI 2014: 17th International Conference, Boston, MA, USA, September 14–18, 2014, Proceedings, Part III 17* (Springer International Publishing, 2014), pp. 305–312
58. H.E. Kim, S. Hwang, *Deconvolutional feature stacking for weakly-supervised semantic segmentation.* arXiv preprint arXiv:1602.04984
59. H.C. Shin, K. Roberts, L. Lu, D. Demner-Fushman, J. Yao, R.M. Summers, Learning to read chest x-rays: Recurrent neural cascade model for automated image annotation. In *Proceedings of the IEEE conference on computer vision and pattern recognition* (2016), pp. 2497–2506
60. M. Gao, Z. Xu, L. Lu, A. Wu, I. Nogues, R.M. Summers, D.J. Mollura, Segmentation label propagation using deep convolutional neural networks and dense conditional random field. in *2016 IEEE 13th international symposium on biomedical imaging (ISBI)* (IEEE, 2016), pp. 1265–1268
61. M. Gao, U. Bagci, L. Lu, A. Wu, M. Buty, H.C. Shin, H. Roth, G.Z. Papadakis, A. Depeursinge, R.M. Summers, Z. Xu, Holistic classification of CT attenuation patterns for interstitial lung diseases via deep convolutional neural networks. Comput. Methods Biomech. Biomed. Eng.: Im-Aging & Vis. **6**(1), 1–6 (2018)
62. A. Rajkomar, S. Lingam, A.G. Taylor, M. Blum, J. Mongan, High-throughput classification of radiographs using deep convolutional neural networks. J. Digit. Imaging **30**, 95–101 (2017)
63. Y. Chen, F. Shi, A.G. Christodoulou, Y. Xie, Z. Zhou, D. Li, Efficient and accurate MRI super-resolution using a generative adversarial network and 3D multi-level densely connected network. in *International conference on medical image computing and computer-assisted intervention* (Springer International Publishing, Cham, 2018), pp. 91–99
64. P. Peng, K. Lekadir, A. Gooya, L. Shao, S.E. Petersen, A.F. Frangi, A review of heart chamber segmentation for structural and functional analysis using cardiac magnetic resonance imaging. Magn. Reson. Mater. Phys., Biol. Med. **29**, 155–195 (2016)
65. S.K. Zhou, H. Greenspan, C. Davatzikos, J.S. Duncan, B. Van Ginneken, A. Madabhushi, J.L. Prince, D. Rueckert, R.M. Summers, A review of deep learning in medical imaging: Imaging traits, technology trends, case studies with progress highlights, and future promises. Proc. IEEE **109**(5), 820–838 (2021)
66. D. Dey, P.J. Slomka, P. Leeson, D. Comaniciu, S. Shrestha, P.P. Sengupta, T.H. Marwick, Artificial intelligence in cardiovascular imaging: JACC state-of-the-art review. J. Am. Coll. Cardiol. **73**(11), 1317–1335 (2019)
67. B. Wu, Y. Fang, X. Lai, Left ventricle automatic segmentation in cardiac MRI using a combined CNN and U-net approach. Comput. Med. Imaging Graph. **82**, 101719 (2020)
68. J. Schlemper, J. Caballero, J.V. Hajnal, A.N. Price, D. Rueckert, A deep cascade of convolutional neural networks for dynamic MR image reconstruction. IEEE Trans. Med. Imaging **37**(2), 491–503 (2017)
69. M.R. Avendi, A. Kheradvar, H. Jafarkhani, A combined deep-learning and deformable-model approach to fully automatic segmentation of the left ventricle in cardiac MRI. Med. Image Anal. **30**, 108–119 (2016)
70. O. Oktay, W. Bai, M. Lee, R. Guerrero, K. Kamnitsas, J. Caballero, A. de Marvao, S. Cook, D. O'Regan, D. Rueckert, Multi-input cardiac image super-resolution using convolutional neural networks. in *Medical Image Computing and Computer-Assisted Intervention-MICCAI 2016: 19th International Conference, Athens, Greece, October 17–21, 2016, Proceedings, Part III 19* (Springer International Publishing, 2016), pp. 246–254
71. N. Lessmann, I. Išgum, A.A. Setio, B.D. de Vos, F. Ciompi, P.A. de Jong, M. Oudkerk, P.T.M. Willem, M.A. Viergever, B. van Ginneken, Deep convolutional neural networks for automatic coronary calcium scoring in a screening study with low-dose chest CT. in *Medical Imaging 2016: Computer-aided diagnosis*, vol. 9785 (SPIE, 2016), pp. 255–260
72. J.M. Wolterink, T. Leiner, B.D. de Vos, R.W. van Hamersvelt, M.A. Viergever, I. Išgum, Automatic coronary artery calcium scoring in cardiac CT angiography using paired convolutional neural networks. Med. Image Anal. **34**, 123–136 (2016)

73. R. Korez, B. Likar, F. Pernuš, T. Vrtovec, Model-based segmentation of vertebral bodies from MR images with 3D CNNs. in *International conference on medical image computing and computer-assisted intervention* (Springer International Publishing, Cham, 2016), pp. 433–441
74. Y. Cai, M. Landis, D.T. Laidley, A. Kornecki, A. Lum, S. Li, Multi-modal vertebrae recognition using transformed deep convolution network. Comput. Med. Imaging Graph. **51**, 11–19 (2016)
75. S. Miao, Z.J. Wang, R. Liao, A CNN regression approach for real-time 2D/3D registration. IEEE Trans. Med. Imaging **35**(5), 1352–1363 (2016)
76. C. Spampinato, S. Palazzo, D. Giordano, M. Aldinucci, R. Leonardi, Deep learning for automated skeletal bone age assessment in X-ray images. Med. Image Anal. **36**, 41–51 (2017)
77. J. Waller, A. O'connor, E. Raafat, A. Amireh, J. Dempsey, C. Martin, M. Umair, Applications and challenges of artificial intelligence in diagnostic and interventional radiology. Pol. J. Radiol. **87**(1), 113–117 (2022)
78. J.W. Gichoya, I. Banerjee, A.R. Bhimireddy, J.L. Burns, L.A. Celi, L.C. Chen, R. Correa, N. Dullerud, M. Ghassemi, S.C. Huang, P.C. Kuo, AI recognition of patient race in medical imaging: a modelling study. Lancet Digit. Health **4**(6), e406–e414 (2022)
79. I.R.I. Alberto, N.R.I. Alberto, A.K. Ghosh, B. Jain, S. Jayakumar, N. Martinez-Martin, N. McCague, D. Moukheiber, L. Moukheiber, M. Moukheiber, S. Moukheiber, The impact of commercial health datasets on medical research and health-care algorithms. Lancet Digit. Health **5**(5), e288–e294 (2023)
80. P. Fahr, J. Buchanan, S. Wordsworth, A review of the challenges of using biomedical big data for economic evaluations of precision medicine. Appl. Health Econ. Health Policy **17**(4), 443–452 (2019)
81. W. Huda, R.B. Abrahams, X-ray-based medical imaging and resolution. Am. J. Roentgenol. **204**(4), W393–W397 (2015)
82. A.K. Dhara, S. Mukhopadhyay, A. Dutta, M. Garg, N. Khandelwal, Content-based image retrieval system for pulmonary nodules: Assisting radiologists in self-learning and diagnosis of lung cancer. J. Digit. Imaging **30**, 63–77 (2017)
83. H.K. Lim, S.C. Hong, W.S. Jung, K.J. Ahn, W.Y. Won, C. Hahn, I.S. Kim, C.U. Lee, Automated segmentation of hippocampal subfields in drug-naïve patients with Alzheimer disease. Am. J. Neuroradiol. **34**(4), 747–751 (2013)
84. D.B. Larson, M.C. Chen, M.P. Lungren, S.S. Halabi, N.V. Stence, C.P. Langlotz, Performance of a deep-learning neural network model in assessing skeletal maturity on pediatric hand radiographs. Radiology **287**(1), 313–322 (2018)
85. X. Ren, T. Li, X. Yang, S. Wang, S. Ahmad, L. Xiang, S.R. Stone, L. Li, Y. Zhan, D. Shen, Q. Wang, Regression convolutional neural network for automated pediatric bone age assessment from hand radiograph. IEEE J. Biomed. Health Inform. **23**(5), 2030–2038 (2018)
86. Q. Shen, Y. Shan, Z. Hu, W. Chen, B. Yang, J. Han, Y. Huang, W. Xu, Z. Feng, Quantitative parameters of CT texture analysis as potential markers for early prediction of spontaneous intracranial hemorrhage enlargement. Eur. Radiol. **28**, 4389–4396 (2018)
87. B. Rathore, Digital transformation 4.0: Integration of artificial intelligence & metaverse in marketing. Eduzone: Int. Peer Rev/Ref. Multidiscip. J., **12**(1), 42–48 (2023)
88. J. Deng, W. Dong, R. Socher, L.J. Li, K. Li, L. Fei-Fei, Imagenet: A large-scale hierarchical image database. In *2009 IEEE conference on computer vision and pattern recognition* (IEEE, 2009), pp. 248–255
89. O. Russakovsky, J. Deng, H. Su, J. Krause, S. Satheesh, S. Ma, Z. Huang, A. Karpathy, A. Khosla, M. Bernstein, A.C. Berg, Imagenet large scale visual recognition challenge. Int. J. Comput. Vision **115**, 211–252 (2015)
90. S. Winklhofer, U. Held, J.M. Burgstaller, T. Finkenstaedt, N. Bolog, N. Ulrich, J. Steurer, G. Andreisek, F. Del Grande, Degenerative lumbar spinal canal stenosis: intra-and inter-reader agreement for magnetic resonance imaging parameters. Eur. Spine J. **26**, 353–361 (2017)
91. J.K. Hoang, W.D. Middleton, A.E. Farjat, S.A. Teefey, N. Abinanti, F.J. Boschini, A.J. Bronner, N. Dahiya, B.S. Hertzberg, J.R. Newman, D. Scanga, Interobserver variability of sonographic features used in the American College of Radiology Thyroid Imaging Reporting and Data System. Am. J. Roentgenol. **211**(1), 162–167 (2018)

92. T.R. Insel, Translating scientific opportunity into public health impact: a strategic plan for research on mental illness. Arch. Gen. Psychiatry **66**(2), 128–133 (2009)
93. L.M. Prevedello, S.S. Halabi, G. Shih, C.C. Wu, M.D. Kohli, F.H. Chokshi, B.J. Erickson, J. Kalpathy-Cramer, K.P. Andriole, A.E. Flanders, Challenges related to artificial intelligence research in medical imaging and the importance of image analysis competitions. Radiol.: Artif. Intell., **1**(1), e180031 (2019)
94. A. Rahman, M.S. Hossain, G. Muhammad, D. Kundu, T. Debnath, M. Rahman, M.S.I. Khan, P. Tiwari, S.S. Band, Federated learning-based AI approaches in smart healthcare: concepts, taxonomies, challenges and open issues. Clust. Comput. **26**(4), 2271–2311 (2023)
95. T. Huang, H. Xu, H. Wang, H. Huang, Y. Xu, B. Li, S. Hong, G. Feng, S. Kui, G. Liu, D. Jiang, Artificial intelligence for medicine: Progress, challenges, and perspectives. Innov. Med., 100030–100031
96. S. Ai, C. Li, X. Li, T. Jiang, M. Grzegorzek, C. Sun, M.M. Rahaman, J. Zhang, Y. Yao, H. Li, A state-of-the-art review for gastric histopathology image analysis approaches and future development. BioMed Res. Int., 2021
97. M. Gidwani, K. Chang, J.B. Patel, K.V. Hoebel, S.R. Ahmed, P. Singh, C.D. Fuller, J. Kalpathy-Cramer, Inconsistent partitioning and unproductive feature associations yield idealized radiomic models. Radiology **307**(1), e220715 (2022)
98. Z.H. Chen, L. Lin, C.F. Wu, C.F. Li, R.H. Xu, Y. Sun, Artificial intelligence for assisting cancer diagnosis and treatment in the era of precision medicine. Cancer Commun. **41**(11), 1100–1115 (2021)
99. Y. Bouchareb, P.M. Khaniabadi, F. Al Kindi, H. Al Dhuhli, I. Shiri, H. Zaidi, A. Rahmim, Artificial intelligence-driven assessment of radiological images for COVID-19. Comput. Biol. Med. **136**, 104665 (2021)
100. F. Binczyk, W. Prazuch, P. Bozek, J. Polanska, Radiomics and artificial intelligence in lung cancer screening. Transl. Lung Cancer Res. **10**(2), 1186 (2021)
101. K.H. Lin, N.H. Lu, T. Okamoto, Y.H. Huang, K.Y. Liu, A. Matsushima, C.C. Chang, T.B. Chen, Fusion-extracted features by deep networks for improved COVID-19 classification with chest X-ray radiography. In *Healthcare* vol. 11, no. 10 (MDPI, 2023), p. 1367
102. S. Liang, R. Zhang, D. Liang, T. Song, T. Ai, C. Xia, L. Xia, Y. Wang, Multimodal 3D DenseNet for IDH genotype prediction in gliomas. Genes **9**(8), 382 (2018)
103. B. Mesko, The role of artificial intelligence in precision medicine. Expert. Rev. Precis. Med. Drug Dev. **2**(5), 239–241 (2017)

Chapter 4
Tracing Historical Connections: The Evolutionary Ties of Artificial Intelligence, Confocal Microscopy, and Marvin Minsky (1927–2016)

Guan-Yu Zhuo, Manikanth Karnati, Mohendra Roy, Nirmal Mazumder, Ankur Gogoi, and Fu-Jen Kao

Abstract The historical convergence of Marvin Minsky's contributions to artificial intelligence (AI) and confocal microscopy (CM) is explored in this contribution, highlighting the intersection of these disciplines. As a luminary in the field of AI, Minsky's foundational work at the Massachusetts Institute of Technology's Artificial Intelligence Laboratory set the stage for neural network research and computational theories of mind. Simultaneously, Minsky's inventive prowess extended to CM, where he introduced the confocal scanning microscope, revolutionizing three-dimensional imaging. This book chapter provides a concise examination of Minsky's groundbreaking legacy, elucidating the profound impact of his work on the evolution of both AI and CM. Minsky's interdisciplinary approach and his ability to seamlessly navigate between AI and advanced microscopy reflected his deep curiosity and commitment to pushing the boundaries of scientific exploration. His impact on AI and CM was not confined to theoretical contributions but extended to practical

Zhou and Karnati: These authors contributed equally to this work.

G.-Y. Zhuo
Institute of Translational Medicine and New Drug Development, China Medical University, Taichung 404328, Taiwan

M. Karnati · N. Mazumder
Department of Biophysics, Manipal School of Life Sciences, Manipal Academy of Higher Education, Manipal, Karnataka 576104, India

M. Roy
Department of Information and Communication Technology, School of Technology, Pandit Deendayal Energy University, Gandhinagar, India

A. Gogoi (✉)
Department of Physics, Jagannath Barooah University, Assam 785001, India
e-mail: ankurgogoi@gmail.com

F.-J. Kao
Institute of Biophotonics, National Yang-Ming Chiao Tung University, Taipei 11221, Taiwan

A. Gogoi and N. Mazumder (eds.), *Biomedical Imaging*, Biological and Medical Physics, Biomedical Engineering, https://doi.org/10.1007/978-981-97-5345-1_4

innovations that continue to shape these fields today. Through a multidisciplinary lens, the chapter aims to underscore Minsky's trailblazing contributions and the enduring interplay between AI and advanced microscopy.

4.1 Introduction

'A picture is worth a thousand words'—a phrase that holds significant importance in the realm of contemporary biomedical imaging, emphasizing the idea that a meticulously captured image can convey a wealth of information. Over the course of centuries, images captured through various optical and other microscopy methods, e.g., magnetic resonance imaging (MRI), computed tomography (CT), X-ray imaging, photoacoustic imaging, linear and nonlinear optical microscopy techniques, etc., have indeed empowered researchers, clinicians, and scientists to precisely visualize the intricacies of biological structures at different spatiotemporal resolutions. This, in turn, has played a crucial role in enabling healthcare professionals to make informed decisions, ultimately leading to improved patient health through accurate diagnoses and effective treatments for a diverse spectrum of diseases [1–3].

Undoubtedly, within the spectrum of advanced biomedical imaging modalities developed to date, optical imaging techniques have gained considerable attention owing to their versatility, expansive spatiotemporal resolution, portability, and ability to acquire real-time and live cell images. Notably, in the realm of optical imaging modalities, the emergence of laser scanning confocal microscopy (CM) in the 1950s stands out as a pivotal breakthrough in far-field optical sectioning microscopy. This innovation facilitated three-dimensional (3D) high-resolution fluorescence imaging in both reflection and transmission modes, minimizing out-of-focus fluorescence signals [4–6]. Since its inception, CM has found widespread applications in clinical disciplines such as dermatology, ophthalmology, oncology, and beyond [7–14]. Furthermore, endoscopic versions of CM have been deployed in diverse biomedical imaging applications, including gastroenterology and bronchoscopy, underscoring its adaptability and significance in various medical fields [15, 16].

Importantly, biomedical images acquired through such advanced imaging modalities often produce extensive datasets rich in information regarding the anatomical and functional aspects of biological structures. This dataset is characterized by its multifaceted nature, incorporating intricate details at various spatial and temporal scales. The inherent complexity of these images, arising from the diverse characteristics of tissues and the nuanced interactions within biological systems, poses a formidable challenge for clinicians to analyze and comprehensively interpret the wealth of information embedded in the data [17, 18]. On top of that, the presence of noise and artifacts, and the need for intricate pattern recognition further complicate the analysis. In this context, the utilization of computational tools and advanced algorithms, especially artificial intelligence (AI)-based techniques, has become imperative for assisting clinicians in obtaining meaningful insights from these intricate

Fig. 4.1 Marvin Minsky in 2008. The photograph is obtained from Wikimedia commons [23] and licensed under the Creative Commons Attribution-Share Alike 2.0 Generic license

biomedical images, ultimately enhancing the accuracy and efficiency of diagnostic and research processes in the fields of biomedical sciences and health care [20–22] (Fig. 4.1).

Interestingly, these two seemingly disparate multidisciplinary scientific domains, i.e., CM and AI, are indeed intertwined through a common figure, Marvin Lee Minsky, who pioneered both of these fields in the 1950s. This visionary scientist, who was born on August 9, 1927 in New York, passed away on January 24, 2016 in Boston at the age of 88 [24]. This book chapter presents a brief overview on the Marvin Minsky's trailblazing legacy in AI and CM.

4.2 Marvin Minsky's Pioneering Work in Artificial Intelligence

The timeframe from 1943 to 1955 is recognized as the gestation period of AI (the term was coined later). Noteworthy figures such as Warren McCulloch, Walter Pitts, Donald Hebb, and Alan Turing significantly contributed to the early development of AI during this era [25, 26]. During this period, Minsky obtained his bachelor's degree in mathematics from Harvard University in 1950. For his bachelor's thesis titled "A Generalization of Kakutani's Fixed-Point Theorem", Minsky used topology to prove a generalization of Kakutani's fixed-point theorem [27, 28].

A pivotal breakthrough occurred in 1951 when Marvin Minsky and Dean Edmonds, both undergraduate students, constructed the first neural network machine, SNARC (Stochastic Neural Analog Reinforcement Calculator) [29]. By utilizing

3000 vacuum tubes and repurposing an excess automatic pilot mechanism from a B-24 bomber, the SNARC simulated a network comprising 40 neurons. Notably, the SNARC demonstrated proficiency in simulating problems analogous to a mouse running in a maze [25, 30, 31]. The SNARC is a pioneering system in the field of artificial intelligence. The network of 40 artificial neurons is interconnected using the principle of Hebb synapses. Hebb synapses are a fundamental concept in neuroscience, where an increase in synaptic efficacy results from the repeated and persistent stimulation of a postsynaptic cell by a presynaptic cell. There is a unique feature known as the probability knob. This knob, which ranges from 0 to 1, represents the probability of signal propagation. If the signal successfully propagates, this event is stored by a capacitor. The system also incorporates a reward mechanism. If the signals align with the desired outcome, a reward is manually given to the system, reinforcing the successful pathways and promoting learning. Each synapse machine in the original SNARC system was equipped with 6 vacuum tubes, a testament to the technology of the time.

Furthering his outstanding academic journey, Marvin Minsky completed his doctorate in mathematics at Princeton University in 1954. His doctoral dissertation, titled "Neural Nets and the Brain Model Problem," encompassed significant contributions to learning theory and the theory of random neural nets [32]. Following the completion of his Ph.D. in 1954, Minsky carried forward his research as a Junior Fellow at Harvard and later at the MIT Lincoln Laboratory. During this period, he authored one of the pioneering papers in the field of artificial intelligence titled "Heuristic Aspects of the Artificial Intelligence Problem" [27, 28, 33].

Notably, Marvin Minsky is one of the four organizers of the 1956 Dartmouth Workshop on Artificial Intelligence, which is recognized as the birthplace of AI. The proposal for this groundbreaking workshop, titled 'A Proposal for the Dartmouth Summer Research Project on Artificial Intelligence,' was submitted by John McCarthy (Dartmouth College), Marvin Minsky (Harvard University), Nathaniel Rochester (IBM), and Claude Shannon (Bell Telephone Laboratories) to the Rockefeller Foundation on August 31, 1955 [34, 35]. This marked the first introduction of the term 'Artificial Intelligence,' with John McCarthy credited for coining this pivotal phrase. The 1956 workshop notably featured the participation of key figures, including Ray Solomonoff, Oliver Selfridge, Trenchard More, Arthur Samuel, Allen Newell and Herbert Simon. It was during this workshop that Newell and Simon unveiled the first computer program, 'Logic Theorist', capable of simulating certain aspects of human problem-solving abilities [25, 36–39].

In 1958, Marvin Minsky joined the MIT faculty, where he, alongside John McCarthy, co-founded the MIT Artificial Intelligence Laboratory in 1959 [25, 26, 31]. In the following years, Minsky authored several pivotal papers on AI, including "Steps Toward Artificial Intelligence", published in 1961. This work conducted a thorough survey and analysis of prior accomplishments, outlining significant challenges for future research in symbolic AI. These challenges encompassed areas such as search, pattern recognition, learning, planning, and induction [29, 40]. In another paper titled "Matter, Mind, and Models," published in 1963, Marvin Minsky explored the connections between physical matter, mental processes, and the application of models to comprehend these phenomena. This work exemplifies his extensive interest in artificial intelligence, cognitive science, and interdisciplinary investigation of the mind [41]. In the late 1960s, Minsky and Seymour Papert collaborated on perceptrons, a type of basic computing device that mimics certain aspects of neural behavior. They explored the capabilities and limitations of perceptrons and jointly authored the book "Perceptrons," [42] in 1969, which became quite influential in the understanding of artificial neural networks. In 1969, Marvin Minsky was honored with the ACM A.M. Turing Award, an annual accolade presented by the Association for Computing Machinery (ACM), in appreciation of his pivotal contributions to the development and progress of the field of artificial intelligence [43].

Later, in 1974, Minsky introduced the concept of 'frames', an AI data structure, in his seminal article "A Framework for Representing Knowledge" [44]. Essentially, frames provide a structured way to organize and represent knowledge, allowing AI systems to efficiently store, process, and retrieve information. In his 1979 essay "K-lines: A Theory of Memory" [45], Marvin Minsky examined fundamental inquiries about the representation, storage, retrieval, and utilization of information in the brain to form memories. In his paper, Minsky proposed that in the memory formation process, a K-line is generated, connecting with the mental agencies engaged in the memorable event. When activated later, it triggers specific mental agencies, leading to a partial mental state that recalls the initial event. Minsky authored several influential essays and books exploring the organization and interpretation of information by minds and machines [46, 47].

Undoubtedly, Minsky was instrumental in the initial phases of AI, enabling subsequent generations to comprehend and emulate human intelligence through computational methods. His enduring legacy remains fundamental, influencing the course of the field and sparking continuous exploration at the forefront of AI and cognitive science.

4.3 Minsky's Contribution to the Development of Confocal Microscopy

Since Zacharias Janssen and his father Hans Janssen invented the compound microscope in 1590 [48, 52], optical microscopy has significantly advanced, playing a crucial role in the development of various scientific fields and enhancing our comprehension of the microscopic world. In fact, optical microscopy has played the most significant role in the advancement of the entire field of biology by enabling scientists to observe and study biological specimens at the cellular and subcellular levels, contributing significantly to our understanding of living organisms. More than 300 years ago, Antonie van Leeuwenhoek employed his self-ground optical lenses to uncover bacteria, marking the inception of microbiology [50]. Similarly, Santiago Ramón y Cajal, a Spanish neuroscientist, made significant contributions to the field of modern neurology in the late nineteenth century. In 1888, he utilized classical optical microscopy to create accurate drawings of neurons in a section of a chick cerebellum stained with an improved Golgi method. Cajal's meticulous observations and illustrations were foundational in advancing our understanding of the nervous system and laid the groundwork for modern neuroscience [50, 51].

Throughout this timeline, significant improvements have been achieved by several scientists, including Robert Hooke, Carl Zeiss, Ernst Abbe, Linnik, Frits Zernike, etc., to enhance magnification, resolution, contrast, noise, optical aberrations, and sensitivity [52]. Notably, Frits Zernike was awarded the Nobel prize for his invention of the phase contrast microscope in 1953 [53]. Notably, following advancements in fluorescence labeling techniques and the emergence of wide-field fluorescence microscopy in the early twentieth century, it became possible to noninvasively image both living and fixed specimens with enhanced chemical and physical specificity and contrast [4, 52, 54, 55]. Such wide-field fluorescence microscopes are valued for their affordability, ease of use, versatility, and rapid imaging capabilities in scientific and biomedical fields. Nonetheless, these methods have inherent limitations, such as increased photobleaching caused by specimen-wide illumination at high excitation power levels, resulting in the degradation of fluorescent signals over time. Moreover, wide-field fluorescence microscopes experience reduced contrast due to the presence of out-of-focus light and lack of clarity when imaging thick biological tissues due to multiple scattering [6].

Marvin Minsky encountered these issues, particularly when attempting to image three-dimensional wiring diagrams (connections) among brain cells to understand how the brain learns. He encountered difficulties in comprehending microscopic brain structures using conventional microscopes due to light scattering in dense neural tissue. This occurred during the period when Minsky completed his Ph.D. and became a Junior Fellow at Harvard. However, his persistent efforts to overcome these obstacles ultimately led to his invention of the confocal scanning microscope in 1955 [29, 33, 56]. The key innovation was the incorporation of one pinhole on the illumination side and another on the detection side, strategically positioned at the conjugate optical planes. This configuration allows light emitted from the illumination pinhole to be directed by a lens onto the specimen at its focal point. Subsequently, the second lens focuses the illuminated sample point onto the second pinhole before reaching the detector. Because of these dual lenses and the use of stage scanners, Minsky referred to his microscopy apparatus as a "double focusing stage scanning system" [4, 5, 56].

In the initial design of the microscope, Minsky used zirconium arcs as the light source, a low noise photomultiplier as the detector and an image constructed on the display of a surplus long-persistence oscilloscope from the military. For scanning, Minsky chose to move the specimen by adjusting the stage directly instead of altering the direction of the light rays. By utilizing a 60 Hz solenoid for vertical platform movement and a solenoid with a lower frequency for horizontal movement, Minsky achieved a frame rate of roughly one image every ten seconds [56, 58]. Minsky obtained a patent (U.S. Patent 3,013,467) for his invention in 1957 [59]. A photograph of the original microscope is shown in Fig. 4.2a.

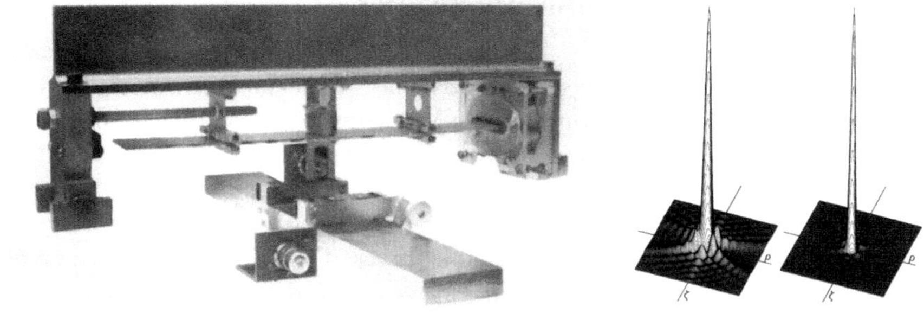

(a) (b)

Fig. 4.2 **a** A photograph of the original confocal microscope is shown above. The figure was reproduced with permission from Minsky [56], Copyright © OFACM, Inc. **b** The point spread function (psf) for a conventional microscopy system, is shown on the left, $p(\zeta, \rho)$, and that for a confocal microscopy system, $p_{conf}(\zeta, \rho) = p(\zeta, \rho) \times p(\zeta, \rho)$, is shown on the right. The sharpening of the effective psf in a confocal microscopy system improves the lateral resolution by a factor of $\sqrt{2}$. The figure was created with permission from IOP Publishing, Ltd., from Webb [57] Copyright (c)1996 IOP Publishing Ltd.; permission conveyed through Copyright Clearance Center, Inc

This ingenious design effectively eliminated out-of-focus light, resulting in sharper and more precise images [57, 60]. Figure 4.2b shows the sharpening of the point spread function (PSF) in confocal microscopy images compared to that obtained by conventional microscopes. In addition, the ability of a confocal microscope to selectively capture optical sections at different depths within a specimen enabled researchers to create detailed three-dimensional reconstructions. Minsky's confocal scanning microscope addressed several challenges faced by contemporary conventional microscopes, such as the degradation of image quality due to scattered light and reduced contrast caused by out-of-focus blur.

Furthermore, integrating adaptive optics into CM has reduced aberrations caused by biological samples or optical components, improving image quality and resolution. CM has been used to explore dynamic biological processes and functional activities within living cells, as well as to visualize cellular architecture. Fluorescence resonance energy transfer (FRET) and fluorescence lifetime imaging microscopy (FLIM) are two approaches that use confocal microscopy to analyze protein–protein interactions, cellular signaling cascades, and metabolic processes in real time [5, 57].

This advancement revolutionized the field of microscopy by providing a powerful tool for visualizing biological samples with unprecedented clarity and detail. Importantly, Minsky's pioneering work laid the foundation for confocal microscopy, and his design principles continue to influence modern confocal microscope configurations and play a pivotal role in advancing microscopy capabilities.

4.4 Conclusion and Future Implications

Notably, Marvin Minsky is considered as one of the pioneers of AI. Several other scientists, such as Alan Turing, John McCarthy, Nathaniel Rochester, Claude Shannon, and others, made substantial contributions to the field before or during Minsky's time [25, 35]. Similarly, in the advancement of confocal microscopy systems, contemporary scientists such as Goldman and Koana have made noteworthy contributions [61–63]. There is no doubt, however, that Minsky's influence on the fields of AI and CM was profound, laying the groundwork for numerous subsequent methodologies and applications that have greatly benefited humankind, especially in the field of biomedical imaging.

In addition to his contributions to AI and confocal microscopy, Marvin Minsky invented several other instruments. These include the first head-mounted graphical display in 1963, the concept of a binary-tree robotic manipulator in the same year, and the serpentine hydraulic robot arm in 1967 [29, 33]. Additionally, Minsky had an interest in science fiction. He acted as a technical consultant for the film 2001: A Space Odyssey and collaborated with Harry Harrison on co-authoring the science fiction novel, The Turing Option [64].

Although historical evidence linking Marvin Minsky's work directly to the integration of AI and CM may be limited, recent research has greatly benefited from leveraging the synergy between AI and cutting-edge imaging methodologies. Notably, Minsky's primary focus was not explicitly on merging AI with confocal microscopy. Nevertheless, his invention indeed served as a foundation for the development of advanced imaging technologies and computational techniques within microscopy. This book chapter is a humble attempt to trace and summarize Minsky's enduring legacy through his distinct contributions to both AI and microscopy, highlighting the convergence of these seemingly distinct fields.

Today, AI has become ubiquitous across various domains, including natural language processing, computer vision, machine learning, and robotics. AI algorithms are increasingly utilized to automate image analysis, improve image quality, and optimize experimental parameters not only in confocal microscopy but also in other imaging modalities. These advancements hold promise for transforming our understanding of complex biological structures and fostering innovation. Moreover, Minsky's exploration of neural networks and computational theories of mind laid the groundwork for neuromorphic computing, which mimics the structure and function of the human brain. Such systems have the potential to enhance AI models, particularly in cognitive computing and machine learning applications. Undoubtedly, Minsky's contributions to confocal microscopy have influenced the development of advanced imaging technologies and computational analysis techniques. His invention of the confocal scanning microscope has paved the way for innovations in imaging technology and computational analysis, shaping the landscape of modern microscopy systems. In the realm of biomedical imaging, the integration of AI has played a crucial role in disease diagnosis, treatment planning, and drug discovery, leading to advancements in personalized medicine and improved patient outcomes.

In summary, Marvin Minsky's visionary ideas have had a lasting impact on scientific and technological progress, shaping the approach to complex problems and fostering innovation across disciplines. Certainly, Minsky's legacy lies not only in his individual contributions to AI and microscopy but also in the bridging of these seemingly disparate disciplines.

Acknowledgements The authors gratefully acknowledge the financial support received from the Science and Engineering Research Board (SERB), Department of Science and Technology (DST), India (Grant No. CRG/2019/004868) and the National Science and Technology Council (NSTC), Taiwan (Grant No. 110-2112-M-039-001-MY3).

References

1. A. Webb, *Introduction to biomedical imaging* (John Wiley & Sons, 2022)
2. M. Obayya, M.K. Saeed, N. Alruwais, S.S. Alotaibi, M. Assiri, A.S. Salama, Hybrid metaheuristics with deep learning based fusion model for biomedical image analysis. IEEE Access **11**, 117149–117158 (2023)
3. K. Heinzmann, L.M. Carter, J.S. Lewis, E.O. Aboagye, Multiplexed imaging for diagnosis and therapy. Nat. Biomed. Eng. **1**(9), 697–713 (2017)
4. M. Renz, Fluorescence microscopy—A historical and technical perspective. Cytometry A **83**(9), 767–779 (2013)
5. A.D. Elliott, Confocal microscopy: principles and modern practices. Curr. Protoc. Cytom. **92**(1), e68 (2020)
6. A. Gogoi, G. Keiser, F.J. Kao, A.S. Chiang, Multiscale and multimodal imaging for connectomics. in *Advanced optical methods for brain imaging* (Springer Singapore, 2019), pp. 3–45
7. S. González, K. Swindells, M. Rajadhyaksha, A. Torres, Changing paradigms in dermatology: confocal microscopy in clinical and surgical dermatology. Clin. Dermatol. **21**(5), 359–369 (2003)
8. R. Jayamohanan, Confocal microscopy–Working principle and applications in dermatology. J. Ski. Sex. Transm. Dis. **5**(2), 81–89 (2023)
9. J.C. Erie, L.W. McLaren, S.V. Sanjay Patel, Confocal microscopy in ophthalmology. Am. J. Ophthalmol., **148**(5), 639–646 (2009)
10. E.C. Ledbetter, Applications of in vivo confocal microscopy in the management of infectious keratitis in veterinary ophthalmology. Vet. Ophthalmol. **25**(S1), 5–16 (2022)
11. R. Shetty, H.S. Dua, L. Tong, G. Kundu, P. Khamar, B. Gorimanipalli, S. D'Souza, Role of in vivo confocal microscopy in dry eye disease and eye pain. Indian J. Ophthalmol. **71**(4), 1099–1104 (2023)
12. M. Ilie, C. Caruntu, M. Lupu, D. Lixandru, M. Tampa, S.R. Georgescu, A. Bastian, C. Constantin, M. Neagu, S. Zurac, D. Boda, Current and future applications of confocal laser scanning microscopy imaging in skin oncology. Oncol. Lett. **17**(5), 4102–4111 (2019)
13. K. Carlson, I. Pavlova, T. Collier, M. Descour, M. Follen, R. Richards-Kortum, Confocal microscopy: Imaging cervical precancerous lesions. Gynecol. Oncol. **99**(3), S84–S88 (2005)
14. C. Longo, S. Guida, M. Mirra, R. Pampena, S. Ciardo, S. Bassoli, A. Casari, F. Rongioletti, M. Spadafora, J. Chester, S. Kaleci, M. Lai, S. Magi, L. Mazzoni, F. Farnetani, I. Stanganelli, G. Pellacani, Dermoscopy and reflectance confocal microscopy for basal cell carcinoma diagnosis and diagnosis prediction score: a prospective and multicenter study on 1005 lesions. J. Am. Acad. Dermatol. (2024). https://doi.org/10.1016/j.jaad.2024.01.035

15. A. Villard, I. Breuskin, O. Casiraghi, S. Asmandar, C. Laplace-Builhe, M. Abbaci, A. Moya Plana, Confocal laser endomicroscopy and confocal microscopy for head and neck cancer imaging: Recent updates and future perspectives. Oral Oncol. **127**, 105826 (2022)
16. I.S. Mamenko, I.V. Vasilev, R.V. Simonov, A.S. Zacharova, P.V. Gavrilov, I.D. Dvorakovskaya, T.A. Novitskaya, P.K. Yablonskiy, Probe-based confocal laser endomicroscopy in diagnosis of diffuse cystic lung disease in Sjögren's syndrome. Monaldi Arch. Chest Dis. (2024). https://doi.org/10.4081/monaldi.2023.2699
17. X.L. Chen, T.Y. Yan, N. Wang, K.M. von Deneen, Rising role of artificial intelligence in image reconstruction for biomedical imaging. Artif. Intell. Med. Imaging **1**(1), 1–5 (2020)
18. A.A. Nawresh, S. Sasikala, Biomedical imaging techniques using AI Systems. in *Handbook of artificial intelligence in biomedical engineering* (Apple Academic Press, 2021), pp. 147–172
19. S. Wang, G. Cao, Y. Wang, S. Liao, Q. Wang, J. Shi, C. Li, D. Shen, Review and prospect: artificial intelligence in advanced medical imaging. Front. Radiol. **1**, 781868 (2021)
20. M.S. Azar, M. Soifer, Z. Zemborain, S. Komai, M.E. Quiroga-Garza, H. Mousa, R.E.R. Lozano, J. Tian, S. Farsiu, V.L. Perez, Deep-learning based analysis of in-vivo confocal microscopy-derived metrics of the corneal nerve whorl in eyes with neuropathic ocular pain and controls. Invest. Ophthalmol. Vis. Sci. **64**(8), 4018 (2023)
21. G. Kundu, R. Shetty, S. D'Souza, P. Khamar, R.M. Nuijts, S. Sethu, A.S. Roy, A novel combination of corneal confocal microscopy, clinical features and artificial intelligence for evaluation of ocular surface pain. PLoS One, **17**(11), 0277086 (2022)
22. J. Selvaraj, T. Prabha, T.D.A. Kumar, S. Palaniappan, Artificial intelligence in biomedical image processing. in *Machine learning and systems biology in genomics and health* (Springer, 2022), pp. 147–188
23. https://commons.wikimedia.org/wiki/File:Marvin_Minsky.jpg. Date Accessed 23rd Mar 2024
24. Marvin Lee Minsky. Weblink: https://history.computer.org/pioneers/minsky.html. Date Accessed 23rd Mar 2024
25. S.J. Russell, P. Norvig, *Artificial intelligence: A modern approach,* 3rd edn (Pearson, 2014)
26. A. Karunananda, *Roadmap of artificial intelligence concepts to reality and future.* Weblink: https://slaai.lk/wp-content/uploads/2021/03/2019-Proceedings/Roadmap-of-Artificial-Intelligence-Concepts-to-Reality-and-Future.pdf
27. M. Minsky, *A generalization of kakutani's fixed-point theorem. Bachelor's Thesis, Department of Mathematics* (Harvard University, Cambridge, MA, 1950)
28. D. Hillis, J. McCarthy, T.M. Mitchell, E.T. Mueller, D. Riecken, A. Sloman, P.H. Winston, In honor of Marvin Minsky's contributions on his 80th birthday. AI Mag. **28**(4), 103–110 (2007)
29. Brief Academic Biography of Marvin Minsky. Weblink: https://web.media.mit.edu/~minsky/minskybiog.html. Date Accessed 23rd March 2024
30. C.D. James, J.B. Aimone, N.E. Miner, C.M. Vineyard, F.H. Rothganger, K.D. Carlson, S.A. Mulder, T.J. Draelos, A. Faust, M.J. Marinella, J.H. Naegle, A historical survey of algorithms and hardware architectures for neural-inspired and neuromorphic computing applications. Biol. Inspired Cogn. Arch. **19**, 49–64 (2017)
31. L.M. Fisher, Marvin Minsky: 1927–2016. Commun. ACM **59**(4), 22–24 (2016)
32. National Research Council, Funding a revolution: Government support for computing research (National Academies Press, 1999)
33. P.H. Winston, Marvin L. Minsky (1927–2016). Nature, **530** (7590), 282 (2016)
34. J. McCarthy, M.L. Minsky, N. Rochester, C.E. Shannon, A proposal for the dartmouth summer research project on artificial intelligence, August 31, 1955. AI Mag. **27**(4), 12–14 (2006)
35. A very short history of artificial intelligence (AI). Weblink: https://www.forbes.com/sites/gilpress/2016/12/30/a-very-short-history-of-artificial-intelligence-ai/?sh=3ba8a7da6fba. Date Accessed 23rd March 2024
36. J. Moor, The dartmouth college artificial intelligence conference: the next 50 years. AI Mag. **27**(4), 87–91 (2006)
37. L. Gugerty, Newell and Simon's logic theorist: historical background and impact on cognitive modeling. in *Proceedings of the human factors and ergonomics society annual meeting*, vol. 50, issue no. 9 (SAGE Publications, Sage, Los Angeles, 2006), pp. 880–884

38. A. Newell, H. Simon, The logic theory machine–A complex information processing system. IRE Trans. Inf. Theory **2**(3), 61–79 (1956)
39. A. Karunananda, Roadmap of artificial intelligence concepts to reality and future. in *SLAAI—International conference on artificial intelligence* (2019), pp. 1–6
40. M. Minsky, Steps toward artificial intelligence. In Proceedings of the IRE **49**(1), 8–30 (1961)
41. M. Minsky, *Matter, mind and models*. Weblink: https://web.media.mit.edu/~minsky/papers/MatterMindModels.html. Date Accessed 23rd March, 2024
42. M. Minsky, S. *Papert, perceptrons: An introduction to computational geometry* (MIT Press, 1969)
43. Weblink: https://amturing.acm.org/award_winners/minsky_7440781.cfm. Date Accessed 23rd March, 2024
44. M. Minsky, *A framework for representing knowledge*. Weblink: https://dspace.mit.edu/bitstream/handle/1721.1/6089/AIM-306.pdf?%2520sequence%3D2. Date Accessed 23rd March 2024
45. M. Minsky, K-lines: A theory of memory. Cogn. Sci. **4**(2), 117–133 (1980)
46. M. Minsky, *The society of mind* (Simon and Schuster, 1987)
47. M. Minsky, *The emotion machine* (Simon and Schuster, 2006)
48. R.L. Haden, The origin of the microscope. Ann. Med. Hist. **1**(1), 30–44 (1939)
50. B. Huang, H. Babcock, X. Zhuang, Breaking the diffraction barrier: super-resolution imaging of cells. Cell **143**(7), 1047–1058 (2010)
51. L.W. Swanson, J.W. Lichtman, From Cajal to connectome and beyond. Annu. Rev. Neurosci. **8**(39), 197–216 (2016)
52. B.R. Masters, *The development of fluorescence microscopy*. Encyclopedia of Life Sciences, (2001)
53. The Nobel Prize in Physics 1953. Weblink: https://www.nobelprize.org/prizes/physics/1953/summary/. Date Accessed 23rd March 2024
54. B.N. Giepmans, S.R. Adams, M.H. Ellisman, R.Y. Tsien, The fluorescent toolbox for assessing protein location and function. Science **312**(5771), 217–224 (2006)
55. F.L. Labarthet, Y.R. Shen, Nonlinear optical microscopy, in *Optical imaging and microscopy*, vol. 87, (Springer, Berlin, Heidelberg, 2003), pp.169–196
56. M. Minsky, Memoir on inventing the confocal scanning microscope. Scanning **10**(4), 128–138 (1988)
57. R.H. Webb, Confocal optical microscopy. Rep. Prog. Phys. **59**(3), 427–471 (1996)
58. R. Leiva-García, J. García-Antón, M.J. Muñoz-Portero, Application of confocal laser scanning microscopy to the in-situ and ex-situ study of corrosion processes. in *Laser scanning-theory and applications* (IntechOpen, 2011), pp. 431–448
59. M. Minsky, *Microscopy apparatus patent* (United States, 1957)
60. Confocal Microscopy. Weblink: https://www.ntmdt-si.com/resources/spm-theory/theoretical-background-of-spm/3-scanning-optical-microscopy/32-confocal-microscopy. Date Accessed 23rd March, 2024
61. C.J.R. Sheppard, *Confocal microscopy*. Weblink: https://analyticalscience.wiley.com/content/article-do/confocal-microscopy
62. H. Naora, Microspectrophotometry and cytochemical analysis of nucleic acids. Science **114**(2959), 279–280 (1951)
63. H. Goldmann, Spaltlampenphotographie und—photometric. Ophthalmologica **98**(5–6), 257–270 (1939)
64. M. Minsky, H. Harrison, *The turing option* (Warner Books, 1992)

Chapter 5
Types of Optical Microscopic Analysis for Cell Death Using Artificial Intelligence

Baishali Das, Sindhoora Kaniyala Melanthota, N. Pooja, Ishita Chakraborty, Indira Govindaraju, Divyani Chatterjee, and Nirmal Mazumder

Abstract Cell death is a biological process that occurs when a cell stops carrying out all its functions. Cell death leads to macroscopic changes in the cell morphology. Coherent anti-Stokes Raman scattering (CARS), Stimulated Raman scattering (SRS) are used as a tool to observe structural and chemical changes at cellular resolution. There are many methods used to detect cell death generally takes time and materials. Spectroscopic and microscopic analyses are commonly used techniques for non-invasive analysis of samples, for cell death. Raman spectroscopy and microscopy techniques detects the inelastic scattering of light due to excitation or de-excitation of the samples. Spectroscopic technique has the advantage of being able to take rapid measurements in real time of in vivo, and in vitro samples, without any undesirable effects on the surrounding living cells and perform analysis regardless of the sample size. Despite the advantages that spectroscopy techniques offer, the complexity of the spectroscopic peaks is a major drawback. This gives rise to the need to develop automated algorithms for analysing high throughput spectroscopic and microscopic value with great efficiency and accuracy. Machine learning, which is a subset of artificial intelligence (AI) may be used to assess and draw conclusions from scientific investigations that give rise to large amounts of data in an automated manner, thereby reducing the scope for manual errors during analysis. Machine learning algorithms is categorized into supervised learning, unsupervised learning, as well as reinforcement learning algorithms. The development of new algorithms like deep learning (DL) has proven to be highly beneficial in classifying large and complex data. When used with fluorescence microscopy, DL has proven to show high specificity and can help in achieving inexpensive and rapid analysis.

B. Das · S. K. Melanthota · N. Pooja · I. Chakraborty · I. Govindaraju · D. Chatterjee · N. Mazumder (✉)
Department of Biophysics, Manipal School of Life Sciences, Manipal Academy of Higher Education, Manipal, Karnataka 576104, India
e-mail: nirmal.mazumder@manipal.edu

A. Gogoi and N. Mazumder (eds.), *Biomedical Imaging*, Biological and Medical Physics, Biomedical Engineering, https://doi.org/10.1007/978-981-97-5345-1_5

5.1 Introduction

When a cell stops performing all its activities, it undergoes through a biological process known as cell death. Cell death leads to macroscopic changes in the cell morphology. Studies on cell death, and its subsequent aftermath has been integral to studies pertaining to various genetic and infectious diseases. Knowledge of cell death, and the various residues of the phenomenon is also crucial for toxicology studies regarding the effect of diverse compounds on cells. Cell death is divided into 3 types– Apoptotic (type I), Autophagic (type II) and Necrotic (type III). Each type of cell death is characterized by a set of marked molecular changes and cascade mechanisms that occur in the cell. Increased caspase activity, condensation of chromatin (pyknosis), fragmentation of nucleus (karyorrhexis), and plasma membrane blebbing, and cytoplasmic deformation are all signs of apoptosis. Apoptotic cells also produce small, seemingly intact vesicles known as apoptotic bodies, which are easily taken up by surrounding cells through phagocytosis and broken down in lysosomes. Autophagy manifests extensive cytoplasmic vacuolization and similarly culminating with phagocytic engulfment of organelles and particles with consequent lysosomal degradation. Necrotic cells do not show any characteristic features of type I or II cell death and terminates with the removal of cell corpses in the absent of apparent lysosomal and phagocytic involvement [16]. It is well known that necrosis causes the cell organelles to enlarge, that results in a loss of plasma and integrity of the cell membrane as well as inflammation of the necrotic tissue. There are many methods used to detect cell death generally takes time and materials. However, minute morphological alterations in the shape and texture of the cell that can serve as the basis for detection occur prior to cell death [10]. Thus, the physiopathological effects generated by various illnesses and therapies on cell viability and cell death are crucial for conducting studies on the therapeutic effects of various compounds and compositions for the treatment of a disease or disorder. Notwithstanding its relevance, the exact molecular components, and the physiological mechanisms involved in the cell death process, and their contribution to the same is yet to be fully understood. Currently, cell death is classified or identified based on the morphology of the cells. In vivo tests however, have been unable to unambiguously report the type of cell death that occurred in a given sample. There are several tests and assays that are in use to determine the cell viability. These tests, however, are known to be destructive, time-consuming, and not very cost-effective, thereby limiting the application of these tests. Moreover, most cytotoxicity assays are dependent on chemicals and biomarkers, which may sometimes interact with cell components, resulting in ambiguity during other pharmacological tests, which does not serve the purpose of the cytotoxicity assays. Kanduc et al., observed that the current assays to identify cell death rely heavily on the post-mortem changes that occur after cell death and not the process of cell death itself, thereby contributing to the ambiguity in the report of cell death [14]. This gave rise to the need for the application of other techniques in biologist's arsenal for the identification and analysis of cell death without causing cytotoxicity and other undesirable effects on the sample. Spectroscopy and microscopy analyses are commonly used techniques for non-invasive analysis of samples, for cell death. Being non-invasive in nature, these techniques definitively overcome the disadvantages of the traditional cytotoxicity assay. Moreover, spectroscopy techniques have the advantage of being able to take rapid measurements in real time of in vivo, and in vitro samples, without any undesirable effects on the surrounding living cells and also perform analysis regardless of the sample size (either a single cell, or a population of cells). These advantages have enabled the application of spectroscopic techniques for the analysis of not only pharmaceuticals [2] and toxins, but also to study their effects on tissues.

Despite the advantages that spectroscopy techniques offer, the complexity of the spectroscopic peaks is a major drawback. The complex nature of the peaks is representative of the components of the cell each of which is responsible for the spectroscopic peaks. Traditionally, peak fitting is performed prior to the data analysis. Peak heights and the concentration of the associated biological component have a linear relationship. Thus, inferences are made on the biophysical parameters associated with the cells' viability and health. The sheer number and complexity of the peaks make the peak fitting process a very arduous task, which paves more way for human error to occur during analyses. This gives rise to the need to develop automated algorithms for analysing high throughput spectroscopic and microscopic value with great efficiency and accuracy. Machine learning, which is a subset of artificial intelligence (AI) may be used to assess and draw conclusions from scientific investigations that give rise to large amounts of data in an automated manner, thereby reducing the scope for manual errors during analysis. Machine learning algorithms can be further categorized based on the nature of datasets used to train the algorithms, into supervised learning, unsupervised learning, as well as reinforcement learning algorithms. Supervised learning algorithms require labelled and annotated datasets to 'learn' to come to the appropriate conclusions. Unsupervised learning algorithms 'learn' to draw conclusions based on an unlabelled dataset, by recognizing patterns in the training dataset. While reinforcement learning is a type of machine learning, wherein the algorithm decides on the suitable course of action to maximise reward in a specific situation. This chapter elucidates the different types of artificial intelligence that are routinely used in the biological fields, and their application in the field of spectroscopy and microscopy for the detection of cell death, and what it means to the healthcare and pharmaceutical sectors.

5.1.1 Types of AI

Artificial intelligence (AI) is one of the most popular academic disciplines in recent years due to the advances in the field of computers and technology. There are numerous definitions for AI. However, AI can generally be defined as a discipline that studies and creates artificial systems/machines/algorithms that can emulate human intelligence and behaviour. Computer vision, natural language processing, and voice recognition are some of the commonly used AI algorithms. Machine learning (ML) however, is one of the most commonly studied and applied algorithms in the sector of AI. The sheer number of ML algorithms is vast enough to make ML a major subset under the umbrella of AI algorithms. ML is a branch or subset of AI, which involves the automation of computer programs such that the program analyses given data after 'learning' from a training dataset, this ability of the algorithm to 'learn' from data has made it incredibly popular and widely used in various scientific and technical sectors. Deep Learning (DL) is a subset of ML, for which the principle of the algorithms is similar. However, DL algorithms differ from their machine learning counterparts in the structure of the learning process. DL algorithms function in a manner similar to that of the human network of nerves, by forming neural networks of many layers. DL algorithms are therefore referred to as neural networks. The application of AI in biology has been discussed in detail by Hassoun et al., wherein the authors emphasize on the ability of artificial intelligence to enable researchers to collect, and correlate different sets of data, thereby drawing conclusions leading to targeted and untargeted discoveries in the field of biology. The predictive capability of AI algorithms makes it popular, thereby finding application in various subdisciplines of sciences [12]. AI is particularly popular in the medical and life sciences sectors due to its ability to reduce human bias and errors that commonly occur during the classification of medical data, thereby aiding the doctor to give an accurate diagnosis [22]. The accuracy of the AI algorithms also makes it reliable for the prediction of patient prognosis, which in turn aids the medical staff to provide patients with better treatment. The sheer number of algorithms makes it important to classify them based on the dataset used for training them necessary, to increase their application in various fields and further update and improve them. The following section discusses the categories of machine learning algorithms.

(A) Supervised learning algorithms: The training dataset for these algorithms includes an input–output pair. The performance and output of the algorithm are monitored through the output or the labelled entries in the training dataset. The algorithm 'learns' to solve various problems of classification and regression, and evaluates its performance based on the labelled training dataset. Regression and classification problems are best solved by supervised training algorithms. Regression tasks involve the prediction of numerical outputs, while classification algorithms are employed to classify items and data into various categories [21]. Supervised machine learning models have often found application in the imaging field, to predict the occurrence of a disease, after studying the features to be given importance. Patel et al., have thoroughly revised the application of supervised machine learning models for the detection and prediction of depression in patients based on MRI imaging data [24].

(B) Unsupervised learning algorithms: These algorithms deal with unlabelled data as their training dataset. This enables the algorithm to make predictions based on the patterns and trends in the training dataset that is 'learned'. Clustering of data points, anomaly detection, and feature extraction, are some of the tasks that are commonly performed by unsupervised learning algorithms [5]. In the cases of multispectral and hyperspectral imagery for earth observations.

In order to quantitatively assess and estimate the geo-biophysical variables, Romero et al. used machine learning technique that was unsupervised rather than supervised machine learning algorithms for the interpretation of hyperspectral pictures. Due to the high dimensionality of the images and a very small amount of labelled data, the authors presented their preference for the unsupervised machine learning algorithms over machine learning algorithms that was supervised [30].

(C) Reinforcement learning algorithms: These algorithms have a reward-based learning system, wherein the algorithm guides the output such that the reward is maximum. Systems with these algorithms are dynamic wherein the algorithm controls and optimizes the system based on some objective. Reinforcement algorithms are generally employed in the field of robotics, and for the objective of driverless cars [31]. Zhou et al., have extensively reviewed the application and benefits of implementing reinforcement learning techniques with that of the deep learning methodology for the analysis of parametric medical images and to solve optimization tasks in the healthcare setup and the medical field, such as neural architecture searching and hyperparameter tuning. The authors also elucidate the advantages of using reinforcement learning algorithms over the traditionally used supervised learning algorithms for the analysis of medical images [38]. Generally, the applications of most machine learning algorithms overlap with each other, and it is the discretion of the user to decide which algorithm would perform the task with the most efficiency and accuracy. The algorithms used for the detection of cell death from microscopy and spectroscopy data, and their optimization are discussed further in this paper.

5.2 Types of Microscopy

5.2.1 Spontaneous Raman Technique

Raman spectroscopy and microscopy techniques detects the inelastic scattering of light due to excitation or de-excitation of the samples. They are utilized as a potent analytical technique to ascertain the structural alterations and fundamental chemical composition more sensitively at the molecular level [35]. Being label-free and non-invasive, this method allows rapid detection and quantification of nucleic acid, various proteins, and other components of the cells [20]. During cell death, a lot of events take place in the cell such as changes in chemical composition, pH, increase or decrease in the level of proteins, etc. which are easily detectable and can be monitored over time using Raman micro-spectroscopy. Activation in T-cell is an important process in immune response and is rigorously regulated by the activation-induced cell death (AICD) pathway. Raman spectroscopy was used to study the activated T-cells as well as the DNA from activated and resting T-cells. The DNA specific peaks were observed at 683 cm^{-1} for guanine and 727 cm^{-1} was assigned adenine, respectively. The DNA backbone's O–P–O, cytosine, and thymine were represented by the broad band at

765–786 cm^{-1}. The Raman spectra was analyzed using PCA analysis and observed that the Raman peak intensities decreased in activated T-cells as compared to resting cells. Thus, showing that Raman spectroscopy insights in monitoring induced cell death [17]. Cytochrome c act as a trigger for apoptosis and is observed to translocate from mitochondria to the cytosol during the same. Raman microscopy was used to study this change in molecular dynamics. Cytochrome c shows Raman scattering intensity at 750 cm^{-1} due to the pyrrole breathing mode and the signal was observed to diffuse into the cytosol after the cell was treated with Actinomycin D (apoptotic inducer). Apart from this, the Raman scattering intensities at 1684, and 2857 cm^{-1} corresponding to the Amide-I band in protein and CH_2 band in lipid were also studied and shown in Fig. 5.1.

It can be inferred that the apoptotic cells showed a larger number of lipid vesicles with no significant difference in the distribution of proteins due to the induction of apoptosis [23]. With the rapid growth of nanotechnology, concerns have also been raised about their toxicological implication in biological systems often leading to cell death. Raman spectroscopy complemented with a support vector machine was used to investigate the particle-cell interaction by performing peak-to-peak analysis. The

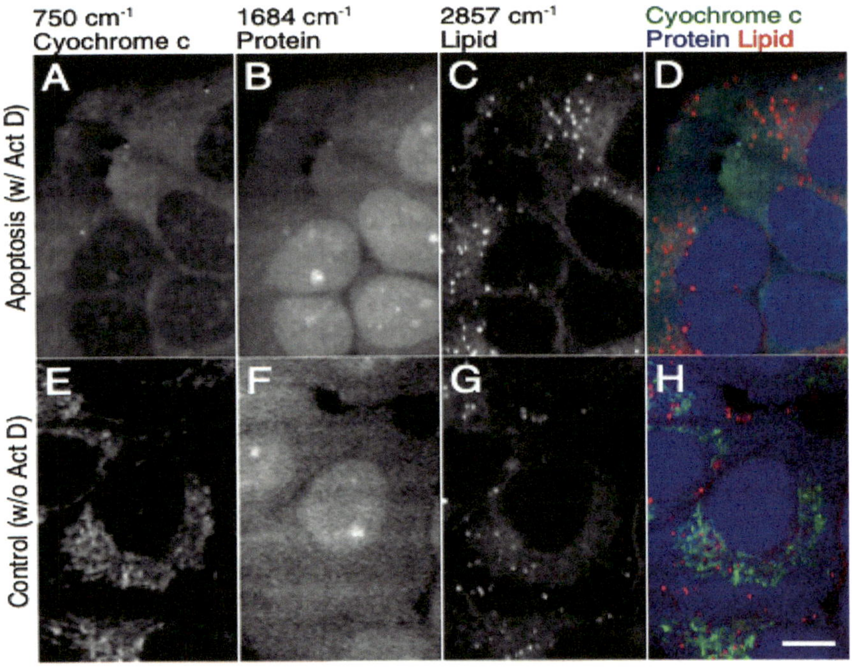

Fig. 5.1 Raman images of apoptotic or nonapoptotic HeLa cells at **a**, **e** 750, **b**, **f** 1684, **c**, **g** 2857 cm^{-1}, showing the distribution of cytochrome c, protein beta sheet, and lipid molecules, respectively. **d**, **h** were constructed by merging images (**a**) through (**c**) or (**e**) through (**g**) with colour channels. The images consist of 154 × 88 pixels. Scale bar, 10 μm. The figure is reproduced with permission from reference [23]

toxicity of TiO2 nanoparticles was analyzed and found that the 788 cm^{-1} peaks corresponding to the O–P–O in the DNA backbone decrease over time when treated with nanoparticles. 1231, 1242, 1258, and 1284 cm^{-1} peaks represent amide III which is the basic component of protein structure and the nanoparticles treated cells showed drastic variation in these peaks compared to the peaks from untreated cells. Showing that Raman spectroscopy can be used for toxicological analysis rapidly with single-cell resolution [26]. Often discrimination of the different cell death in vitro is challenging and needs to rely on various biochemical assays for discrimination. Raman microspectroscopy was used to analyze apoptotic and necrotic cell death events in cells complemented with PCA analysis. Raman spectra of the early apoptotic cell showed a band at 1375 cm^{-1} which became prominent in the late apoptotic stage and was absent in control Saos-2 cells. The late apoptotic cells showed a decrease in intensity at 1003 cm^{-1} whereas the narcotic cells showed increased relative intensity at the same band. The amide I peak at 1658 cm^{-1} in viable and apoptotic Saos-2 cells showed a shift in narcotic cells. Further, the peak at 1052 $cm^{-1}$was found to be absent in both type of cell death as seen in Fig. 5.2 [3]. In conclusion, spontaneous Raman micro and spectroscopy provide label free, rapid detection and monitoring of cell death with increased specificity and sensitivity. The methods complemented with analytical approaches make the whole process more efficient.

5.2.2 Fluorescence Techniques

The fluorescence-based method, unlike other approaches like electron microscopy and mass spectrometry, eliminates the need for sample preparation and cell fixation because fluorescent probes can be applied to living cells and the in-situ apoptosis process may be seen. The high sensitivity methods and capability to utilize a large number of fluorescent probes to assay apoptosis at varied time frames make fluorescence-based methods an ideal choice over their peers. Several assays developed in the past decades to analyze cell death, utilized features such as DNA fragmentation, protein modifications, and certain other morphological changes like cell detachment and shrinkage which is peculiar to cell death. The introduction of AI has created an enormous wave of innovation in the analysis and interpretation of data from these techniques. The development of new algorithms like deep learning (DL) has proven to be highly beneficial in classifying large and complex data [10]. When used with fluorescence microscopy, DL has proven to show high specificity and can help in achieving inexpensive and rapid analysis. The fluorescent methodology was able to detect apoptotic probes with the optimal trade-offs for its outstanding mixture of signal-to-noise ratios, spatial and temporal resolution. There are different modes of fluorescence spectroscopy and microscopy. Quiroz et al. [27] used Fluorescence Resonance Energy Transfer (FRET) to examine the interaction between proteins. This method involves a peptide linker which couples the donor and acceptor pair. The caspase activities are mainly studied using the FRET technique. Smith et al. (2019) used Fluorescence Lifetime Imaging Microscopy (FLIM), an ideal technique

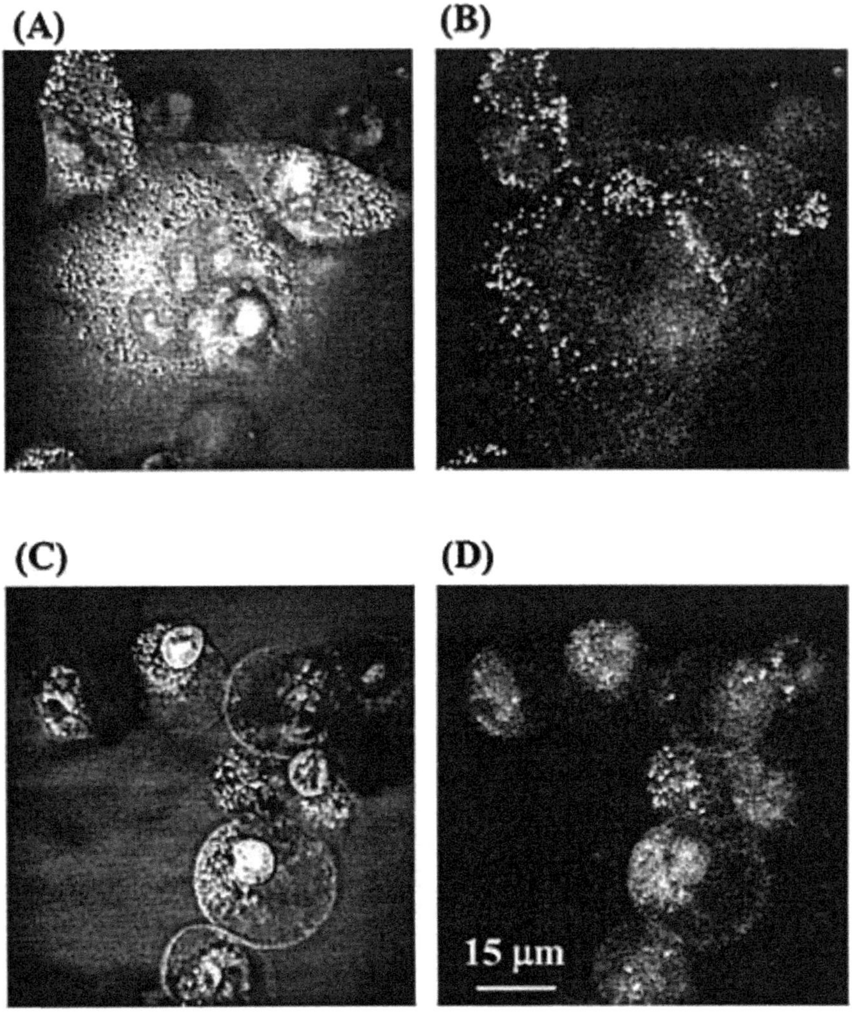

Fig. 5.2 Stages of apoptosis using CARS microscopy: early stages (**a**, **b**); later stages (**c**, **d**). The figure is reproduced with permission from reference [4]

exploited in various fields of molecular biology and biomedical research. The technique involves quantitative imaging of cellular micro-environments which can segregate two fluorophores with the similar fluorescence emissions. However, it strictly depends on intricate data-fitting practices to obtain the desired outcome. Hence deep neural network (DNN) architecture was combined with the FLIM which was named as fluorescence lifetime imaging network (FLI-Net). This FLI-Net utilizes the formation of FLI image and DL for the quantification of fluorescence decays over the complete image at a higher speed. FLI-Net can serve as a real-time, user-oriented appeal in analyzing intricate fluorescence lifetime images of cell death and different stages of apoptosis. The large-scale detection of apoptosis was quite challenging without fast signal processing algorithms and high-end computers. Recently, a phasor approach is being applied to analyze FLIM data. It delivers a fit-free and model-less approach to time-resolved data that is flexible for visual inspection. The method uses a phasor plot where the fluorescence decay data provided by each pixel is converted to a point in the plot. The phasor approach provides a faster analysis when compared to the traditional fitting methods [29, 33]. Aftab et al. [1] successfully designed a deep learning algorithm for large-scale detection of chemical-induced apoptosis using label-free time-lapse microscopy. A fluorescence-based assay was conducted to determine the apoptosis specific caspase activity. A set of results obtained from deep learning dependent detection of apoptosis was compared and confirmed with the fluorescence-based assay. On validation, 2866 compounds were screened on a large scale at a time to detect their role in apoptosis, if any. Traditional optical imaging technologies have certain limitations such as the need for skilled technicians, difficulties in staining procedures, etc. These limitations can be overcome by using two photon fluorescence microscopy (TPM). TPM has become a promising tool in deep tissue imaging. The inclusion of DL algorithms in TPM image processing can be helpful in the rapid analysis of acquired images. Injured neuron detection becomes a prerequisite for determining the degree of ischemic brain damage. Ischemic neuronal death can be due to both neuronal necrosis and apoptosis. Wang et al. [34] showed the capability of TPM in label-free detection of neuronal injury in a rat cerebral ischemia–reperfusion paradigm. The identification and evaluation of neuronal damage areas on TPM pictures were accomplished using a deep learning model that utilised a convolutional neural network.

Computational identification of fluorescence microscopy picture is quite laborious due to the sparse cell distribution. Du et al. [6] designed a segmentation algorithm to solve this issue by fusing a range filter and the marching square approach to differentiate between normal and apoptotic cells. This can be used as a pre-filtering step that provides approximate cell position coordinates in the two-dimensional matrix where the image is stored. The cell boundaries are further studied by the active contours without edges method. The morphological data is then fed to a support vector machine (SVM) classifier that will further differentiate between normal and apoptotic cells. The manual methodology involved in multi-channel fluorescence image investigation is a laborious process. To process huge image datasets, an automatic analytical method is required. Nuquantus (Nuclei quantification utility software), a cutting-edge machine learning-based technique introduced by Gross et al. [11], allows to examine the morphological characteristics of tissue under diverse pathological circumstances. The software was able to identify, quantify, and categorise nuclei in cells without discernible cell boundaries. The detection of cell proliferation and cell death in several cell classes using Nuquantus proved successful.

5.2.3 Coherent Raman Scattering (CARS and SRS) Microscopy

Microscopy including bright field microscopy and fluorescence microscopy with both vital and exclusion dyes are efficient and quick methods to detect cell death based on changes in cell morphology. Recently, nonlinear optical microscopy (NOM), primarily coherent anti-Stokes Raman scattering (CARS), Stimulated Raman scattering (SRS), has gained widespread popularity as a tool to observe structural and chemical changes at cellular resolution [13]. Also, these techniques are non-destructive and do not require labelling to visualize [23]. Therefore, these techniques are appropriate for studying the cell death process. A third order nonlinear optical technique, CARS that can be used to follow local biochemical changes during the apoptotic process and selectively image the key biomolecules present in a cell, including nucleic acids, lipids, proteins, and saccharides. The first report of CARS used excitation of aliphatic C-H vibrations in the lipid membrane to monitor the apoptotic NIH 3T3 cells. With progress, CARS microscopy was used to the progression of apoptosis in cells. At an early stage, the nucleolar reorganization was studied with protein molecule accumulation at the nucleolus periphery. With apoptosis progress, the proteins are seen to abandon the nucleolus, resulting in low signal intensity. Further, at the final stages, a hollow disintegrated nucleolus was observed. In contrast, the nucleoplasm exhibits a protein redistribution into a less uniform pattern. The CARS signal from lipids in cytoplasmic masses arising from the compaction of cytoplasmic organelles was comparatively high [4, 25]. Figure 5.2a and b depict early stages of apoptosis with flat cells and intact nuclei as observed under CARS microscopy. Figure 5.2c and d exhibit later stages of apoptosis with cytoplasmic masses of lipid from degrading cell organelles.

Further, CARS was used to observe intracellular molecular dynamics occurring during apoptosis or programmed cell death. It was used to track the movement of cytochrome c from the mitochondria to cytosol which acts as a trigger of caspase cascade activation ref. CARS microscopy has been used to observe hepatocyte necrosis associated with liver diseases like fibrosis or steatosis in bile duct ligated (BDL) rats. After bile duct ablation, the morphology of hepatocyte become incredibly erratic as a result of significant collagen fibril and hepatic lipid droplet accumulations brought on by necrosis [18]. In the case of glioblastoma, the tumour detection was based on CARS signal intensity and the Extensive necrosis which is most likely the result of chemotherapy treatment for the primary tumour is explained by the buildup of foam cells. CARS image analysis extensively needs image processing and feature engineering [8]. The primary flaw with CARS is the non-resonant background that results from electronic contributions even when vibrational resonance is absent. Analysis of CARS images can be done using a deep learning algorithm. Transfer learning (TL), a machine learning (ML) research issue, applies the acquired knowledge on one task to a different domain [15]. A statistical technique called cross-validation is used to evaluate the expertise of machine learning models. For instance, CARS images are analyzed by deep learning using a nine-fold cross-validation approach.

The bottleneck values are cached and calculated for each image, which is the information stored in the penultimate layer. Google's TensorFlow (version r1.1) deep learning framework56 was used to conduct this investigation on an Amazon Web Services (AWS) EC2 p2.xlarge system [36].

Like CARS, SRS is a nonlinear optical method that uses two ultrafast laser pulses known as the strokes pulse and the pump pulse to excite molecules into vibrationally excited states. SRS microscopy is used for quantitative protein mapping, and lipid and nucleic acid distributions as depicted in Fig. 5.3. Deep learning is used for denoising the images which significantly removes the noise. Without significantly blurring any characteristics, deep learning deionizing recovers the predicted axons in the channel lipid and improves nucleus from the background in the protein channel. This assists in observing carefully the changes taking place in the cell during cell death. The signals from protein and lipid channels show how shallow tissue imaging using a deep learning approach can enhance images taken at a much greater depth [19]. Differences in signals arising from heterochromatin and euchromatin within cells were observed, which assisted in determining whether a cell is dividing. SRS detects a nucleolus present in non-dividing cells during the interphase and the increased levels of DNA condensation that occur in dividing cells. This technique is beneficial in identifying cells undergoing cell division or apoptosis within living tissue, making it possible to study diseases such as cancer [37].

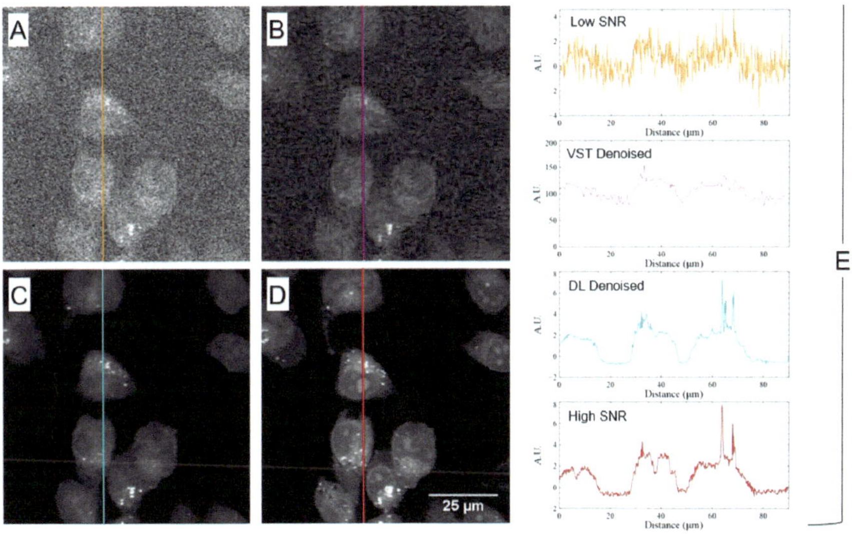

Fig. 5.3 **a-d** SRS images at 2920 cm^{-1} of fixed HeLa cells where **a** was imaged using 1mW Stokes and 20 mW pump (low power) then subsequently denoised using VST (**b**) and the trained deep learning algorithm (**c**); **d** shows the same field of view as (**a–c**) imaged at 20 mW Stokes and pump (high power). Corresponding pixel value plots along the shown line are demonstrated in (**e**). The figure is reproduced with permission from reference [19]

5.3 Conclusion

The present work summarizes the spectroscopic and microscopic analysis for cell death using artificial intelligence. Several distinct molecular alterations and cascade mechanisms that take place inside the cell define cell death. Apoptosis is characterised by deformation of cytoplasm, nucleus fragment, increase in caspase activity, condensation of chromatin, blebbing in plasma membrane. The vitality of the cells is assessed using a variety of procedures and assays. However, the use of these tests is constrained by their reputation for being time-consuming, destructive, and inefficient. CARS is used to detect regional biochemical alterations brought on by apoptosis and to focus on specific biomolecules within a cell, such as proteins, lipids, nucleic acids, and saccharides whereas quantitative protein mapping, nucleic acid and lipid distributions are all performed using SRS microscopy. Methods for non-invasively analysing samples for cell death include spectroscopic and microscopic investigations. When samples are excited or de-excited, inelastic light scattering is detected using Raman spectroscopy and microscopy techniques. In vivo and in vitro samples can be quickly measured in real time using spectroscopic techniques, without negatively affecting the surrounding living cells, and analysis can be done on any size sample. The intricacy of the spectroscopic peaks is a significant disadvantage of spectroscopy techniques despite the benefits they provide. This necessitates the development of automated methods for very efficient and accurate analysis of high throughput spectroscopic and microscopic values. Machine learning can be used to automatically evaluate and derive conclusions from scientific studies that generate enormous amounts of data, minimising the possibility of human mistake during assessment. The convergence of machine learning like deep learning (DL) with fluorescence imaging techniques holds potential for rapid cell death detection and analysis and might also help in distinguishing different types of cell death in the near future.

Acknowledgements We thank the Department of Science and Technology (DST), Government of India, India (Project Number-GITA/DST/TWN/P-95/2021) and Indian Council of Medical Research, Government of India, India (Project Number-ITR/Ad-hoc/43/2020-21) for the financial support. We thank Manipal School of Life Sciences, Manipal Academy of Higher Education, Manipal, Karnataka, India for providing the infrastructure and facilities.

Conflict of Interest The authors do not have any conflict of interest.

References

1. O. Aftab, M. Nazir, M. Fryknäs, U. Hammerling, R. Larsson, M.G. Gustafsson, Label free high throughput screening for apoptosis inducing chemicals using time-lapse microscopy signal processing. Apoptosis **19**(9), 1411–1418 (2014)
2. M.T. Alula, Z.T. Mengesha, E. Mwenesongole, Advances in surface-enhanced Raman spectroscopy for analysis of pharmaceuticals: A review. Vib. Spectrosc. **98**, 50–63 (2018)

3. E. Brauchle, S. Thude, S.Y. Brucker, K. Schenke-Layland, Cell death stages in single apoptotic and necrotic cells monitored by Raman microspectroscopy. Sci. Rep. **4**(1), 1–9 (2014)
4. J.X. Cheng, Y.K. Jia, G. Zheng, X.S. Xie, Laser-scanning coherent anti-Stokes Raman scattering microscopy and applications to cell biology. Biophys. J . **83**(1), 502–509 (2002)
5. M.C. Cooper, V.S. Jacob, P.A. Lewis, A study of the classification capabilities of neural networks using unsupervised learning: A comparison withK-means clustering. Psychometrika **59**(4), 509–525 (1994)
6. Y. Du, H.M. Budman, T.A. Duever, Segmentation and quantitative analysis of normal and apoptotic cells from fluorescence microscopy images. IFAC-PapersOnLine **49**(7), 603–608 (2016)
7. J. Fang, A. Swain, R. Unni, Y. Zheng, Decoding optical data with machine learning. Laser Photonics Rev. **15**(2), 2000422 (2021)
8. R. Galli, O. Uckermann, A. Temme, E. Leipnitz, M. Meinhardt, E. Koch, M. Kirsch, Assessing the efficacy of coherent anti-Stokes Raman scattering microscopy for the detection of infiltrating glioblastoma in fresh brain samples. J. Biophotonics **10**(3), 404–414 (2017)
9. L. Galluzzi, I. Vitale, J.M. Abrams, E.S. Alnemri, E.H. Baehrecke, M.V. Blagosklonny, G. Kroemer, Molecular definitions of cell death subroutines: recommendations of the Nomenclature Committee on Cell Death 2012. Cell Death Differ. **19**(1), 107–120 (2012)
10. A.L. Greca, N. Pérez, P.M. Milone, S. Castaneda, M.A. Scarafia, A.M. Möbbs, S.G. Miriuka, *Celldeath: a tool for simple detection of cell death in transmitted light microscopy images by visual deep learning analysis* (2020). bioRxiv, 2020–03
11. P. Gross, N. Honnorat, E. Varol, M. Wallner, D.M. Trappanese, T.E. Sharp, S.R. Houser, Nuquantus: Machine learning software for the characterization and quantification of cell nuclei in complex immunofluorescent tissue images. Sci. Rep. **6**(1), 1–12 (2016)
12. S. Hassoun, F. Jefferson, X. Shi, B. Stucky, J. Wang, E. Rosa Jr., Artificial intelligence for biology. Integr. Comp. Biol. **61**(6), 2267–2275 (2021)
13. W. Hu, G. Zhao, C. Wang, J. Zhang, L. Fu, Nonlinear optical microscopy for histology of fresh normal and cancerous pancreatic tissues. PLoS ONE **7**(5), e37962 (2012)
14. D. Kanduc, A. Mittelman, R.O.S.A.R.I.O. Serpico, E.B.E.R.T.A. Sinigaglia, A.A. Sinha, C. Natale, E.M.A.N.U.E.L. Farber, Cell death: apoptosis versus necrosis. Int. J. Oncol. **21**(1), 165–170 (2002)
15. S. Khan, N. Islam, Z. Jan, I.U. Din, J.J.C. Rodrigues, A novel deep learning based framework for the detection and classification of breast cancer using transfer learning. Pattern Recogn. Lett. **125**, 1–6 (2019)
16. G. Kroemer, L. Galluzzi, P. Vandenabeele, J. Abrams, E.S. Alnemri, E.H. Baehrecke, G. Melino, Classification of cell death: recommendations of the Nomenclature Committee on Cell Death 2009. Cell Death Differ. **16**(1), 3–11 (2009)
17. Y.J. Lee, H.J. Ahn, G.J. Lee, G.B. Jung, G. Lee, D. Kim, H.K. Park, Investigation of biochemical property changes in activation-induced CD8+ T cell apoptosis using Raman spectroscopy. J. Biomed. Opt. **20**(7), 075001 (2015)
18. J. Lin, F. Lu, W. Zheng, Z. Huang, S. Xu, D.C. Tai, H. Yu, Assessment of liver steatosis and fibrosis in rats using integrated coherent anti-Stokes Raman scattering and multiphoton imaging technique. J. Biomed. Opt. **16**(11), 116024 (2011)
19. B. Manifold, E. Thomas, A.T. Francis, A.H. Hill, D. Fu, Denoising of stimulated Raman scattering microscopy images via deep learning. Biomed. Opt. Express **10**(8), 3860–3874 (2019)
20. M.M. Mariani, P.J. Day, V. Deckert, Applications of modern micro-Raman spectroscopy for cell analyses. Integr. Biol. **2**(2–3), 94–101 (2010)
21. N. Murali, A. Kucukkaya, A. Petukhova, J. Onofrey, J. Chapiro, Supervised machine learning in oncology: a clinician's guide. Dig. Dis. Interv. **4**(01), 073–081 (2020)
22. J. Neng, Q. Zhang, P. Sun, Application of surface-enhanced Raman spectroscopy in fast detection of toxic and harmful substances in food. Biosens. Bioelectron. **167**, 112480 (2020)
23. M. Okada, N.I. Smith, A.F. Palonpon, H. Endo, S. Kawata, M. Sodeoka, K. Fujita, Label-free Raman observation of cytochrome c dynamics during apoptosis. Proc. Natl. Acad. Sci. **109**(1), 28–32 (2012)

24. M.J. Patel, A. Khalaf, H.J. Aizenstein, Studying depression using imaging and machine learning methods. NeuroImage: Clinical, **10**, 115–123 (2016)
25. A. Pliss, A.N. Kuzmin, A.V. Kachynski, P.N. Prasad, Biophotonic probing of macromolecular transformations during apoptosis. Proc. Natl. Acad. Sci. **107**(29), 12771–12776 (2010)
26. G. Pyrgiotakis, O.E. Kundakcioglu, P.M. Pardalos, B.M. Moudgil, Raman spectroscopy and support vector machines for quick toxicological evaluation of titania nanoparticles. J. Raman Spectrosc. **42**(6), 1222–1231 (2011)
27. M.M. Quiroz, G. Khanal, D. Pappas, Apoptosis (Programed Cell Death) studied by fluorescence spectroscopy. Encycl. Anal. Chem.: Appl., Theory Instrum., 1–15 (2006)
28. S. Rangan, H.G. Schulze, M.Z. Vardaki, M.W. Blades, J.M. Piret, R.F. Turner, Applications of Raman spectroscopy in the development of cell therapies: State of the art and future perspectives. Analyst **145**(6), 2070–2105 (2020)
29. S. Ranjit, K. Henriksen, A. Dvornikov, M. Delsante, A. Rosenberg, M. Levi, E. Gratton, Phasor approach to autofluorescence lifetime imaging FLIM can be a quantitative biomarker of chronic renal parenchymal injury. Kidney Int. **98**(5), 1341–1346 (2020)
30. A. Romero, C. Gatta, G. Camps-Valls, Unsupervised deep feature extraction for remote sensing image classification. IEEE Trans. Geosci. Remote Sens. **54**(3), 1349–1362 (2015)
31. A.E. Sallab, M. Abdou, E. Perot, S. Yogamani, Deep reinforcement learning framework for autonomous driving. Electron. Imaging **2017**(19), 70–76 (2017)
32. J.T. Smith, R. Yao, N. Sinsuebphon, A. Rudkouskaya, N. Un, J. Mazurkiewicz, X. Intes, Fast fit-free analysis of fluorescence lifetime imaging via deep learning. Proc. Natl. Acad. Sci. **116**(48), 24019–24030 (2019)
33. M. Štefl, N.G. James, J.A. Ross, D.M. Jameson, Applications of phasors to in vitro time-resolved fluorescence measurements. Anal. Biochem. **410**(1), 62–69 (2011)
34. S. Wang, B. Lin, G. Lin, R. Lin, F. Huang, W. Liu, J. Chen, Automated label-free detection of injured neuron with deep learning by two-photon microscopy. J. Biophotonics **13**(1), e201960062 (2020)
35. M. Wegener, *Extreme nonlinear optics: an introduction* (Springer Science & Business Media, 2005)
36. S. Weng, X. Xu, J. Li, S.T. Wong, Combining deep learning and coherent anti-Stokes Raman scattering imaging for automated differential diagnosis of lung cancer. J. Biomed. Opt. **22**(10), 106017 (2017)
37. X. Zhang, M.B. Roeffaers, S. Basu, J.R. Daniele, D. Fu, C.W. Freudiger, X.S. Xie, Label-free live-cell imaging of nucleic acids using stimulated Raman scattering microscopy. ChemPhysChem **13**(4), 1054–1059 (2012)
38. S.K. Zhou, H.N. Le, K. Luu, H.V. Nguyen, N. Ayache, Deep reinforcement learning in medical imaging: A literature review. Med. Image Anal. **73**, 102193 (2021)

Chapter 6
Applications of Artificial Intelligence in the Analysis of Images of the Oral Cavity for Cancer Detection

Adarsh Kudva, Vidya Kudva, Vathsala Patil, Neha Mohan, Preethi R. Shenoy, and Archi Rai

Abstract This chapter explores the transformative role of artificial intelligence (AI) in the analysis of images of the oral cavity for cancer detection. AI-driven technologies have revolutionized healthcare, particularly in the realm of medical imaging analysis. In the context of oral cavity cancer detection, AI systems leverage advanced algorithms to analyze digital images captured using smartphones or digital cameras. These images serve as visual records of the oral cavity, allowing clinicians to document suspicious lesions or abnormalities and monitor changes over time. AI-powered image analysis offers several advantages, including early detection of cancerous growth or progression and facilitating remote consultations with specialists. The chapter discusses key AI methodologies and strategies employed in image analysis for oral cavity cancer detection, such as convolutional neural networks (CNNs), deep learning models, and attention-based mechanisms. It also addresses challenges and limitations, such as ensuring image quality, addressing algorithmic biases, and the importance of the human–computer interface. Additionally, ethical considerations related to data privacy, informed consent, and regulatory frameworks in deploying AI systems in healthcare are highlighted.

A. Kudva (✉)
Department of Oral and Maxillofacial Surgery, Manipal College of Dental Sciences, Manipal. Manipal Academy of Higher Education (MAHE). India, Manipal, Karnataka 576104, India
e-mail: adarsh.kudva@manipal.edu

V. Kudva
NITTE (Deemed to be University), Department of Electronics and Communication Engineering, NMAM Institute of Technology, Nitte, Karnataka 574110, India

V. Patil
Department of Oral Medicine and Radiology, Manipal College of Dental Sciences, Manipal, Manipal Academy of Higher Education, Manipal, Karnataka 576104, India

N. Mohan · P. R. Shenoy · A. Rai
Manipal College of Dental Sciences, Manipal Academy of Higher Education, Manipal, Karnataka 576104, India

A. Gogoi and N. Mazumder (eds.), *Biomedical Imaging*, Biological and Medical Physics, Biomedical Engineering, https://doi.org/10.1007/978-981-97-5345-1_6

6.1 Introduction

Technological advancements in the area of artificial intelligence (AI) and its subfields like machine learning (ML) and deep learning (DL), have brought in a paradigm shift in the healthcare sector. AI, with its ability to mimic human behavior through the design and training of computers, has opened up new possibilities in medical applications [1]. For instance, AI-powered systems can analyze vast amounts of medical data to aid in the early diagnosis of diseases, interpret complex medical images such as MRI scans and X-rays, assist in creating personalized treatment plans based on patient-specific data, and even predict patient outcomes and prognosis [2–5].

Machine learning, as a subset of AI, plays a crucial role in processing these large datasets and developing decision-making capabilities. ML algorithms can identify patterns and trends in medical data that may not be readily apparent to human observers, leading to more accurate diagnoses and treatment recommendations. Additionally, deep learning algorithms, a more advanced form of ML, can learn from and adapt to new information, further enhancing their effectiveness in healthcare applications [6].

The integration of AI and ML technologies into healthcare systems holds immense promise for improving patient care. These technologies can help healthcare providers make more informed decisions, reduce diagnostic errors, streamline administrative tasks, and ultimately enhance the overall quality and efficiency of healthcare delivery. Moreover, AI-driven innovations have the potential to drive down healthcare costs by optimizing resource utilization and promoting preventive care strategies. AI is revolutionizing healthcare by enabling smarter, more data-driven approaches to medical diagnosis, treatment, and patient management. The continued advancement and adoption of these technologies are expected to lead to significant improvements in healthcare outcomes and patient experiences in the years to come [7].

Healthcare facilities that utilize AI and ML technology primarily aim to enhance patient care. Among the many advantages that these technologies provide to healthcare providers are [8]:

(a) ***Improved Decision-Making***: Clinical decision-making is aided by the insights provided by AI and ML algorithms, which can analyze massive amounts of patient data. It has the potential to improve patient outcomes by leading to more precise diagnosis and treatment programs.

(b) ***Fewer Diagnostic Errors***: Healthcare professionals may lower the potential of diagnostic errors and increase the probability that patients will obtain accurate diagnoses promptly by using diagnostic tools driven by artificial intelligence.

(c) ***Easier Administration***: Automated administrative processes like patient scheduling, billing, and record-keeping are possible with AI-driven solutions, freeing up healthcare workers to concentrate on direct patient care.

(d) ***Better Healthcare Overall***: Patients can anticipate better healthcare experiences overall as an outcome of more efficient and tailored healthcare delivery enabled by AI and ML integration.

The emergence of artificial intelligence (AI) and machine learning (ML) in healthcare has ushered in a new era of data-driven medical diagnosis, treatment, and patient management [3]. These technologies are transforming healthcare practices, offering the potential for significant improvements in healthcare outcomes and patient experiences in the future. Oral cancer, ranked as the 6th most common malignancy globally with high mortality and morbidity rates, presents formidable challenges in early detection and accurate diagnosis [3]. The efficacy of non-invasive techniques such as autofluorescence, chemiluminescence, and narrow-band imaging used for early detection remains uncertain with biopsy and histopathological analysis remaining the gold standard diagnostic approach. However, biopsies are not without limitations. They are invasive, prone to sampling errors, and challenging to identify representative regions due to the non-uniform appearance of lesions [9]. Furthermore, the intratumor heterogeneity in oral squamous cell carcinoma often necessitates evaluation by experienced pathologists, leading to limited access and healthcare resource constraints, particularly in developing nations [3].

Early diagnosis plays a pivotal role in improving outcomes and survival rates in oral cancer. Nevertheless, patient preferences for non-invasive investigations over invasive procedures can lead to delayed diagnosis and compromised prognosis. This underscores the urgent need for the development of efficient and non-invasive diagnostic tools. In recent years, AI and ML have demonstrated remarkable progress in the realm of oral cancer. They have shown promising results in the early screening of potentially malignant oral lesions, diagnosing oral cancer, and deciphering the molecular signaling pathways underlying cancer pathogenesis [3]. These technological advancements provide healthcare professionals with invaluable diagnostic aids, enhancing their capabilities in patient care and subsequent investigative procedures [10].

6.2 AI Applications in Oral Cancer

Artificial intelligence has shown its efficacy in various fields of cancer [11].

(a) Recognizing complex patterns of images (oral images, histological micro pictograph). This technique makes image interpretation less subjective and reproducible.
(b) Complement clinical decisions using images that are not detectable by humans.
(c) Enable the aggregation of multiple data streams into powerful integrated diagnostic systems spanning radiographic images, genomics, pathology, electronic health records, and social networks.
(d) Helps in the detection, characterization, and monitoring of tumors.

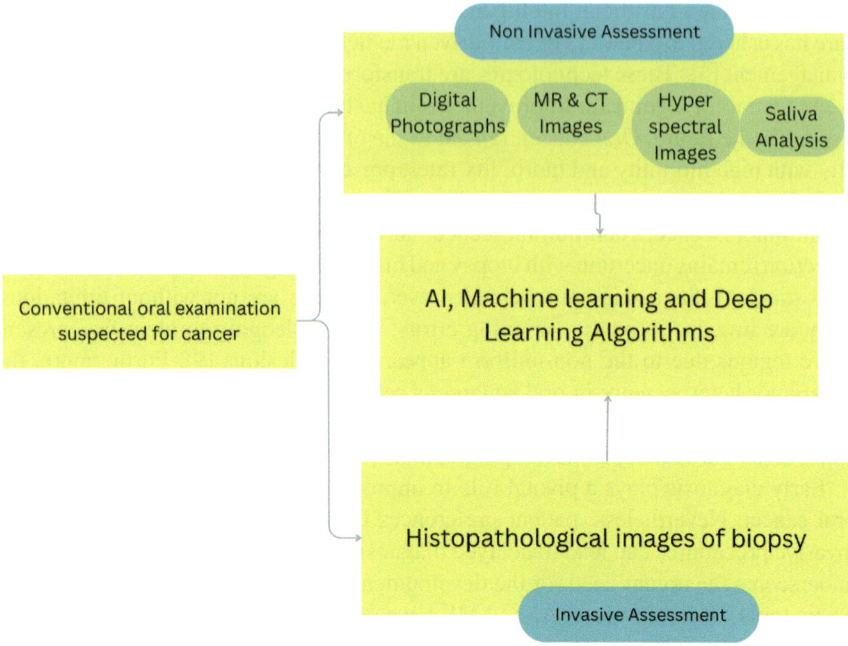

Fig. 6.1 Different imaging modalities for screening of oral cavity cancer [12]

6.2.1 Imaging Modalities or Datasets for Screening of Oral Cavity Cancer

Researchers have developed AI models capable of accurately detecting and segmenting cancerous regions from oral cavity images obtained through various imaging modalities. Different imaging modalities for which AI algorithms have been developed are illustrated in Fig. 6.1.

6.2.2 Imaging Data Employed for Detection of Oral Cancer Through Artificial Intelligence Technique

Various imaging modalities are employed for the detection of oral cavity cancer. These modalities play a crucial role in the diagnosis, staging, and treatment planning for patients. Some common imaging modalities used in oral cavity cancer detection include:

Digital photographs: These images of the oral cavity, captured using smartphones or digital cameras, serve as valuable visual records that allow clinicians to document suspicious lesions or abnormalities. They play a crucial role in monitoring changes over time, aiding in the early detection of cancerous growth or progression. Moreover, these digital images enable remote consultations with specialists, particularly in areas with limited access to healthcare services, as they can be captured and transmitted to healthcare providers for evaluation and diagnosis. However, the quality of these images is a critical factor that can influence the accuracy of diagnostic outcomes, as highlighted in studies such as [13]. To address this, several strategies have been proposed to establish the region of interest (ROI) in images automatically. Techniques like GRID-Cam, 72U-Net architecture with EfficientNet-b7 and YoloV5, and segmentation masking have shown significant impacts on image quality and performance in convolutional neural network (CNN) procedures.

Additionally, using uniform cameras, maintaining a consistent distance during image capture, ensuring dry oral mucosa before imaging, and focusing on patterns with the lens's optical axis perpendicular to the lesion being investigated are essential for optimal image quality and diagnostic accuracy. In machine learning-based healthcare applications, increasing the size and diversity of samples within training datasets is a straightforward approach to bolster diagnostic accuracy. However, obtaining reliable expert annotations for these datasets remains a major limiting factor, highlighting the ongoing challenges in AI-driven healthcare advancements.

Computed Tomography (CT): CT scans provide detailed cross-sectional images of the oral cavity and its associated facial parts, allowing for the visualization of tumors, lymph node involvement, and adjacent structures. It helps in staging the cancer and determining the extent of its spread.

Magnetic Resonance Imaging (MRI): MRI utilizes magnetic fields and radio waves to produce detailed images of the oral cavity. It is particularly useful for assessing soft tissue involvement, nerve invasion, and evaluating the extent of tumor spread.

Hyperspectral Imaging (HSI): HSI captures images across a wide range of wavelengths, from the visible to the near-infrared spectrum. Each pixel in the image contains spectral information, allowing for detailed analysis of tissue characteristics. HSI enables visualization of subtle biochemical and structural differences in tissues that may not be apparent with conventional imaging techniques.

Histopathological images: Micro pictographs of biopsies from the oral cavity can provide critical information for diagnosing and characterizing oral cavity lesions, including benign and malignant tumors. These images are obtained by examining thin slices of tissue under a microscope after staining with various dyes to highlight cellular structures.

AI Approaches used for analysis of oral cavity images

AI approaches used for the analysis of oral cavity images can be classified into traditional machine learning approaches and deep learning approaches as depicted in Fig. 6.2.

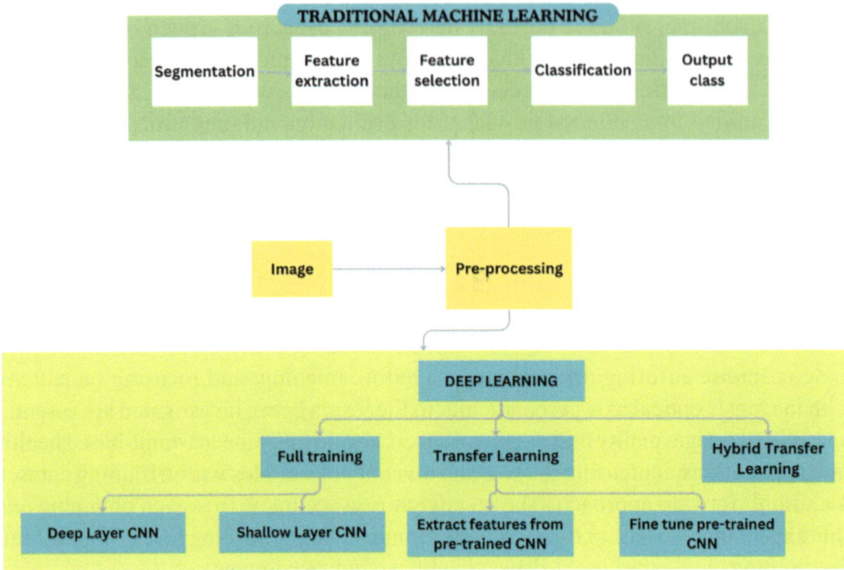

Fig. 6.2 AI approaches used for analysis of oral cavity images

6.2.3 *Traditional Image Processing Approach*

The traditional image processing approach involves a sequence of steps such as pre-processing, segmentation, feature extraction, feature selection, and classification.

Pre-processing is used to enhance the quality and usability of the images for subsequent analysis. To improve the brightness and contrast of images, methods such as Histogram Equalization, Adaptive local Histogram equalization technique [12], and Control Enhancement Adaptive local Histogram equalization (CLAHE) [14] are used. Different filters such as Savitzky–Golay filter [15], Median filter [16], Gabor filter [17, 18], Adaptive median filter [19], and Anisotropic diffusion filter [20] are used for noise removal.

Segmentation of the region of interest (ROI) in oral cavity images is crucial for isolating the relevant anatomical structures or lesions from the background. Thresholding is a simple and widely used technique for image segmentation [21]. In oral cavity images, thresholding can be applied to separate the oral cavity region from the background based on intensity values. This method works well when there is a clear contrast between the oral cavity and surrounding tissues. A few variants of thresholding are entropy threshold, Ostu's threshold, and k-means clustering [22, 23]. A few researchers used a region-based method for segmenting regions with similar characteristics based on seed points. In oral cavity images, region growing can be initiated from seed points within the oral cavity and expanded iteratively by adding neighboring pixels that meet certain similarity criteria, such as intensity or texture [24]. Edge detection algorithms, such as the Canny edge detector or the Sobel operator, can be used to identify boundaries between the oral cavity and surrounding tissues. Once edges are detected, region-filling techniques can be applied to delineate the oral cavity region [25]. Machine learning algorithms, such as convolutional neural networks (CNNs), can be trained to automatically segment the oral cavity region from images. Training data typically consists of manually annotated images, and the CNN learns to predict the segmentation mask based on image features [26].

The feature extraction step aims to extract the unique attributes present in images of different classes. The commonly used features are texture, shape, and color [27]. Gray level cooccurrence matrix (GLCM), a feature of transfer learning [28], and Hu's moment extraction [29] are also used. Deep learning is very efficient in automating the feature extraction process [30].

The feature selection step identifies the most relevant features for classification. Based on the selected features, a classifier detects the cancerous cases. Wide range of classifiers such as Support Vector Machine (SVM), Random Forest (RF), Logistic Regression (LR), K-Nearest Neighbors (KNN), and Artificial Neural Networks (ANN) [31]. These classifiers can be trained on labeled datasets with extracted features from oral cavity images and evaluated using performance metrics such as accuracy, sensitivity, specificity, and area under the receiver operating characteristic curve (AUC-ROC). Choosing the most appropriate classifier depends on factors such as dataset size, feature dimensionality, computational resources, and the desired balance between interpretability and predictive accuracy. The researchers [32, 33] provide an overview of various traditional machine learning algorithms used in oral cancer detection from medical images, discussing their strengths, weaknesses, and future research directions.

6.2.4 Deep Learning Approach

Deep learning techniques, particularly convolutional neural networks (CNNs), have shown remarkable performance in analyzing medical images related to the oral cavity. In the deep learning approach, Convolutional Neural Networks (CNNs) are used for image analysis. CNNs can be employed for image classification in two ways. The first approach is to train the CNN from scratch which is called full training. Depending on the number of layers, CNNs can be classified as deep-layer CNN and shallow-layer CNN. Full training of CNNs with deep architecture requires a large amount of annotated data which is usually not available in the medical domain. Hence, the most popular deep CNNs [34–36] are trained with ImageNet data [37, 38]. Deep CNNs are capable of obtaining a full representation of the training data. Image analysis in the medical domain usually uses shallow-layer CNNs for full training. However, images in the medical domain have high feature variability.

To exploit the advantages of deep architecture on a limited dataset, another approach known as transfer learning is utilized. This approach facilitates the use of the knowledge acquired by pre-trained CNNs which are trained using large amounts of ImageNet data. Transfer learning facilitates the direct use of CNN's learned weights and architecture to perform a new task. In transfer learning, attributes can be extracted from different layers of the CNN and can be used to train a new classifier to accomplish the appropriate task. It is also possible to fine-tune the pretrained CNN until optimum performance is achieved. Disha et al. [39] performed transfer learning using pre-trained CNNs MobileNet, VGG19, VGG16, InceptionV3, and ResNet50 for identifying oral pre-cancerous and cancerous lesions and to differentiate them from normal mucosa using a dataset of clinically annotated photographic images. Rahaman et al. [40] performed transfer learning using AlexNet to extract rank features from oral squamous cell carcinoma (OSCC) biopsy images [41].

Santisudha et al. [41] used fine-tuning of pre-trained models VGG16, VGG19, ResNet50, InceptionV3, and MobileNet using Oral Squamous Cell Carcinoma (OSCC) histopathology images for classification. However, fine-tuning a deep-layered CNN requires extensive computational resources. Therefore, in low data and low computational resource situations, the solution is to use a shallow-layered CNN with filter weights initialized with the weights of appropriate pretrained CNN filters. This method is called Hybrid transfer learning [42]. Such a scenario is shown in Fig. 6.3.

For oral squamous cell carcinoma, popular CNNs such as U-Net and Inceptionv4 have demonstrated considerable potential in detecting and localizing areas of malignancy on digital slides. In addition, these pathology informatics studies suggest that the combination of deep learning models and human clinical judgment has the greatest potential to increase diagnostic accuracy and decrease total time to diagnosis [43, 44].

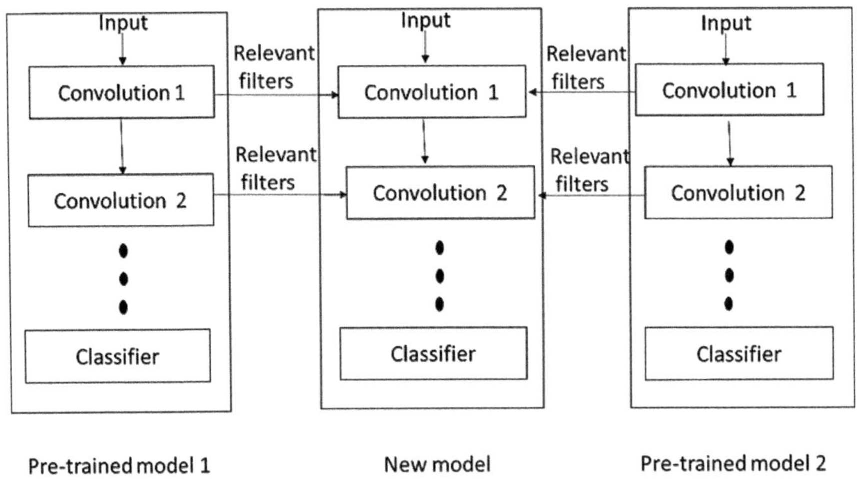

Fig. 6.3 Hybrid transfer learning

6.3 Clinical Applications of AI in Oral Cancer

AI-assisted models have shown remarkable success in various aspects of oral squamous cell carcinoma (OSCC) diagnosis, treatment, and prognosis. For instance, Rahman et al. [45] developed a model that achieved an impressive accuracy of 99.78% in classifying OSCC using morphological and textural features [45]. In OSCC resection surgeries, AI-assisted methods, as demonstrated by Trajanovski et al. [46], have shown excellent results in segmenting tumor tissue [46]. One critical aspect where AI shines is in predicting lymph node metastasis, a common occurrence in OSCC. Shan et al. [47] reported on the efficacy of AI in this area, given the frequent rate of metastasis to cervical lymph nodes in OSCC cases [47]. Furthermore, AI technologies have been instrumental in predicting the 5-year overall survival rate and molecular features of head and neck carcinoma, as discussed in studies by Alkhadar et al. [48], Chen et al. [49] and Zhao et al. [50].

The AI applications not only support diagnosis and prognosis but are also being integrated into treatment planning. Howard et al. [51] found that machine learning methods can identify individuals at high risk who would benefit from chemotherapy and radiation, aiding in personalized treatment strategies [51]. In terms of detection and localization of malignancy in OSCC, popular convolutional neural networks (CNNs) such as UNet and Inceptionv4 have shown promise, as highlighted by Halicek et al. [43] and Mavuduru et al. [44]. Transfer learning, as discussed by Kim et al. [52], is another approach that leverages pretrained neural networks for new classification tasks, enhancing diagnostic capabilities [52].

Tanriver et al. [53] proposed a deep learning-based model for automatic real-time detection and classification of oral lesions, showcasing the feasibility of AI-driven approaches in clinical settings [53]. A systematic review and meta-analysis by Elmakaty et al. [54] further confirmed the high accuracy of AI-assisted models in differentiating healthy oral tissue from OSCC [54]. The attention-based mechanisms, discussed by Amor et al. [55], play a crucial role by selectively focusing on relevant regions of input images for predictive outputs, contributing to improved diagnostic accuracy and efficiency in AI-driven healthcare applications for OSCC [55].

6.4 Ethical Concerns of AI

The ethical concerns surrounding the deployment of AI in healthcare are indeed significant and require careful consideration. Issues such as data privacy and confidentiality, informed consent, and algorithmic biases are crucial to address in order to ensure the responsible and ethical use of AI systems. These concerns highlight the need for a multidisciplinary approach involving experts from various fields to develop ethical and legal guidelines that can guide the successful implementation of AI systems in healthcare, earning the trust of both patients and healthcare professionals [56].

There are certain limitations of AI and its clinical applications that deserve attention. These include unregulated training set algorithms, unsupervised data sets, and the challenge of classification based on numerous types of cancers. These limitations underscore the importance of the human–computer interface and the need for continuous monitoring and evaluation to ensure the accuracy, reliability, and ethical use of AI technologies in healthcare.

6.5 Conclusions

The advancements in AI technology collectively hold promise for improving the early detection and management of oral cavity cancer, ultimately leading to better patient outcomes. However, continued research, validation, and collaboration between multidisciplinary teams are crucial to realizing the full potential of AI and ML in this domain. Integrating information from multiple imaging modalities can improve the accuracy and reliability of oral cavity cancer detection. Combining data from modalities like MRI, CT, PET, and optical imaging techniques allows for a more comprehensive assessment of tissue morphology, physiology, and molecular characteristics, leading to better diagnostic outcomes. AI-based systems can automate the detection and classification of oral cavity lesions. These systems can assist clinicians in identifying suspicious regions in images, reducing the burden of manual inspection and potentially improving early detection rates. AI-based tools are capable of providing real-time diagnostic support during endoscopic procedures or point-of-care examinations. AI avoids emotional problems, cultural and moral beliefs, and fatigue. These tools aid clinicians in making immediate decisions regarding the presence of cancerous lesions, thereby facilitating timely interventions and improving patient outcomes. As AI-based algorithms for oral cavity cancer detection continue to mature, there is increasing emphasis on their clinical validation and integration into routine clinical workflows. Collaborative efforts involving clinicians, researchers, and industry partners are essential to ensure the safety, efficacy, and regulatory compliance of these technologies before widespread clinical adoption.

References

1. A. Kolokythas, Can Artificial Intelligence (AI) assist in the diagnosis of oral mucosal lesions and/or oral cancer? in *Oral surgery, oral medicine, oral pathology and oral radiology*, vol. 134, issue 4 (Elsevier Inc., 2022), pp. 413–414. https://doi.org/10.1016/j.oooo.2022.07.004
2. R. Sathishkumar, M. Govindarajan, A comprehensive study on artificial intelligence techniques for oral cancer diagnosis: challenges and opportunities. in *2023 International Conference on System, Computation, Automation and Networking (ICSCAN)* (IEEE, 2023) (pp. 1–5)
3. B. Ilhan, K. Lin, P. Guneri, P. Wilder-Smith, Improving oral cancer outcomes with imaging and artificial intelligence. J. Dent. Res. **99**(3), 241–248 (2020)

4. C. Kavyashree, H.S. Vimala, J. Shreyas, A systematic review of artificial intelligence techniques for oral cancer detection. Healthc. Anal., 100304 (2024)
5. S. Hegde, V. Ajila, W. Zhu, C. Zeng, Artificial intelligence in early diagnosis and prevention of oral cancer. Asia Pac. J. Oncol. Nurs. **9**(12), 100133 (2022)
6. S. Chakravorty, B.K. Aulakh, M. Shil, M. Nepale, R. Puthenkandathil, W. Syed, Role of Artificial Intelligence (AI) in dentistry: A literature review. J. Pharm. Bioallied Sci. **16**(Suppl 1), S14–S16 (2024)
7. F. Schwendicke, L.T. Arsiwala-Scheppach, J. Krois, Artificial intelligence: What it is and what it can do for dentists. Dent. Update **50**(4), 245–249 (2023)
8. A. Raut, S. Samantaray, R.R. Sahu, Digital technology and artificial intelligence in dentistry: recent applications and imminent perspectives. Cogn. Inform. Soft Comput.: Proceeding CISC **2021**, 417–430 (2022)
9. Yakob et al., (2014)
10. B.Z. Hameed, N. Naik, S. Ibrahim, N.S. Tatkar, M.J. Shah, D. Prasad, B.K. Somani, Breaking barriers: Unveiling factors influencing the adoption of artificial intelligence by healthcare providers. Big Data Cogn. Comput. **7**(2), 105 (2023)
11. W.L. Bi, A. Hosny, M.B. Schabath, M.L. Giger, N.J. Birkbak, A. Mehrtash, T. Allison, O. Arnaout, C. Abbosh, I.F. Dunn, R.H. Mak, Artificial intelligence in cancer imaging: clinical challenges and applications. CA: Cancer J. Clin., **69**(2), 127–157 (2019)
12. M. Badawy, H.M. Balaha, A.S. Maklad, A.M. Almars, M.A. Elhosseini, Revolutionizing oral cancer detection: an approach using aquila and gorilla algorithms optimized transfer learning-based CNNs. Biomimetics (Basel). **8**(6), 499 (2023). https://doi.org/10.3390/biomimetics8060499.PMID:37887629;PMCID:PMC10604828
13. J.M. Brown, J.P. Campbell, A. Beers, K. Chang, S. Ostmo, R.V.P. Chan, J. Dy, D. Erdogmus, S. Ioannidis, J. Kalpathy-Cramer, M.F. Chiang, Automated diagnosis of plus disease in retinopathy of prematurity using deep convolutional neural networks. JAMA Ophthalmol. **136**(7), 803 (2018). https://doi.org/10.1001/jamaophthalmol.2018.1934
14. M.P. Kirubabai, G. Arumugam, Deep learning classification method to detect and diagnose the cancer regions in oral MRI images Medico Legal. Update **21**(1), 462–468 (2021)
15. M. Chakraborty, R.D. Gupta, S. Mukhopadhyay, N. Anjum, S. Patsa, J. Ray, An introductory analysis of digital infrared thermal imaging guided oral cancer detection using ultiresolution rotation invariant texture features. in Medical imaging 2017: Computer-aided diagnosis, vol. 10134 (International Society for Optics and Photonics, 2017), p. 101343D
16. M.-J. Jeng, M. Sharma, L. Sharma, T.-Y. Chao, S.-F. Huang, L.-B. Chang, S.-L. Wu, L. Chow, Raman spectroscopy analysis for optical diagnosis of oral cancer detection. J. Clin. Med. **8**(9), 1313 (2019)
17. R.O. Alabi, O. Youssef, M. Pirinen, M. Elmusrati, A.A. Mäkitie, I. Leivo, A. Almangush, Machine learning in oral squamous cell carcinoma: Current status, clinical concerns and prospects for future—A systematic review. Artif. Intell. Med., **115** (2021)
18. S. Panigrahi, T. Swarnkar, Automated classification of oral cancer histopathology images using convolutional neural network.in *2019 IEEE International Conference on Bioinformatics and Biomedicine, BIBM* (IEEE, 2019), pp. 1232–1234
19. A.A. Nawandhar, L. Yamujala, N. Kumar, Performance analysis of image segmentation for oral tissue. in 2017 ninth international conference on advances in pattern recognition, ICAPR (IEEE, 2017), pp. 1–6
20. D.P. Pragna, S. Dandu, M. Meenakzshi, C. Jyotsna, J. Amudha, Health alert system to detect oral cancer. in 2017 *international conference on inventive communication and computational technologies, ICICCT* (IEEE, 2017), pp. 258–262
21. S. Harnale, D. Maktedar, Oral cancer detection: Feature extraction & SVM classification. Int. J. Adv. Netw. Appl. **11**(3), 4294–4297 (2019)
22. A.A. Nawandhar, L. Yamujala, N. Kumar, Image segmentation using thresholding for cell nuclei detection of colon tissue. in *2015 International conference on advances in computing, communications and informatics, ICACCI* (IEEE, 2015), pp. 1199–1203

23. B. Ilhan, P. Guneri, P. Wilder-Smith, The contribution of artificial intelligence to reducing the diagnostic delay in oral cancer. Oral Oncol., **116** (2021). Article 105254
24. G. Arumugam, M.P. Kirubabai, Detection and segmentation of cancer regions in oral MRI images using ANFIS classification method Int. J. Recent Technol. Eng. **8**(3), 6376–6380 (2019)
25. T.Y. Rahman, L.B. Mahanta, A.K. Das, J.D. Sarma, Automated oral squamous cell carcinoma identification using shape, texture and color features of whole image strips . Tissue Cell, **63** (2020a). Article 101322
45. T.Y. Rahman, L.B. Mahanta, H. Choudhury, A.K. Das, J.D. Sarma, Study of morphological and textural features for classification of oral squamous cell carcinoma by traditional machine learning techniques. Cancer Reports, **3**(6), (2020b). https://doi.org/10.1002/cnr2.1293
26. X. Wang, X. Meng, S. Yan, Deep learning-based image segmentation of cone-beam computed tomography images for oral lesion detection. J. Healthc. Eng., **2021**, 7 (2021). Article ID 4603475. https://doi.org/10.1155/2021/4603475
27. R. Kumar, R. Srivastava, S. Srivastava, Detection and classification of cancer from microscopic biopsy images using clinically significant and biologically interpretable features. J. Med. Eng., **2015** (2015)
28. A. Elnakib, H.M. Amer, F.E. Abou-Chadi, Computer aided detection system for early cancerous pulmonary nodules by optimizing deep learning features, in *Proceedings of the 2019 8th international conference on software and information engineering* (2019), pp. 75–79
29. D.K. Das, S. Koley C. Chakraborty A.K. Maiti, Automated segmentation of mitotic cells for in vitro histological evaluation of oral squamous cell carcinoma. in *2014 IEEE international symposium on signal processing and information technology, ISSPIT* (IEEE 2014), pp. 000354–000357
30. F. Martino, D.D. Bloisi, A. Pennisi, M. Fawakherji, G. Ilardi, D. Russo, D. Nardi, S. Staibano, F. Merolla, Deep learning-based pixel-wise lesion segmentation on oral squamous cell carcinoma images. Appl. Sci. **10**(22), 8285 (2020)
31. K. Kourou, T.P. Exarchos, K.P. Exarchos, M.V. Karamouzis, D.I. Fotiadis, Machine learning applications in cancer prognosis and prediction. Comput. Struct. Biotechnol. J., **13** (2015), 8–17, https://doi.org/10.1016/j.csbj.2014.11.005. https://www.sciencedirect.com/science/article/pii/S2001037014000464
32. S. Dixit, A. Kumar, K. Srinivasan, A current review of machine learning and deep learning models in oral cancer diagnosis: recent technologies, open challenges, and future research directions. Diagnostics. **13**(7), 1353 (2023). https://doi.org/10.3390/diagnostics13071353
33. C. Kavyashree, H.S. Vimala, J. Shreyas, A systematic review of artificial intelligence techniques for oral cancer detection. Healthc. Anal., **5**, (2024)
34. Y. LeCun, L. Bottou, Y. Bengio, P. Haner, Gradient-based learning applied to document recognition. Proc. IEEE, **86**(11), 2278–2324 (1998)
35. A. Krizhevsky, I. Sutskever, G.E. Hinton, ImageNet classification with deep convolutional neural networks. in Proc 25th international conference on neural information processing systems, vol. 1 (Lake Tahoe, Nevada, 2012), pp. 1097–1105
36. C. Szegedy, W. Liu, Y. Jia, P. Sermanet, S. Reed, D. Anguelov, D. Erhan, V. Vanhoucke, A. Rabinovich, Going deeper with convolutions. in *Proc IEEE conference on computer vision and pattern recognition* (Boston, MA, USA, 2015), pp. 1–9
37. J. Deng, W. Dong, R. Socher, L.J. Li, K. Li, F.F. Li, ImageNet: A large-scale hierarchical image database. in *Proc IEEE conference on computer vision and pattern recognition* (Miami, USA, 2009), pp. 248–255
38. O. Russakovsky, J. Deng, H. Su, J. Krause, S. Satheesh, S. Ma, Z. Huang, A. Karpathy, A. Khosla, M. Bernstein, C.B. Alexander, F.F. Li, ImageNet large scale visual recognition challenge. Int. J. Comput. Vision **115**(3), 211–252 (2015)
39. D. Sharma, V. Kudva, V. Patil, A. Kudva, R.S. Bhat, A convolutional neural network based deep learning algorithm for identification of oral precancerous and cancerous lesion and differentiation from normal mucosa: A retrospective study. Eng. Sci., **18**, 278–287 (2022)
40. A.U. Rahman, A. Alqahtani, N. Aldhafferi, M.U. Nasir, M.F. Khan, M.A. Khan, A. Mosavi, Histopathologic oral cancer prediction using oral squamous cell carcinoma biopsy empowered

with transfer learning. Sensors (Basel). **22**(10), 3833 (2022). https://doi.org/10.3390/s22103833.PMID:35632242;PMCID:PMC9146317
41. P. Santisudha, S.N. Bhabani, B. Ruchi, K. Kundan, G. Susmita, S. Tripti, *Classifying histopathological images of oral squamous cell carcinoma using deep transfer learning*, vol. 9, Issue 3 (Heliyon, 2023)
42. V. Kudva, K. Prasad, S. Guruvare, Hybrid transfer learning for classification of uterine cervix images for cervical cancer screening. J. Digit. Imaging **33**(3), 619–631 (2020). https://doi.org/10.1007/s10278-019-00269-1.PMID:31848896;PMCID:PMC7256135
43. M. Halicek, M. Shahedi, J.V. Little, A.Y. Chen, L.L. Myers, B.D. Sumer, B. Fei, Detection of squamous cell carcinoma in digitized histological images from the head and neck using convolutional neural networks. in *Medical imaging 2019: Digital pathology*, eds. by J.E. Tomaszewski, A.D. Ward (SPIE, 2019), p. 18. https://doi.org/10.1117/12.2512570
44. A. Mavuduru, M. Halicek, M. Shahedi, J.V. Little, A.Y. Chen, L.L. Myers, B. Fei, Using a 22-layer U-Net to perform segmentation of squamous cell carcinoma on digitized head and neck histological images. in *Medical Imaging 2020: Digital Pathology*, eds. by J. E. Tomaszewski, A.D. Ward (SPIE, 2020), p. 35. https://doi.org/10.1117/12.2549061
46. S. Trajanovski, C. Shan, P.J.C. Weijtmans, S.G.B. de Koning, T.J.M. Ruers, Tongue tumor detection in hyperspectral images using deep learning semantic segmentation. IEEE Trans. Biomed. Eng. **68**(4), 1330–1340 (2021). https://doi.org/10.1109/TBME.2020.3026683
47. J. Shan, R. Jiang, X. Chen, Y. Zhong, W. Zhang, L. Xie, J. Cheng, H. Jiang, Machine learning predicts lymph node metastasis in early-stage oral tongue squamous cell carcinoma. J. Oral Maxillofac. Surg. **78**(12), 2208–2218 (2020). https://doi.org/10.1016/j.joms.2020.06.015
48. H. Alkhadar, M. Macluskey, S. White, I. Ellis, A. Gardner, Comparison of machine learning algorithms for the prediction of five-year survival in oral squamous cell carcinoma. J. Oral Pathol. Med. **50**(4), 378–384 (2021). https://doi.org/10.1111/jop.13135
49. L. Chen, H. Zeng, M. Zhang, Y. Luo, X. Ma, Histopathological image and gene expression pattern analysis for predicting molecular features and prognosis of head and neck squamous cell carcinoma. Cancer Med. **10**(13), 4615–4628 (2021). https://doi.org/10.1002/cam4.3965
50. Z. Zhao, Y. Li, Y. Wu, R. Chen, Deep learning-based model for predicting progression in patients with head and neck squamous cell carcinoma. Cancer Biomark. **27**(1), 19–28 (2019). https://doi.org/10.3233/CBM-190380
51. F.M. Howard, S. Kochanny, M. Koshy, M. Spiotto, A.T. Pearson, Machine learning-guided adjuvant treatment of head and neck cancer. JAMA Netw. Open **3**(11), e2025881 (2020). https://doi.org/10.1001/jamanetworkopen.2020.25881
52. H.E. Kim, A. Cosa-Linan, N. Santhanam, M. Jannesari, M.E. Maros, T. Ganslandt, Transfer learning for medical image classification: a literature review. BMC Med. Imaging **22**(1), 69 (2022). https://doi.org/10.1186/s12880-022-00793-7
53. G. Tanriver, M. Soluk Tekkesin, O. Ergen, Automated detection and classification of oral lesions using deep learning to detect oral potentially malignant disorders. Cancers **13**(11), 2766 (2021). https://doi.org/10.3390/cancers13112766
54. I. Elmakaty, M. Elmarasi, A. Amarah, R. Abdo, M.I. Malki, Accuracy of artificial intelligence-assisted detection of oral squamous cell carcinoma: A systematic review and meta-analysis. in *Critical reviews in oncology/hematology*, vol. 178 (Elsevier Ireland Ltd., 2022). https://doi.org/10.1016/j.critrevonc.2022.103777
55. R. del Amor, L. Launet, A. Colomer, A. Moscardó, A. Mosquera-Zamudio, C. Monteagudo, V. Naranjo, An attention-based weakly supervised framework for spitzoid melanocytic lesion diagnosis in whole slide images. Artif. Intell. Med. **121**, 102197 (2021). https://doi.org/10.1016/j.artmed.2021.102197
56. S. Gerke, T. Minssen, G. Cohen, Ethical and legal challenges of artificial intelligence-driven healthcare. in *Artificial intelligence in healthcare* (Elsevier, 2020), pp. 295–336. https://doi.org/10.1016/B978-0-12-818438-7.00012-5
57. N. Naik, B.M. Hameed, D.K. Shetty, D. Swain, M. Shah, R. Paul, B.K. Somani, Legal and ethical consideration in artificial intelligence in healthcare: who takes responsibility? Front. Surg. **9**, 266 (2022)

Chapter 7
Leveraging Machine Learning for Advanced Biomedical Imaging: Insights from Speckle Pattern Analysis

Zeev Kalyuzhner and Zeev Zalevsky

Abstract In this chapter we present the usage of back scattered laser light for biomedical imaging and sensing. The back scattered laser light forms a self-interference pattern called secondary speckle pattern. By applying various machine learning approaches to process the spatial and the temporal variations in the distribution of those patterns, various biomedical modalities are extracted. Specifically, we give several applicable examples: we analyze the temporal fluctuations of the speckle patterns collected from the sclera to perform remote sensing of blood oxygen saturation. Then, we use the spatial distribution of the speckle patterns to perform biometric authentication and identification. Finally, we use the spatial–temporal variations of those patterns collected from different regions of the brain in order to detect stimulation of different cortexes associated with human senses. We end the chapter by addressing the challenges and the future perspectives related to the proposed novel sensing methodology and then conclude the chapter.

7.1 Introduction

In recent years, the rapid evolution of machine learning (ML) and artificial intelligence (AI) has ushered in a transformative era within the realm of biomedical imaging. The integration of these advanced technologies has not only catalyzed the automation of intricate image analysis processes but has also paved the way for unprecedented insights into the complexities of biological structures and functions. This paradigm shift has positioned ML and AI as indispensable tools in the pursuit of refining diagnostic and prognostic approaches in healthcare [1–4]. Biomedical imaging, a cornerstone of modern medical diagnostics, has benefited immensely from the application of ML and AI. The capacity of these technologies to discern intricate patterns within voluminous and intricate imaging datasets has empowered

Z. Kalyuzhner · Z. Zalevsky (✉)
Faculty of Engineering and The Nano-Technology Center, Bar-Ilan University, 52900 Ramat-Gan, Israel
e-mail: Zeev.Zalevsky@biu.ac.il

A. Gogoi and N. Mazumder (eds.), *Biomedical Imaging*, Biological and Medical Physics, Biomedical Engineering, https://doi.org/10.1007/978-981-97-5345-1_7

clinicians and researchers alike with an unparalleled ability to extract nuanced information. This newfound capability transcends the limitations of traditional analytical methodologies, offering enhanced diagnostic accuracy, sensitivity, and specificity. As a result, the landscape of biomedical imaging has undergone a revolutionary transformation, with the amalgamation of ML and AI serving as a catalyst for innovative applications and methodologies [1, 3]. This chapter specifically dives into the utilization of ML in the analysis of speckle patterns within the context of optical biomedical imaging. Optical imaging techniques, characterized by their non-invasive nature and high-resolution capabilities, have garnered attention for their potential in monitoring various physiological parameters and disease states. The intricate analysis of speckle patterns, an integral component of optical imaging, emerges as a focal point for exploration, underscoring its substantial promise in unravelling crucial insights into biological phenomena [5, 6]. The significance of speckle pattern analysis lies in its potential to offer non-invasive monitoring solutions that transcend conventional methodologies. This technique holds promise in elucidating dynamic physiological processes and detecting subtle alterations indicative of pathological conditions. As such, the integration of ML in the analysis of speckle patterns contributes to a burgeoning field with far-reaching implications for personalized medicine, early disease detection, and treatment optimization [7, 8].

7.1.1 Speckle Pattern Analysis in Biomedical Imaging

Speckle patterns, essentially random interference patterns produced by the scattering of coherent light (such as a laser), contain rich information about the scattering medium. In biomedical contexts, these patterns can be analyzed to extract vital information about biological tissues [8].

7.1.2 Advancements in Optical Recognition Through Speckle Pattern Analysis

The deployment of optical methods for human recognition has been widespread, with iris recognition [9–11] and facial features recognition [12–14] representing prominent examples. Despite the well-established nature of the speckle phenomenon and its extensive investigation, its practical application in this domain remains relatively limited. In essence, a speckle is a stochastic pattern resulting from the superposition of light waves with distinct phases and amplitudes of the same frequency, a manifestation that is intricately linked to the properties of the light scattering medium [8, 15].

The following exploration focuses broadly on speckle patterns generated in the far field, particularly those resulting from the reflection of monochromatic laser light on a surface. These patterns encapsulate valuable information about surface deformations, providing a means to image dynamic or static changes in the illuminated area [8].

The work by Zalevsky et al. [8] pioneers the measurement of far-field speckle patterns to extract various biomedical parameters through correlation-based operations. This involves capturing the tilting motion and converting it into translational movement of the speckle patterns. Mathematically, this process is expressed as:

$$T_m(x_o, y_o) = \iint \exp\left[i\phi(x, y)\right] \exp\left[\frac{\pi i}{\lambda Z_1}\left((x - x_o)^2 + (y - y_o)^2\right)\right] dxdy$$
$$= A_m(x_0, y_0) \exp\left[i\psi(x_0, y_0)\right]$$

where (x, y) are the coordinates of the transversal plane while the axial axis are denoted by Z. λ is the optical wavelength, ϕ—random phase created by surface roughness, and Z_1—distance between the object and plane captured by imaging system, with the intensity of the obtained speckle image being:

The resulting speckle image's intensity, denoted as (x_s, y_s), is determined by the spatial impulse response, magnification inverse, and sensor plane coordinates [8]:

$$I(x_s, y_s) = \left| \iint T_m(x_o, y_o) h(x_o - Mx_s, y_o - My_s) \right|^2$$

where h is the spatial impulse response, M is the inverse of the magnification of the imaging system, and (x_s, y_s) is the sensor plane coordinates' set.

Zalevsky et al.'s [8] methodology employs this computational approach to authenticate individuals based on heartbeats extracted from speckle recordings and pulse measurements. This authentication relies on the dynamic properties of sensed speckle patterns, specifically the unique vibrational signature of the heartbeat. The application of speckle pattern analysis was extended to various use cases beyond biomedical parameters. Instead of adhering to theoretical frameworks and domain-specific feature extraction, we adopt deep neural network models [16] tailored for visual data analysis to perform classification tasks. These findings demonstrate the repeatability of person authentication under diverse conditions, contributing to the broader landscape of optical recognition research and showcasing the adaptability of deep learning models for processing static speckle patterns in various contexts [16].

7.1.3 Machine Learning in Speckle Analysis

Addressing the intricacies of modelling the relationship between identity and indirect measurements, such as speckle patterns represented in visual data format, inherently aligns with the domain of machine learning algorithms. Notably, Deep Learning (DL) models have exhibited remarkable efficacy in handling visual data, as exemplified by their exemplary performance in challenges like the ImageNet Challenge [17]. The foundational architectures of these models are rooted in neural networks, drawing inspiration from the structure of biological neurons [18]. Neural networks, fundamental to DL models, are constructed with layers comprising individual "neurons." Each neuron, acting as a unit, applies a nonlinear function (such as Rectified Linear Unit, ReLU [19]) to the received data. Subsequently, these neurons collaboratively compose and pass the transformed data to the next layer, creating a hierarchical representation. This process continues until the data reaches the final layer, where it is evaluated by a loss function. Adjustments to the model are then propagated backward through the network, enabling the refinement of the learning process [16]. The iterative nature of this learning process ensures that each layer learns to transform its input into a progressively more intricate representation, with each subsequent layer fine-tuning these representations. The term "depth" in neural networks refers to the number of layers they incorporate.

7.1.4 Applications in Biomedical Imaging

The three research papers [20–22] explored in this chapter further underscore the transformative impact of ML on speckle pattern analysis. In the realm of biomedical imaging, these studies dive into diverse applications, ranging from authentication based on static speckle patterns on the forehead to the classification and detection of basic senses using laser-induced speckle patterns. Each research paper presents a unique perspective, collectively expanding the horizons of this interdisciplinary field. With applications in smart homes, smart cars, and potential contributions to the understanding of neurological disorders, the synthesis of speckle pattern analysis and ML emerges as a pivotal force in the evolution of medical diagnostics and sensing technology.

7.2 Remote Photonic Sensing of Blood Oxygen Saturation

7.2.1 Motivation and Experimenting

A low level of oxygen in the blood, termed hypoxemia, can lead to hypoxia, resulting in medical complications and even death [23]. Various causes of hypoxia include lung disease, birth defects, and chronic obstructive pulmonary disease [23]. The phenomenon of silent hypoxemia is causing confusion among physicians treating COVID-19 patients, where individuals appear well despite dangerously low blood oxygen saturation levels [23–26]. This situation challenges conventional understanding and prompts questions about COVID-19's impact on the lungs and the need for more effective diagnostic and treatment methods.

The most common method for measuring blood oxygen saturation is pulse oximetry, which requires direct contact with the patient's finger [27]. This paper proposes a novel approach using laser illumination, similar to non-contact body temperature measurement, to remotely test blood oxygen saturation without physical contact [27]. The method focuses on sampling micro-saccades oscillations in the patient's eye, leveraging machine learning to associate these oscillations with blood oxygen saturation levels [28, 29]. Micro-saccades, tiny eye sclera movements, reflect physiological traits and can indicate blood saturation and short-term hypobaric hypoxia [20].

While speckle-based remote sensing has been explored in biomedical applications [8, 20, 30, 31], this paper advances the concept by integrating speckle-based technology with an artificial intelligence algorithm. This innovative approach aims to remotely sense blood oxygen levels by tracking random eye movements, addressing the challenges associated with traditional methods for relating eye tracker information to blood saturation levels.

The motivation behind the study lies in the optical characteristics of the human sclera, which, when illuminated by a laser, produces secondary speckle patterns. These patterns, captured by a fast camera with defocused optics, allow for the analysis of temporal changes related to nano-movements in the illuminated surface, specifically the eye sclera [20]. In an experimental setting, data was collected from eight healthy human subjects in a controlled laboratory environment. The room was shuttered to eliminate background noises, and each subject was positioned 50 cm from a camera-equipped sensor. The subject's head was restrained to minimize disturbances, and the eye sclera was illuminated by a safe laser (770 μW at 1550 nm). Blood oxygen saturation levels were concurrently recorded using a pulse oximeter [20].

The experiment involved recording two scenarios for each subject: (i) normal blood oxygen saturation level with normal breathing, and (ii) relatively low oxygen saturation level induced by prolonged respiratory arrests. The oximeter had an accuracy tolerance of $\pm$ 2 digits (about 1%), with a threshold set at $\geq$ 96% for normal blood oxygen saturation. Video recordings of the illuminated eye movements were captured at 500 frames per second for 10 s per sample, using a defocused camera controlled by MATLAB. The video data, comprising about 100 videos (2 GB) with 2 million frames, was collected on different dates. To prevent data leakage during the training process, videos were divided into training and test sets, with six subjects in the former and two in the latter [8].

Notably, the study emphasizes the simplicity of the optics despite the advanced sensor, highlighting that precise eye movement sensing can be achieved at large distances without the need for high-quality optics. The analysis focused on back-reflected speckle patterns rather than capturing direct images of the eye. To introduce variation and mitigate biases, data gathering occurred on separate dates, with each subject recorded multiple times in one continuous session [8]. The experimental setup, data collection, and division into training and test sets lay the foundation for the subsequent development and testing of the deep learning model.

7.2.2 Data Exploration and Pre-Processing

The data exploration and pre-processing stages are crucial in constructing a classifier, particularly in machine learning. The initial data source for this study consisted of laboratory-recorded videos capturing temporal variations in speckle patterns. Videos were divided into training and testing sets, with each video assigned to one of these groups. In the pre-processing phase, frames were extracted from each video based on its recording rate (FPS) and saved as PNG files to preserve image quality [32]. During data exploration, efforts were made to establish a correlation between frames within the same section of each video. A 2-dimensional cross-correlational function was employed on consecutive frames to detect differences in their motion direction, simulating micro-saccadic movements (Fig. 7.1a). Unlike ordinary eye images with irrelevant information and noise, the speckle signal image revealed relevant information through a cross-correlation operation between successive frames. Each frame was labelled based on its known recording time, corresponding to the blood oxygen saturation value measured by the oximeter. The correlation between tiny movements (micro-saccades) represented by camera pixels was visually examined, revealing short, consistent, and varied movements under normal blood oxygen saturation conditions (Fig. 7.1b). In contrast, during relatively low blood oxygen saturation levels (Fig. 7.1c), the micro-saccades were more scattered with smaller deviations.

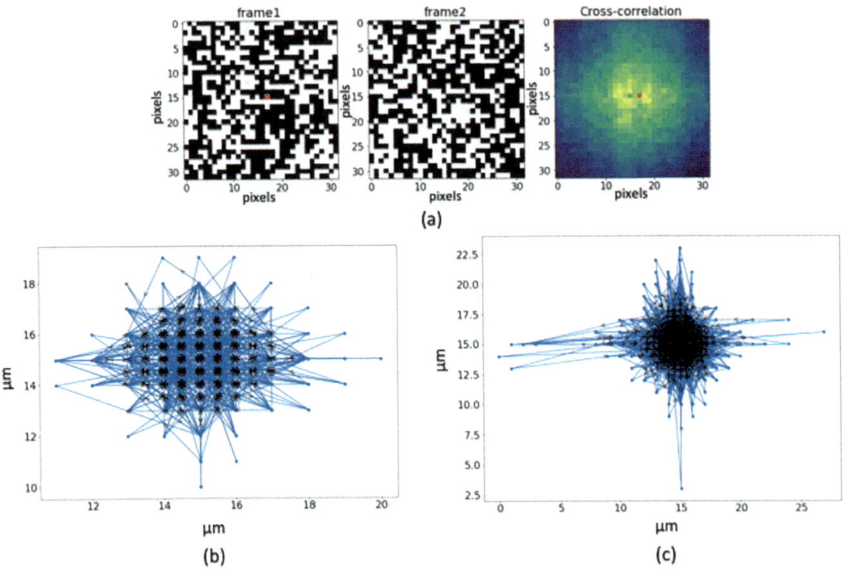

Fig. 7.1 Calculation of micro-saccades movement correlation for 10 s. recordings: **a** finding 2D correlation between two consecutive frames; **b** micro-saccades movement during normal oxygen levels in the blood; **c** micro-saccades movement during relatively low oxygen levels in the blood [21]

The camera sensor translated inputs into pixels of 4.8 × 4.8 μm size, allowing for the observation of eye movements in the range of about 4–5 pixels or 20–25 μm (Figs. 7.1b and 7.3c). After analyzing various data videos, significant differences in micro-saccade patterns were identified between normal and relatively low blood oxygen saturation levels. These differences in speckle patterns suggested a problem suitable for a deep learning model. The selection of the algorithm for building the classification model was based on the hypothesis that deep learning could effectively handle the identified patterns in micro-saccadic movements.

7.2.3 *Method Overview*

The research employed deep learning methodology to address the problem of blood oxygen saturation level classification. The choice of this approach was driven by the nature of the data, particularly the expression of the speckle signal through images and deep neural networks, specifically convolution-based neural networks (CNN) [20]. Deep learning models, known for handling noisy data and learning repetitive patterns, were deemed advantageous for biomedical indices that exhibit interruptions by various noises over time.

A small convnet, illustrated in Fig. 7.2a, was utilized with a few layers and filters per layer, incorporating a data dropout function to reduce overfitting [17, 33]. The dropout function was represented by Eq. (7.1), introducing a fraction of weights during each training iteration.

$$y = \left(t - \sum_{i=1}^{n} \partial_i \mathrm{w}_i I_i\right)^2 \tag{7.1}$$

I_i is a single training input.

t is the true label for this input.

w_i is the weight of the input.

∂ is the dropout rate, where ∂~Bernoulli (p). This means that ∂ is equal to 1 with probability p and 0 otherwise.

The model architecture comprised three 2D convolution layers with ReLU activation (Eq. 7.2) to simplify gradient computation.

$$f(x) = \max(0, x) \tag{7.2}$$

ReLU is followed by a max-pooling layer for variance reduction, computation complexity, and extraction of low-level features, aiding in identifying changes in micro-saccade movement [34, 35].

Two fully connected layers were attached on top of the model, leading to a single unit with sigmoid activation suitable for binary classification, aligning with the problem definition of classifying normal and relatively low blood oxygen saturation

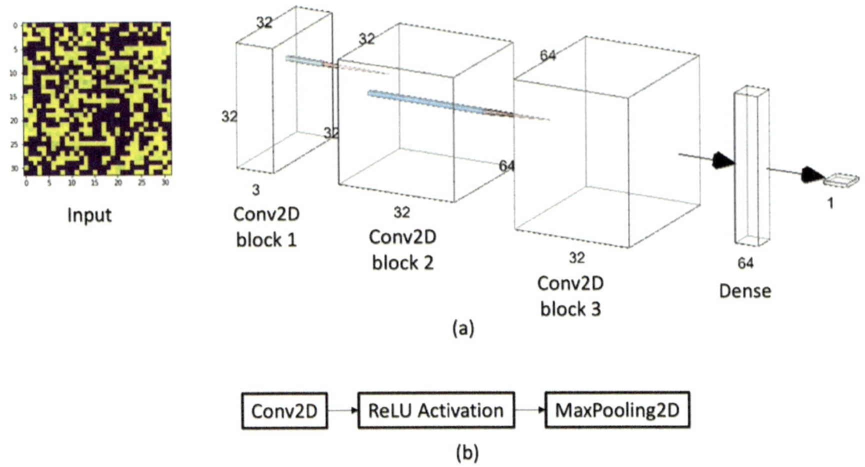

Fig. 7.2 Classifier architecture: **a** deep learning model; **b** convolutional block with Conv2D layer, activation and pooling functions [21]

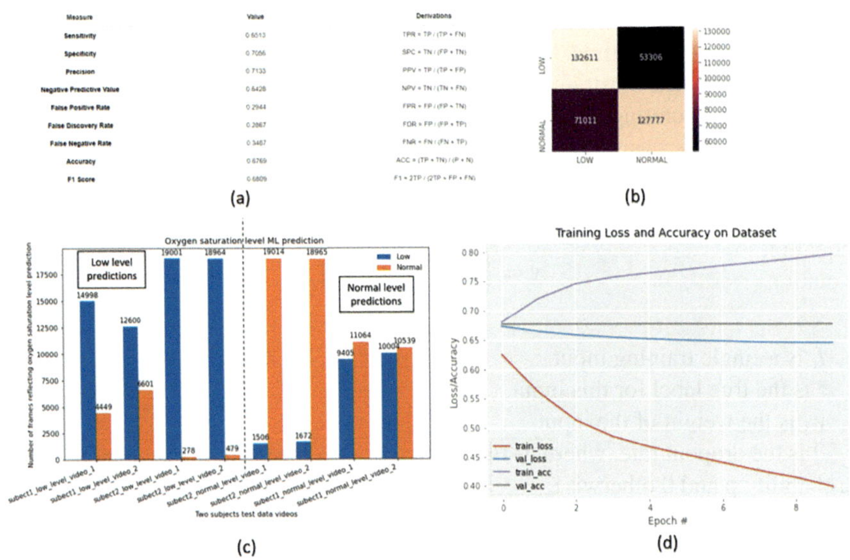

Fig. 7.3 Model training graph and results: **a** model results on test set; **b** confusion matrix of the best model and the test set; **c** ML oxygen level saturation predictions on two subjects' test data; **d** model training process graph including training and validation cost functions and their accuracy [21]

levels. The binary cross-entropy loss function (Eq. 7.3) and RMSprop optimizer [36, 37] were employed for model training. Equation (7.3) represents the binary cross-entropy loss function, calculating the probability of the input being a "relatively low oxygen level" case, while Eq. (7.4) represents the root mean square (RMSprop) for updating the mathematical equations during training. The overall architecture and choices in the algorithm were guided by the hypothesis that deep learning models could effectively handle the identified patterns in micro-saccadic movements for blood oxygen saturation level classification.

$$H_p(q) = -\frac{1}{N}\sum_{i=1}^{N} y_i * \log(p(y_i)) + (1 - y_i) * \log(1 - p(y_i)) \tag{7.3}$$

$$E\left[g^2\right]_t = \beta E\left[g^2\right]_{t-1} + (1-\beta)\left(\frac{\delta C}{\delta \mathrm{w}}\right)^2, \mathrm{w}_t = \mathrm{w}_{t-1} - \frac{\mu}{\sqrt{E\left[g^2\right]_t}}\frac{\delta C}{\delta \mathrm{w}} \tag{7.4}$$

Equation (7.3) represents the binary cross-entropy loss function, where:

y is the label.

p(y) is the predicted probability of the input being the "relatively low oxygen level" case for all N points.

log(p(y)) is the log probability of y being the "relatively low oxygen level" case.

log(1–p(y)) is the log probability of y being the "normal oxygen level" case.

Equation (7.4) represents the root mean square (RMSprop) of the updated math equations where:

E[g] is the moving average of squared gradients.

δC/δw is the gradient of the cost function with respect to the weight.

μ is the learning rate.

β is the moving average parameter.

7.2.4 Experimental Results

The trained model, based on data from six tested subjects, underwent evaluation on a test set containing samples from the two remaining subjects excluded from the training process. The best model achieved an accuracy of 68% on the test set for frame classification (Fig. 7.3a) and a perfect accuracy of 100% on the test set for video classification. Examining the model's success for each frame individually, the confusion matrix of classification results on the frames test set is presented in Fig. 7.3b.

In testing the model's success for each recorded video, the study predicted the oxygen saturation level of each subject's video frames (Fig. 7.3c). The method involved grouping model predictions on video frames, dividing them into columns for each case, and checking if most frames were classified correctly according to the video's label corresponding to blood oxygen saturation level recordings. This approach reflected the desired solution and represented a remote blood oxygen saturation level test for a patient.

The study deduced that the difference in micro-saccades motility between the two oxygen saturation classes played a crucial role in the model's success, particularly for the 2D convolution layer with its specialized cross-correlation. This layer had the highest impact on the model output as it learned patterns of different inputs. Figure 7.3a demonstrated the model's performance using well-known metrics for testing success in the field of deep learning.

The low levels of false positive and false negative rates indicated the reliability of the model in handling various data noises. However, Fig. 7.3c revealed a small difference between normal and relatively low oxygen saturation columns for two subjects with normal oxygen saturation. The discrepancy was attributed to the unique physiological behaviour of micro-saccades during stress caused by relatively low blood oxygen saturation levels, introducing inconsistency throughout the measurement [38].

Figure 7.3d showcased the duration of the model training process. The test accuracy curve reached a plateau relatively early in the training process, suggesting the potential for improvement with actions such as adding Long Short-Term Memory layers, convolution layers, enriching the dataset, etc. Notably, the training process was halted after achieving satisfactory results, and the test-set accuracy curve surpassed a pre-set threshold to conserve computing resources and maintain environmental sustainability.

7.2.5 *Findings and Implications*

This paper introduces a further advancement in a novel approach linking micro-saccades with reduced blood oxygen saturation levels, potentially leading to hypoxia. The experimental validation, conducted on a small predefined number of cases, demonstrated, for the first time, the remote sensing and analysis of anomalies in micro-saccades found in recorded speckle patterns. The system utilized a laser to illuminate the eye sclera, a defocused infrared digital camera capturing speckle patterns, and a computer equipped with MATLAB and machine learning algorithms.

Testing normal and relatively low blood oxygen levels on eight healthy subjects revealed that micro-saccades anomalies, analyzed based on dynamic speckles, exhibited highly accurate performance in distinguishing between different oxygen saturation cases. While the input data showed a reduction in hypoxia validation common to all participants, it still provided a clear indication of the physiological condition. The authors suggest extending the tests to include subjects with silent hypoxemia and updating them. Additionally, different groups of participants should be tested and compared to eliminate the effects of stress conditions during breath restriction, as micro-saccades may also indicate the presence of complementary diseases.

The proposed method holds applicability in medical institutions and public spaces where contactless monitoring of medical metrics is required, such as during the COVID-19 pandemic. Combining this method with remote temperature measurement devices could offer a more reliable module for early tracking of COVID-19 patients.

The distinct difference observed between normal and relatively low blood oxygen saturation levels in this research provides a foundation for future higher-resolution studies. These studies could involve additional cases corresponding to numerical grading, akin to a reference pulse oximeter device. The conclusions emphasize the potential of the proposed method and underscore the need for future investigations and refinements, especially in the context of broader applications and advanced studies.

7.3 Human Identification via Speckle Patterns

7.3.1 Motivational Pursuits

Optical methods have long been employed in human recognition, with iris and facial recognition techniques being common. While the speckle phenomenon, a random pattern resulting from the self-interference of light waves, has been known, its practical application in this domain is less widespread. Speckle patterns, specifically those in the far field resulting from reflection by surfaces illuminated with monochromatic laser light, can carry information about surface deformations. Zalevsky et al.'s work proposed measuring far-field speckle for extracting biomedical parameters, where tilting motion is translated into translational movement of speckle patterns [8].

In contrast, the current approach focuses on processing static speckle patterns refracted from human skin in the forehead area. Instead of theoretical methods, deep neural network models are employed for a classification task to identify individuals from a predefined set. The paper demonstrates that person authentication remains repeatable under different conditions, including sweating and wetting the forehead. This approach offers a novel perspective on speckle-based person identification using deep learning techniques [16].

7.3.2 Experimental Setup

The data collection process involved five human subjects in a controlled laboratory environment. To minimize background noises, the room where speckle measurements were taken was darkened. Each subject was seated 50 cm across from a sensor equipped with a camera recorder, and their face was restrained in a headset to record the forehead area with minimal disturbance and mitigate involuntary head movements. A laser with an output power of 770 μW at a wavelength of 1550 nm (invisible eye-safe laser) illuminated the subject, as shown in Fig. 7.4.

The camera, controlled with MATLAB, produced a video of 100 frames per second quality, focused on the area of interest, for each 10-s sample. Each sample was taken in three resolutions: 32 × 32, 64 × 64, and 128 × 128, with the subject remaining static

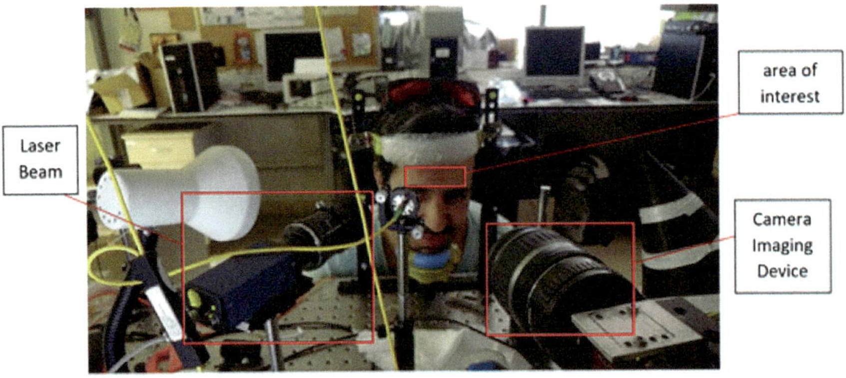

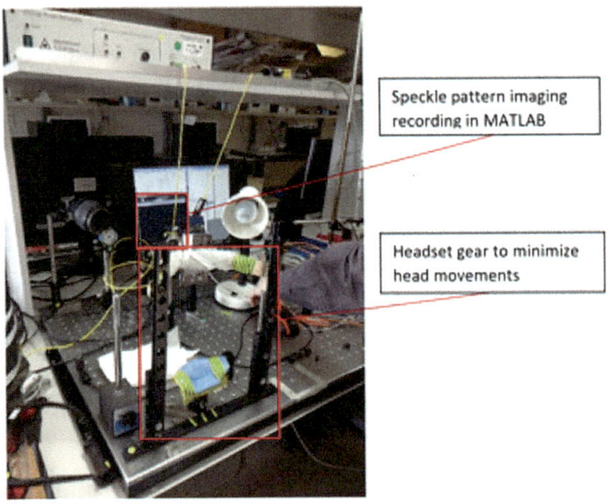

Fig. 7.4 Collecting the data. **a** recording speckle patterns of a person. **b** lab equipment and headset [20]

during recording. To introduce variability in the data and avoid biases inherent in the experiment procedure, data gathering occurred on separate dates. In one continuous session, each subject was recorded in seven different physical circumstances and facial expressions, including neutral, angry, smiling, eyebrows arched upward, sad, water sprayed over the forehead, and after physical exercise. Each situation was recorded in the mentioned resolutions.

Unlike dynamic speckle processing in a reference paper [10], the study did not attempt to correlate between frames of the recorded video. The video was solely used to extract a large number of different static speckle patterns, and no temporal variations analysis was performed. This comprehensive experimental design aimed to capture diverse static speckle patterns under various conditions, providing a rich dataset for the subsequent application of deep learning models in the classification task.

7.3.3 *Method Overview*

The problem of modelling the relationship between identity and indirect measurements, such as speckle patterns presented as visual data, is addressed using machine learning algorithms, with a focus on Deep Learning (DL) models known for their efficacy in handling visual data, as demonstrated in the ImageNet Challenge [17]. These DL models are based on neural networks inspired by biological neurons, organized in layers with each layer containing neurons that apply nonlinear functions (e.g., ReLU) to the received data. The data is transformed through layers, and the network learns to recognize relevant features from the input data.

For the current application, speckle patterns are treated as frames derived from experimental subjects. The output layer of the network produces a vector of five values, each corresponding to a specific class (person in the experiment), representing the probability of the input data belonging to that class. Models for Speckle Patterns Analysis:

7.3.3.1 Simple Neural Networks

As an initial attempt, simple neural networks with 1–2 hidden layers are constructed, where the layers are fully connected. This architecture consists of fully connected layers performing the bulk of feature processing. To mitigate overfitting, a known disadvantage of fully connected layers, the dropout technique is employed, removing individual neurons from the network during training.

7.3.3.2 Convolutional Neural Networks (CNN)

CNNs, known for their success in visual data tasks, are adopted for speckle patterns analysis. CNNs leverage convolution layers, applying a two-dimensional filter across the input data to compute dot products and extract visual features. Down-sampling is used to compress convolved data for faster operation. The hierarchical order includes several convolutional layers, pooling layers for non-linear down-sampling, and a fully connected layer before the output level.

These models are designed to process speckle patterns, extracting meaningful features for person identification. The choice of neural network architectures aims to evaluate different approaches, from simpler models to more complex ones like CNNs, to assess their effectiveness in handling the unique characteristics of speckle data.

7.3.4 Experimental Results

In testing the models, the best-performing model achieved an accuracy of 80.1%. Further analysis of this top-performing model provides insights into its classification performance and areas of potential improvement. Two graphs are presented to illustrate key aspects:

7.3.4.1 Overall Accuracy

The best model achieved an overall accuracy of 80.1%, indicating its ability to correctly identify individuals based on static speckle patterns. This accuracy metric reflects the model's performance across the entire dataset.

7.3.4.2 Intra-class Identification Sensitivity

Figure 7.5 illustrates the intra-class identification sensitivity of distinct subcategories (physical situations) within each class (person). This analysis dives into how well the model performs in distinguishing between different physical circumstances and facial expressions within each individual. Understanding the model's sensitivity to various conditions provides valuable information for refining its robustness.

These results offer a comprehensive view of the model's performance, shedding light on both its strengths and weaknesses. The intra-class identification sensitivity analysis contributes to a nuanced understanding of the model's capabilities in

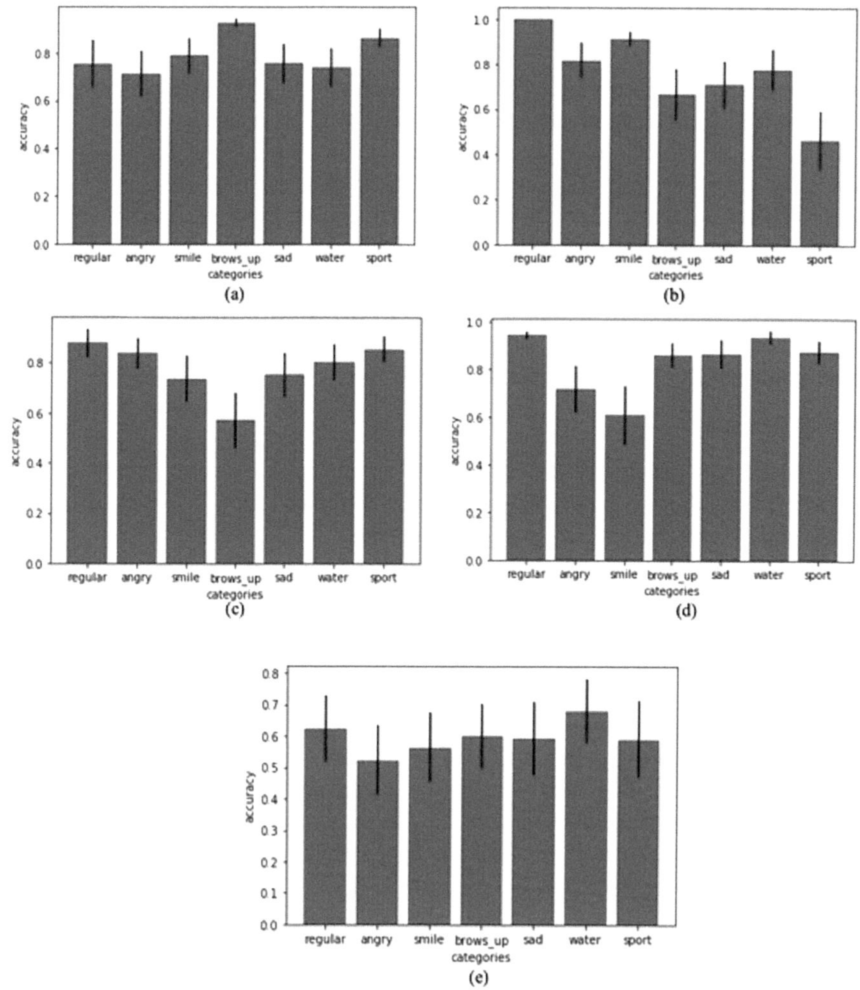

Fig. 7.5 Accuracy rates of classifying correctly each class within every physical subset of it. The bar represents a mean accuracy rate based on model accuracy each time on a randomly chosen subsets of the data. Straight lines indicate the variance of accuracies. **a**—**e**. represent the five different subjects [20]

handling diverse scenarios and expressions. This information can guide future iterations and improvements to enhance the model's accuracy and reliability in real-world applications.

7.3.5 Insights and Recommendations

In this paper, we introduced a pioneering approach to photonic speckle pattern-based authentication, showcasing experimental validation on a predefined set of classes. The authentication system holds promising applications in smart homes and smart cars, demonstrating high accuracy in class separation even under diverse measurement conditions. The primary strength of the proposed model lies in its ability to unveil insights in scenarios where human discernment is challenging. By leveraging neural networks to decipher intricate speckle patterns, this research opens new avenues for exploration with broad applications yet to be integrated. The model's success, achieved without the need for an intricately tailored solution and lacking prior knowledge of speckle properties, signifies a promising breakthrough.

Key Takeaways:

- Versatile Applicability: The demonstrated authentication system's adaptability to different conditions marks it as a potential solution for varied real-world scenarios, especially in the realms of smart environments.
- Neural Networks' Potential: The utilization of neural networks in decoding imperceptible speckle patterns underscores their potential as tools for addressing complex problems with a wide range of applications.
- Open-Source Implementation: The accessibility of the full code, available on GitHub, encourages further exploration and experimentation. Researchers and practitioners can engage with the codebase, run custom training on their data, and tailor configurations to their specific needs.
- Future Prospects: The success achieved in this study paves the way for future investigations into identifying additional facial features linked to sentiments and advanced physical conditions. This suggests a roadmap for extending the model's capabilities and applications.

In conclusion, this research not only presents a robust authentication model but also sets the stage for continued exploration and advancements in leveraging photonic speckle patterns for diverse technological applications.

7.4 Brain Activity Sensing

7.4.1 Inspiration and Exploration

In recent years, the exploration of neural activity and its connection to sensory information and behaviour control has gained significant attention. However, existing methods for sensing brain activity often involve expensive equipment and close physical proximity to the subject. The study of human sensory systems and their link to brain activity has become increasingly popular, with a focus on the five basic senses: sight [39], touch [40], hearing [41, 42], smell [43–46], and taste [47]. Electroencephalography (EEG) has been a prominent tool for analyzing brain activity, and optical techniques, including laser speckle imaging, have been employed for monitoring brain activity using image contrast analysis [48, 49] and cross-correlation based analysis [50].

The temple area in front of the cerebral cortex, while not optically ideal, presents an opportunity for speckle-based remote sensing. Illumination by a laser beam produces secondary speckle patterns, visible with a digital camera using defocused optics. This study proposes using a recurrent neural network (RNN) architecture to classify speckle patterns, considering their temporal dependency.

The research introduces a method for classifying and detecting three basic senses—smell, taste, and hearing—by projecting a laser beam onto a specific area of the human head associated with cerebral cortex activity. The recorded speckle patterns are then analyzed using deep neural networks (DNN). The approach's reliability is validated by comparing results with simultaneously recorded EEG data [39], a recognized method for detecting brain activity related to human senses.

The study holds significance for patients with conditions like stroke or cancer [51–54], especially those experiencing irregularities in basic senses such as taste and smell. Additionally, it explores the potential application of predicting sensory loss associated with COVID-19 [55, 56], contributing to the identification of virus carriers. In summary, this research aims to advance the understanding of neural activity through a novel speckle-based remote sensing approach, with potential clinical applications and relevance to challenges like sensory loss, particularly in the context of COVID-19.

7.4.2 Experiment Setup

The experimental setup involved a laser (770 μW, 1550 nm) and a digital camera with defocused optics to generate and capture speckle patterns. Simultaneously, EEG electrodes from an OpenBCI EEG headband with a Ganglion bio-sensing 4-channels board were synchronized to record brain signals [39]. The study included data from eight healthy participants aged 29–73. The laboratory environment was controlled to prevent background noises during testing, with each participant seated 50 cm from the setup. To restrict head movements, a headset with protective gear was employed, and each subject's sense-related brain activity was recorded for two cases: at rest and under sensory stimulation.

For sensory stimulation, participants engaged in activities relevant to smell, taste, and hearing. For example, the sense of smell was stimulated by smelling an alcohol-soaked cotton ball, taste by consuming sweet chocolate, and hearing by exposure to a continuous constant-frequency noise. High-speed cameras with defocused optics recorded temporal changes in speckle images at a frame rate of 750 FPS and a spatial resolution of 32×32 pixels. Each sample was recorded for 10 s. The dataset, containing approximately 150,000 frames, included unique identification for each subject, specifying the subject's ID, measurement duration, sense type, and binary sign indicating activity or inactivity of the sense.

Data collection occurred on separate dates, with each subject recorded multiple times in continuous sessions. The dataset was subdivided into training and test datasets, ensuring no mixing between them to prevent biases. To assess the proposed method quantitatively, metrics such as True Positive (TP), True Negative (TN), False Positive (FP), and False Negative (FN) were calculated pixelwise. Metrics such as accuracy, precision, recall, and F1-score were computed using these values [39].

7.4.3 Method Overview

The proposed approach involves processing sequential speckle images using a 2D ConvLSTM layer, which consists of an LSTM layer with internal matrix multiplications along with 2D convolution operations. This layer retains the original input dimensions rather than projecting onto a 1D feature vector. The ConvLSTM layer's equations are expressed as follows:

$$
\begin{aligned}
i_t &= \sigma(W_{xi}{*}X_t + W_{hi}{*}H_{t-1} + W_{ci} \circ C_{t-1} + b_i) \\
f_t &= \sigma\left(W_{xf}{*}X_t + W_{hf}{*}H_{t-1} + W_{cf} \circ C_{t-1} + b_f\right) \\
C_t &= f_t \circ C_{t-1} + i_t \circ \tanh(W_{xc}{*}X_t + W_{hc}{*}H_{t-1} + b_c) \\
o_t &= \sigma(W_{xo}{*}X_t + W_{ho}{*}H_{t-1} + W_{co} \circ C_t + b_o) \\
H_t &= o_t \circ \tanh(c_t)
\end{aligned}
$$

where $*$ denotes the convolution operator, $\circ$ is the Hadamard product, $X_1 \ldots X_\mathrm{t}$ represent the model input, $C_1 \ldots C_\mathrm{t}$ are ConvLSTM cell outputs, $H_1 \ldots H_\mathrm{t}$ are hidden states, and o_t, f_t, i_t are 3D tensors of the ConvLSTM layer. Padding is applied before the convolution operation to ensure consistency in rows and columns.

For the ConvLSTM model, a patch size of 1×1 is set, representing each 32×32 frame with a $32 \times 32 \times 3$ tensor. The single-layer ConvLSTM network contains 64 hidden states, input-to-state for 64 input frames, and state-to-state kernels of size 3×3. The output feeds into a fully connected binary classification head with 256 units and ReLU activations. Training, implemented in TensorFlow, includes binary cross-entropy as a loss function, dropout to reduce overfitting, and Adam optimizer with a learning rate of 0.001.

To verify the method, EEG data is recorded simultaneously with speckle patterns. EEG signal classification employs a CNN with three 1D-Conv layers, ReLU activation, and 1D-MaxPooling followed by two fully connected layers. Binary cross-entropy is used as a loss function, and training with Adam optimizer is conducted on a single 1080Ti GPU, converging in approximately 10 epochs [57, 58].

7.4.4 Research Summary

In this study, we introduced a novel approach to photonic speckle patterns-based authentication, focusing on a limited number of predefined classes with potential applications in smart homes and smart cars. The authentication process, relying on static speckles, exhibited high accuracy across various measurement conditions. The significance of the proposed model lies in its ability to provide valuable insights into challenging questions, particularly when dealing with speckle patterns that are imperceptible to the human eye. Neural networks, as employed in this research, open new avenues for exploration with a lower entry barrier and diverse applications. Interested users can leverage this repository to conduct custom training on their own data, utilizing different configurations and hyperparameters.

The achieved promising results without the need for an intricately tailored model and the absence of prior knowledge about speckle properties underscore the potential of this approach. Future investigations could dive deeper into identifying facial features associated with sentiments and advanced physical conditions, further expanding the scope of applications for this technology.

7.5 Challenges and Future Perspectives

In the realm of biomedical imaging, leveraging artificial intelligence and machine learning for speckle pattern analysis offers exciting possibilities. Future research could dive into higher-resolution studies, allowing for more nuanced classifications corresponding to numerical grading. The method's application in diverse scenarios, such as medical institutions and public spaces during pandemics, suggests potential for broader use. Further exploration of combining this approach with remote temperature measurement devices for early COVID-19 tracking showcases a promising interdisciplinary avenue [21].

While the research on linking micro-saccades with blood oxygen saturation presents promising findings for non-invasive health monitoring, several challenges and future directions should be considered. Expanding the participant pool to include individuals with silent hypoxemia and exploring the potential correlation with complementary diseases would enhance the method's robustness. Additionally, investigating the impact of stress conditions during breath restriction and examining the possibility of detecting other medical conditions through micro-saccades could contribute to a more comprehensive understanding of the technology's applications [59].

The exploration of neural activity and sensory connections through laser speckle imaging presents a novel approach with applications in detecting sensory irregularities. Challenges include refining the methodology to enhance accuracy and reliability across diverse sensory experiences. Future research should focus on scaling up the study to encompass a broader range of senses and expanding the participant demographics. Additionally, exploring the potential application in identifying sensory loss associated with health conditions like stroke, cancer, and diseases such as COVID-19 marks a significant avenue for future investigation [60].

One common challenge across these studies is the need for integration and interdisciplinary collaboration. Combining insights from micro-saccades, speckle pattern analysis, and sensory activity detection could lead to innovative approaches in health monitoring and diagnostic technologies. Collaborations between researchers specializing in ophthalmology, biomedical imaging, and neural activity could bridge gaps and foster a more comprehensive understanding of the human body's interconnected systems.

As these technologies advance, ethical considerations become paramount. Issues related to privacy, consent, and potential misuse of health data need careful attention. Establishing clear ethical guidelines and ensuring adherence to regulations are essential to the responsible development and deployment of these innovative technologies.

In conclusion, overcoming challenges and embracing interdisciplinary collaboration will be key to unlocking the full potential of these cutting-edge technologies in advancing healthcare and diagnostic capabilities. The journey from research insights to practical applications requires ongoing exploration, refinement, and a commitment to ethical practices.

7.6 Conclusions

The amalgamation of Machine Learning (ML) with speckle pattern analysis in biomedical imaging marks a dynamic frontier, brimming with transformative potential. The ongoing evolution of ML algorithms in deciphering intricate speckle patterns promises a paradigm shift in non-invasive diagnostic methodologies. This synergy not only enhances those diagnostic capabilities but also unveils unprecedented insights into diverse biological processes and states of health and disease. The current landscape showcases a glimpse of the immense possibilities that lie ahead. ML algorithms, with their capacity to navigate the complexities of speckle patterns, are poised to redefine the boundaries of medical diagnostics. The trajectory of research in this domain points toward a future characterized by increasingly sophisticated models and algorithms. These advancements will undoubtedly contribute to heightened precision, expanded applicability, and the realization of novel diagnostic techniques.

As the scientific community continues to unravel the intricacies of speckle patterns and harness the prowess of ML, the synergy between these realms is destined to bring about revolutionary advancements in biomedical imaging. The journey ahead holds the promise of not only refining those diagnostic tools but also offering a deeper comprehension of the subtle nuances within biological systems.

In conclusion, the integration of ML into speckle pattern analysis represents an exciting chapter in the ongoing narrative of medical innovation. As the field matures, we can anticipate a cascade of innovations that will redefine the landscape of diagnostics and propel us toward a future where precision, accessibility, and insights converge to elevate healthcare to unprecedented heights.

References

1. A. Barragán-Montero et al., Artificial intelligence and machine learning for medical imaging: A technology review. Physica Med. **83**, 242–256 (2021)
2. I. Castiglioni et al., AI applications to medical images: From machine learning to deep learning. Physica Med. **83**, 9–24 (2021)
3. G. Varoquaux, V. Cheplygina, Machine learning for medical imaging: methodological failures and recommendations for the future. NPJ Digit Med **5**, 48 (2022)
4. A. Husnain, S. Rasool, A. Saeed, A. Yousaf Gill, H. Khawar Hussain, AI'S healing touch: examining machine learning's transformative effects on healthcare. J. World Sci. **2**, 1681–1695 (2023)
5. H. Zhu, S.O. Isikman, O. Mudanyali, A. Greenbaum, A. Ozcan, Optical imaging techniques for point-of-care diagnostics. Lab Chip **13**, 51–67 (2013)
6. S.A. Boppart, R. Richards-Kortum, Point-of-care and point-of-procedure optical imaging technologies for primary care and global health. Sci Transl Med. **6**, (2014)
7. J. Park, J.B. Kang, J.H. Chang, Y. Yoo, Speckle reduction techniques in medical ultrasound imaging. Biomed. Eng. Lett. **4**, 32–40 (2014)
8. Z. Zalevsky et al., Simultaneous remote extraction of multiple speech sources and heart beats from secondary speckles pattern. Opt. Express **17**, 21566 (2009)
9. J.G. Daugman, High confidence visual recognition of persons by a test of statistical independence. IEEE Trans. Pattern Anal. Mach. Intell. **15**, 1148–1161 (1993)
10. Y. Zhu, T. Tan Y. Wang, Biometric personal identification based on iris patterns. in *Proceedings 15th International Conference on Pattern Recognition. ICPR-2000* (IEEE Comput. Soc), pp. 801–804. https://doi.org/10.1109/ICPR.2000.906197
11. Z. He et al., Boosting ordinal features for accurate and fast iris recognition. in *2008 IEEE conference on computer vision and pattern recognition* (IEEE, 2008), pp. 1–8. https://doi.org/10.1109/CVPR.2008.4587645
12. R. Brunelli, T. Poggio, Face recognition: features versus templates. IEEE Trans. Pattern Anal. Mach. Intell. **15**, 1042–1052 (1993)
13. A.M. Bronstein, M.M. Bronstein, R. Kimmel, Three-dimensional face recognition. Int. J. Comput. Vis. **64**, 5–30 (2005)
14. D.A. Socolinsky, A. Selinger, Thermal face recognition in an operational scenario. in *Proceedings of the 2004 IEEE computer society conference on computer vision and pattern recognition, 2004. CVPR 2004* (IEEE), pp. 1012–1019. https://doi.org/10.1109/CVPR.2004.1315275
15. *Laser Speckle and Related Phenomena*. vol. 9 (Springer Berlin Heidelberg, 1975)
16. J. Schmidhuber, Deep learning in neural networks: An overview. Neural Netw. **61**, 85–117 (2015)
17. A. Krizhevsky, I. Sutskever, G.E. Hinton, ImageNet classification with deep convolutional neural networks. Commun. ACM **60**, 84–90 (2017)
18. S.C. Kleene, Representation of events in nerve nets and finite automata. in *Automata Studies. (AM-34)* (Princeton University Press, 1956), pp. 3–42. https://doi.org/10.1515/9781400882618-002
19. V. Nair, G. Hinton, Rectified linear units improve restricted boltzmann machines vinod nair. Proceedings of ICML **27**, 807–814 (2010)

20. Z. Kalyzhner, O. Levitas, F. Kalichman, R. Jacobson, Z. Zalevsky, Photonic human identification based on deep learning of back scattered laser speckle patterns. Opt. Express **27**, 36002 (2019)
21. Z. Kalyuzhner, S. Agdarov, A. Bennett, Y. Beiderman, Z. Zalevsky, Remote photonic sensing of blood oxygen saturation via tracking of anomalies in micro-saccades patterns. Opt. Express **29**, 3386 (2021)
22. Z. Kalyuzhner et al., Remote photonic detection of human senses using secondary speckle patterns. Sci Rep **12**, (2022)
23. W. McNicholas, Kent, Mitchell, Hypoxemia in patients with COPD: cause, effects, and disease progression. Int J Chron Obstruct Pulmon Dis. **199**, (2011) https://doi.org/10.2147/COPD.S10611
24. M.J. Tobin, F. Laghi, A. Jubran, Why COVID-19 silent hypoxemia is baffling to physicians. Am. J. Respir. Crit. Care Med. **202**, 356–360 (2020)
25. M. Nouri-Vaskeh, A. Sharifi, N. Khalili, R. Zand, A. Sharifi, Dyspneic and non-dyspneic (silent) hypoxemia in COVID-19: Possible neurological mechanism. Clin. Neurol. Neurosurg. **198**, 106217 (2020)
26. J. Couzin-Frankel, The mystery of the pandemic's 'happy hypoxia.' Science **1979**(368), 455–456 (2020)
27. M. Nitzan, A. Romem, R. Koppel, Pulse oximetry: fundamentals and technology update. Med. Devices: Evid. Res. **231**, (2014) https://doi.org/10.2147/MDER.S47319
28. I. Margalit et al., New method for remote and repeatable monitoring of intraocular pressure variations. J. Biomed. Opt. **19**, 027002 (2014)
29. A. Shapira et al., Increased gamma band activity for lateral interactions in humans. PLoS ONE **12**, e0187520 (2017)
30. B. Lengenfelder et al., Remote photoacoustic sensing using speckle-analysis. Sci. Rep. **9**, 1057 (2019)
31. A. Bennett et al., Corneal thickness measurement by secondary speckle tracking and image processing using machine-learning algorithms. J. Biomed. Opt. **24**, 1 (2019)
32. G. Roelofs, R. Koman, *PNG: The definitive guide.* (O'Reilly & Associates, Inc., 1999)
33. G.E. Hinton, A. Krizhevsky, I. Sutskever, System and method for addressing overfitting in a neural network. **1**, 4–6 (2016)
34. A.F.M. Agarap, *Deep learning using Rectified Linear Units (ReLU)* (2018). ArXiv 2–8
35. H. Wu, X. Gu, Max-pooling dropout for regularization of convolutional neural networks. in (2015), pp. 46–54. https://doi.org/10.1007/978-3-319-26532-2_6
36. Z. Zhang, M. Sabuncu, Generalized cross entropy loss for training deep neural networks with noisy labels. in *Advances in neural information processing systems,* vol. 31, eds. by S. Bengio et al. (Curran Associates, Inc., 2018)
37. S. Ruder, *An overview of gradient descent optimization algorithms* (2016)
38. L.L. Di Stasi et al., Intersaccadic drift velocity is sensitive to short-term hypobaric hypoxia. Eur. J. Neurosci. **39**, 1384–1390 (2014)
39. S.M. Alarcão, M.J. Fonseca, Emotions recognition using EEG signals: A survey. IEEE Trans. Affect. Comput. **10**, 374–393 (2019)
40. T. Nakamura, Y. Tomita, Y. Mitsukura, A method of obtaining sense of touch by using EEG. Information (Japan) **14**, 621–632 (2011)
41. C.B. Christensen, J.M. Harte, T. Lunner, P. Kidmose, Ear-EEG-based objective hearing threshold estimation evaluated on normal hearing subjects. IEEE Trans. Biomed. Eng. **65**, 1026–1034 (2018)
42. B. Zoefel, R. VanRullen, EEG oscillations entrain their phase to high-level features of speech sound. Neuroimage **124**, 16–23 (2016)
43. T.S. Lorig, The application of electroencephalographic techniques to the study of human olfaction: A review and tutorial. Int. J. Psychophysiol. **36**, 91–104 (2000)
44. G.N. Martin, Human electroencephalographic (EEG) response to olfactory stimulation: Two experiments using the aroma of food. Int. J. Psychophysiol. **30**, 287–302 (1998)

45. A. Saha, A. Konar, A. Chatterjee, A. Ralescu, A.K. Nagar, EEG analysis for olfactory perceptual-ability measurement using a recurrent neural classifier. IEEE Trans Hum Mach Syst **44**, 717–730 (2014)
46. A. Saha, A. Konar, P. Rakshit, A.L. Ralescu, A.K. Nagar, Olfaction recognition by EEG analysis using differential evolution induced Hopfield neural net. in *Proceedings of the international joint conference on neural networks* (2013) https://doi.org/10.1109/IJCNN.2013.6706874
47. C. Park, D. Looney, D.P. Mandic, Estimating human response to taste using EEG. in *Proceedings of the annual international conference of the IEEE Engineering in Medicine and Biology Society, EMBS* (2011), pp. 6331–6332 https://doi.org/10.1109/IEMBS.2011.6091563
48. D.A. Boas, A.K. Dunn, Laser speckle contrast imaging in biomedical optics. J. Biomed. Opt. **15**, 011109 (2010)
49. M. Jiang et al., Dynamic imaging of cerebral blood flow using laser speckle during epileptic events. Biomedical Optics, BIOMED **2012**, 195–201 (2012). https://doi.org/10.1364/biomed.2012.btu3a.45
50. N. Ozana et al., Remote photonic sensing of cerebral hemodynamic changes via temporal spatial analysis of acoustic vibrations. J. Biophotonics **13**, 1–12 (2020)
51. T.K. Banerjee, M.K. Roy, K.K. Bhoi, Is stroke increasing in India—Preventive measures that need to be implemented. J. Indian Med. Assoc. **103** 162, 164, 166 passim–162, 164, 166 passim (2005). Preprint at https://europepmc.org/article/med/16173293
52. T.L. Green, L.D. McGregor, K.M. King, Smell and taste dysfunction following minor stroke: a case report. Can. J. Neurosci. Nurs. **30**, 10–13 (2008)
53. W.H.T.R.J.P. Penry, J. Kiffin, Complex partial seizures Clinical characteristics and differential diagnosis. Handb. Parkinson's Dis., Fifth Ed. **33**, 11515–11515 (1983)
54. M.Q. Nguyen, N.J.P. Ryba, A smell that causes seizure. PLoS ONE **7**, 1–10 (2012)
55. J.Y. Tong, A. Wong, D. Zhu, J.H. Fastenberg, T. Tham, The prevalence of olfactory and gustatory dysfunction in COVID-19 Patients: A systematic review and meta-analysis. Otolaryngol.-Head Neck Surg. (U. S.) **163**, 3–11 (2020)
56. A. Walker, G. Pottinger, A. Scott, C. Hopkins, Anosmia and loss of smell in the era of covid-19. The BMJ **370**, 1–4 (2020)
57. X. Shi et al., Convolutional LSTM network: A machine learning approach for precipitation nowcasting. Adv Neural Inf Process Syst. **2015-Janua**, 802–810 (2015)
58. A. Craik, Y. He, J.L. Contreras-Vidal, Deep learning for electroencephalogram (EEG) classification tasks: A review. J Neural Eng. **16**, (2019)
59. Y. Beiderman et al., Remote estimation of blood pulse pressure via temporal tracking of reflected secondary speckles pattern. J. Biomed. Opt. **15**, 061707 (2010)
60. K.W. Cooper et al., COVID-19 and the chemical senses: Supporting players take center stage. Neuron **107**, 219–233 (2020)

Chapter 8
Two Photon Fluorescence Integrated Machine Learning for Data Analysis and Interpretation

Gagan Raju, Manikanth Karnati, Yury V. Kistenev, and Nirmal Mazumder

Abstract The integration of two-photon fluorescence excitation microscopy with machine learning (ML) heralds a new era in biomedical research, empowering scientists to tackle complex biological questions with unprecedented precision. This chapter explores the transformative potential of this integrated approach across multiple domains, particularly in cancer biology, stem cell research, and brain science. By harnessing the high-resolution imaging capabilities of two-photon microscopy and the analytical prowess of ML algorithms, researchers can dissect the intricacies of tumour microenvironments, predict tumour behaviour, and identify therapeutic targets with enhanced accuracy. Case studies exemplifying the automated classification of breast cancer histologic grade and the quantification of tumour microenvironment heterogeneity underscore the profound impact of this integration on cancer diagnosis and treatment. Similarly, in stem cell research, the combination of two-photon microscopy and ML enables quantitative analysis of stem cell dynamics, lineage commitment, and differentiation trajectories in complex 3D microenvironments, facilitating advancements in regenerative medicine and drug discovery. Moreover, in brain research, this approach elucidates dynamic neuronal processes, synaptic connectivity, and neural circuitry, offering insights into neurological disorders and brain function. Ultimately, the integration of two-photon fluorescence microscopy with ML holds immense promise for advancing biomedical research, driving innovations in diagnostics, therapeutics, and personalized medicine, and ultimately improving healthcare outcomes for patients worldwide.

G. Raju · M. Karnati · N. Mazumder (✉)
Department of Biophysics, Manipal School of Life Sciences, Manipal Academy of Higher Education, Manipal 576104, India
e-mail: nirmal.mazumder@manipal.edu

Y. V. Kistenev
Laboratory LMIML, Tomsk State University, Tomsk 634050, Russia

A. Gogoi and N. Mazumder (eds.), *Biomedical Imaging*, Biological and Medical Physics, Biomedical Engineering, https://doi.org/10.1007/978-981-97-5345-1_8

8.1 Introduction to Two-Photon Excitation Process

Two-photon excitation (TPE) is a non-linear process where a molecule is excited by the absorption of two photons instead of one. This phenomenon occurs when two photons of lower energy simultaneously interact with a molecule to produce an electronic transition to a higher energy state [1]. TPE requires high-intensity light sources, such as pulsed lasers, and is commonly used in fluorescence microscopy and multiphoton imaging techniques [2]. The probability of TPE is determined by the intensity of the incident light. The higher the intensity, the more likely it is that two photons will be absorbed by the molecule simultaneously. However, TPE is a non-linear process, meaning that the probability of excitation is proportional to the square of the incident light intensity [3]. This is because the probability of two photons arriving at the same time at the same location is proportional to the square of their individual probabilities. The advantage of TPE is that it allows for deeper imaging of biological tissues with reduced photodamage, as the excitation occurs only in the focal plane of the laser beam [4]. Single photon excitation involves the absorption of a single photon of sufficient energy to promote the molecule to an excited state. In contrast, TPE involves the absorption of two lower-energy photons that combine their energies to excite the molecule to the same energy level as a single higher-energy photon [5]. Another significant difference between the two phenomena is that TPE is a nonlinear process, this means that the probability of TPE is proportional to the square of the incident light intensity, whereas single-photon excitation probability is directly proportional to the intensity.

8.2 TPM Implementation and Data Acquisition

The use of artificial intelligence (AI) in image processing has been particularly beneficial in the field of microscopy. Microscopic images often contain a large amount of data that can be difficult to analyse manually. By implementing AI techniques, researchers and scientists can extract meaningful information from these images more efficiently and accurately [6]. Object detection is one of the primary applications of AI in microscopic image processing. With the use of deep learning algorithms, AI can accurately identify objects within an image, such as cells or tissues. This is particularly useful in medical research, where identifying and tracking specific cells can help diagnose diseases and develop new treatments [7]. Image segmentation is another important task in image processing that can be automated using AI. Image segmentation involves dividing an image into different regions, based on similarities in colour, texture, or other features [8]. This can be a time-consuming and complex task for humans to perform manually, but AI algorithms can quickly and accurately segment an image. This is particularly useful in microscopy, where researchers need to identify specific regions of interest within an image. Further, AI algorithms can be trained to classify different types of cells or tissues based on their characteristics.

This can help identify abnormalities or diseases in samples and aid in the development of new treatment strategies [9]. AI can also be used for restoration in image processing that involves removing noise or other distortions from an image, which can improve its quality and accuracy. AI algorithms can be trained to restore images, making it easier to analyse and interpret them.

Overall, the use of AI in image processing has revolutionized the way researchers and scientists analyse and interpret microscopic images. By automating tasks such as object detection, segmentation, classification, and restoration, AI has made it easier and more efficient to extract meaningful information from these images [10]. As AI technology continues to advance, it is likely that it will play an increasingly important role in the field of microscopy and other areas of scientific research. The field of image processing has seen significant advancements through the use of deep learning, a subset of machine learning. Convolutional neural networks (CNNs), a type of deep learning algorithm, have been effective in recognizing complex features in images [11]. CNNs have been successfully utilized for image classification, object detection, and segmentation, often achieving state-of-the-art results. Additionally, AI has been employed in image restoration tasks, such as denoising and deblurring, where AI algorithms are trained to identify and remove noise or blurs from images, leading to clearer and more useful images. The implementation of AI in image processing has the potential to automate and improve various tasks, resulting in reduced time and effort needed for image analysis, as well as increased accuracy and reliability of results [12].

8.3 TPM Data Preprocessing

The main purpose of using machine learning (ML) is to create a data model, which allows making a prediction of new experimental data to a definite class, for example, corresponding to absence or presence of a target disease. Creation of such a model is conducted by the several data processing steps implementation called ML pipeline. Usually, ML pipeline includes preprocessing, feature extraction/selection, prediction model construction, and its validation (see Fig. 8.1) [13].

The right choice of algorithms and their parameters used at a definite step of the ML pipeline should be validated. Some criteria of the optimal choice can be applied immediately, but the most general way is to do it at the final stage of the prediction model validation and efficiency evaluation. The goal of applying immediately (particular) criteria is to make enable the further data analysis. The goal of the general criteria is increasing the prediction accuracy of a classifier. It means that ML pipeline implementation is not a linear step-by-step process but embedding several iterative cycles to optimize the parameters of each step of the ML pipeline. It also means that the first steps are very crucial because a wrong decision at this stage influences on the efficiency of algorithms used on next steps and also prediction model in the whole. These recommendations are quite general and implemented through universal methods. Sometimes, more specific methods oriented on peculiarities of

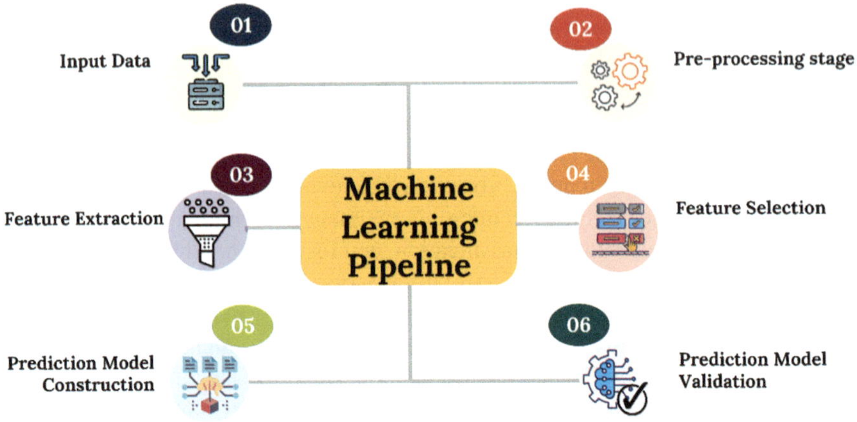

Fig. 8.1 A typical machine learning pipeline [13]

TPM data are necessary. The examples of similar methods are presented below. Two-photon endo-microscopy integrated with a gradient refractive index (GRIN) lens is widely used for brain imaging. A method of adaptive optics based on using the two-photon excited fluorescence signal as the intrinsic guide star inside biological tissues to compensate the intrinsic aberrations of the GRIN lens was developed [14]. 3D image reconstruction of a hippocampal CA1 pyramidal neurons in Thy1-GFP mice model with standard aberration correction only and with full aberration correction and subsequent deconvolution is shown in Fig. 8.2.

In-vivo two-photon microscopy (TPM) images often suffer from complicated physiological noise. A cyclic descent algorithm based on maximum likelihood temporal fluorescence signal estimation was developed to separate stimulus-evoked fluorescence response from background activity and noise [15]. Raw data is often irrelevant, redundant, and incomplete and requires pre-processing [16, 17]. There is an opinion that data pre-processing consumes 50–80% of analysis time [18]. An outlier is data sample differing significantly from the major data. The outlier detection and removal can be conducted through the statistical-based approaches like a parametric Z-score [19], isolation forest [20], density-based spatial clustering [21]. Z-score characterizes the width of random value distribution in terms of the relation of its standard deviation to the mean value. The isolation forest algorithm generates recursively random partitions on the samples in a dataset to isolate samples in a poorly populated area of the dataset spatial distribution considered as outlets. The density-based spatial clustering algorithm uses estimating density spatial distribution of the dataset, marking as outliers points that are lonely in areas with low density.

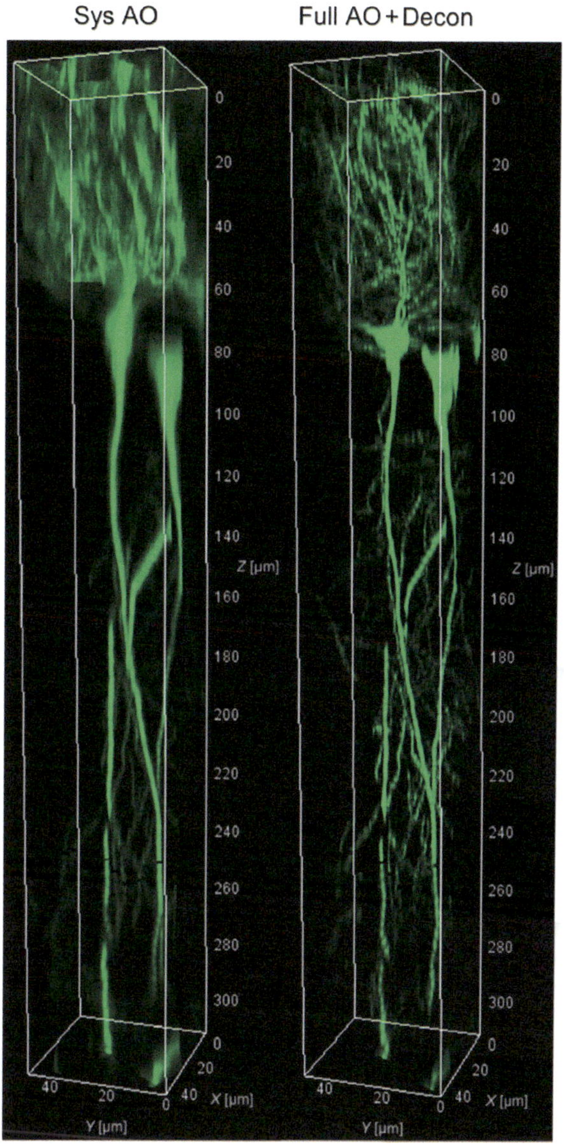

Fig. 8.2 3D image reconstruction of a hippocampal CA1 pyramidal neurons in Thy1-GFP mice. Here, "Sys AO" stands for standard aberration correction, "Full AO + Decon" stands for full aberration correction and subsequent deconvolution. Figure reproduced from [14] with permission from science

8.4 TPM Data Informative Features Extraction

Original TPM data are related to 2D images. A standard way to find regions of interest (ROI) in visual data is based on a textural analysis. Here, the texture is a part of an image with homogeneous parameters inside it. A preliminary image preprocessing by gradient methods allows emphasizing the image structure before finding textures [22, 23]. Texture features can be extracted by the following groups of methods.

Structural methods allow describing textures in terms of visual primitives extracted from graphical data by the mathematical morphology methods [24, 25]. Spectral methods are based on an image convolution with a special mask through application of Fourier, Gabor, or wavelet transforms [26–29]. Model methods describe the texture characteristics it terms of a mathematical model [23]. A combination of spectral and model approaches is based on extracting basic graphical primitives in original image and, after that, presenting the image as a composition of these primitives. The illustrations of a such approach are Bag-of-words [30] and convolutional neural networks [31]. Statistical methods describe a texture in terms of the first or the second order statistics [18]. The first-order statistics (FOSs) describe statistical properties of image pixels independently of each other, for example, the image mean brightness, contrast, and statistical moments. The second-order statistics (SOSs) describe an image in terms of statistical characteristics of various pixel pairs, for example, the gray-level co-occurrence matrix [32]. Cells are specific ROIs in TPM tissue images. Process of such ROIs identification requires frame-by-frame analysis that is very computer-consuming. A convolutional neural network (CNN)-based algorithm named "Cell Identification and Trace Extraction Online" for fast automatic cell identification, segmentation, tracking, and trace extraction in TPM calcium imaging data was developed [33]. An original 3D CNN named STNeuroNet was developed and applied to identify, and segment active neurons regarding their spatial density in TPM images [34]. The STNeuroNet architecture is shown in Fig. 8.3. The CNN was trained on a cortical layer TPM images. The example of application of the STNeuroNet is shown in Fig. 8.4.

The network transforms TPM data into three feature maps with different resolutions using dense feature stacks combined with convolutional layers. A final bilinear upsampling layer transforms the feature maps to the initial TPM data resolution. A max-pooling layer summarizes the features along the time dimension. After that, 2D convolutional layers generate segmentation logits.

TPM data classification using ML is based on using algorithms. A suitable supervised ML algorithm depends on used informative features character. When original TPM image data are transformed into a set of quantitative features like FOSs or

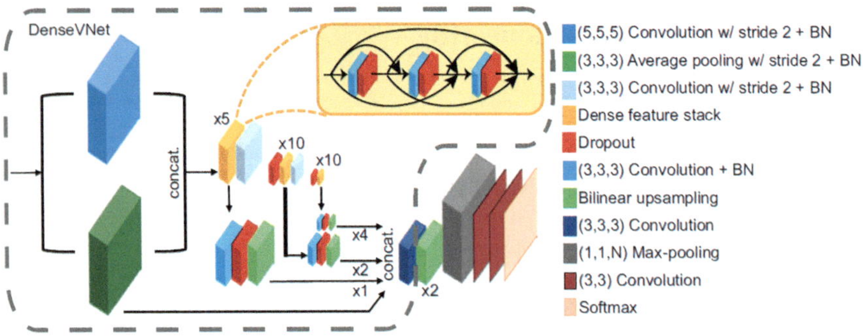

Fig. 8.3 The STNeuroNet architecture. Figure reproduced from [34] with permission from PNAS

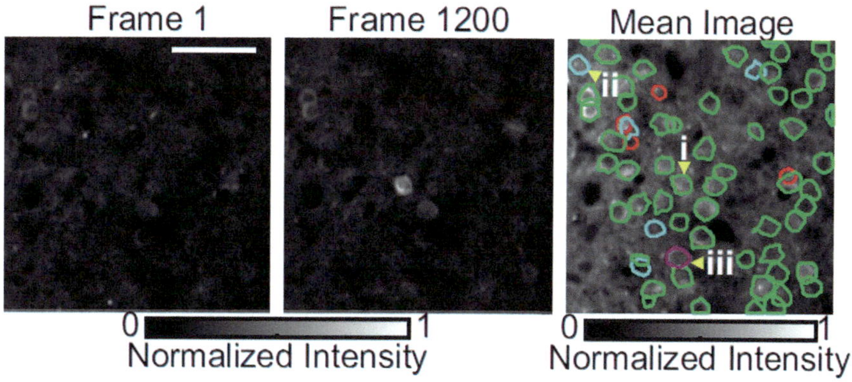

Fig. 8.4 The active neurons detection in TPM images using STNeuroNet. Figure reproduced from [34] with permission from PNAS. The analysis was performed over 1200 2D frames. The illustrations include the first and the last frames, normalized mean image, and correlation image

SOSs, universal ML algorithms can be used. For example, we used Support vector machine to differentiate TMP images of lymphedema skin from healthy skin [22]. The informative features were parameters of Histogram-oriented-gradients (HOG) feature vector. HOG is an effective method of analysis of image brightness gradients. When feature vector is an image or graphical data, then computer vision, image recognition methods should be used to classify similar data. This field is very popular and is developing fast. The effectiveness of concrete image recognition methods depends on visual data complexity and volume of training dataset.

8.5 Applications

The integration of TPM with ML has yielded transformative applications in various domains of cellular imaging, revolutionizing our understanding of complex biological processes. Through the synergy of TPM's high-resolution imaging capabilities and ML's data analysis prowess, researchers have made significant strides in elucidating cellular dynamics and unraveling the mysteries of cellular biology.

Neuronal Morphology Analysis

One of the prominent applications of TPM-ML integration lies in the analysis of neuronal morphology, offering insights into the intricate architecture of the brain. Studies have utilized TPM to capture high-resolution images of neuronal structures, including dendrites, axons, and synapses, providing unprecedented detail of neuronal morphology [35]. ML algorithms, such as convolutional neural networks (CNNs) and graph-based approaches, have been employed to automatically segment and classify neuronal structures from TPM images [36]. These approaches enable the quantification of morphological features and the reconstruction of neuronal networks with remarkable accuracy, facilitating neuroscience research aimed at understanding brain function and dysfunction. In the dynamic world of neuroscience, where the brain's complexities often elude our grasp, a series of groundbreaking studies embarked on a mission to unveil the secrets of neuronal morphology. Armed with two-photon fluorescence microscopy (TPM) and the potent capabilities of machine learning (ML), these studies delved into the intricate realm of dendritic spines and synaptic connections (Fig. 8.5).

Chen et al. [37], who sought to map the terrain of single spines in cortical neurons. Through TPM's lens, they observed the delicate movements of dendritic spines in real-time, unveiling their diverse shapes and sizes. With ML's assistance, they meticulously analyzed and categorized each spine with unprecedented precision, shedding light on the mechanisms governing synaptic plasticity and cognitive function [37]. Following this, Xu et al. [38] endeavoured to decipher the language of calcium transients in neuronal populations. Using TPM to capture neuronal activity, they witnessed the mesmerizing patterns of neural firing across networks. Integrating ML techniques, they decoded these patterns, revealing hidden insights into neural circuitry organization [38]. Their findings paved the way for advancements in brain-machine interfaces and neurorehabilitation. Continuing the quest, Ghosh et al. [39] explored the intricacies of synaptic connectivity within dendritic spines. Through TPM's capabilities, they observed the ever-changing landscape of synaptic contacts. Employing ML algorithms, they decoded the structural features of spines, providing crucial insights into synaptic plasticity and memory formation [39]. Finally, Runge et al. [40] ventured into the dynamic world of dendritic spines in the mouse barrel cortex. Using TPM, they captured the fluctuations in spine dynamics over time, revealing the brain's ability to adapt and rewire. Leveraging ML techniques, they quantified spine turnover rates, unraveling the mechanisms underlying sensory processing and learning [40]. In summary, the discoveries from studying neuronal morphology send a clear message to scientists: by combining TPM and ML, we can uncover hidden secrets of the brain. This not only helps us understand neuroscience better but also opens doors for new breakthroughs in healthcare and science.

Cancer Biology

In the field of cancer biology, TPM-ML integration has emerged as a powerful tool for characterizing tumour microenvironments and elucidating cancer progression mechanisms. By leveraging TPM's ability to visualize cellular dynamics within intact tissues, researchers can study tumour heterogeneity, cell–cell interactions, and tumour-stroma crosstalk in real-time [41]. ML algorithms, such as support vector machines (SVMs) and random forests, have been applied to classify different cell types within the tumour microenvironment based on their spatial and morphological features extracted from TPM images [42]. These approaches enable the identification of prognostic biomarkers, prediction of treatment responses, and development of personalized therapeutic strategies for cancer patients. Xi et al. aimed to improve breast cancer histologic grade classification using multiphoton microscopy and generative adversarial networks (GANs) [43]. They utilized multiphoton microscopy

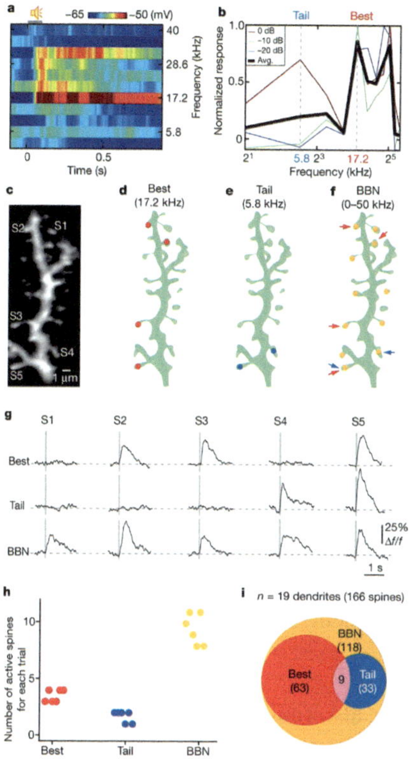

Fig. 8.5 **a**, Colour map showing somatic electrical responses to pure tones at different frequencies (2 kHz to 40 kHz; 10 dB attenuation) in a layer 2/3 neuron. **b**, Summary of electrical responses to pure tones at different frequencies and at different sound levels (0–20 dB attenuations). Two dashed lines mark the tail frequency (light blue, 5.8 kHz) and best frequency (red, 17.2 kHz) used in the following panels. **c**, Two-photon image (average of 6,250 frames) of a dendrite from the same neuron as in panels **a** and **b**. **d**–**f**, Cartoons of the same dendrite as in **c**, showing the location of active spines during three different stimuli: best frequency (**d**, red), tail frequency (**e**, blue) and broadband noise (BBN, **f**, yellow). Red arrows and blue arrows point to spines tuned to the best frequency and the tail frequency, respectively. **g**, Examples of calcium transients in response to the best frequency, the tail frequency and broadband noise from five spines indicated in **c**. Each trace is an average of five individual trials. **h**, Plot of the number of active spines in the dendrite shown in **c**–**f** during six consecutive trials, in response to the best frequency, the tail frequency and broadband noise. **i**, Venn diagram summarizing the total number of active spines for each stimulus (n = 19 dendrites, 8 neurons, 166 visible spines). The number of active spines for each stimulus is indicated in each area. The pink area indicates the spines that responded to both best and tail frequencies. Figure reproduced from [37] with permission from Nature

to capture high-resolution images of breast tissue samples, revealing important morphological features indicative of cancer grade. To automate the classification process, they employed GANs, a type of deep learning algorithm. The GANs were trained to generate synthetic images resembling real breast tissue samples corresponding to different histologic grades. Through iterative training, the GANs learned to discriminate between histologic grades with high accuracy. The automated approach significantly improved classification performance compared to traditional methods, reducing subjectivity and variability in grading.

This innovative method offers a rapid and reliable way to assess tumour pathology, aiding clinicians in making informed treatment decisions and potentially improving patient outcomes. By streamlining the grading process, healthcare providers can tailor therapies to individual patients more effectively. Furthermore, the integration of multiphoton microscopy and machine learning represents a significant advancement in cancer biology research, with implications for precision medicine and personalized cancer care. In another study, Gomes et al. [44] utilized multiphoton imaging, automated image analysis, and machine learning to classify prostate cancer tissue, showcasing the potential of this integrated approach in addressing challenges in cancer biology [44, 45]. By harnessing the power of two-photon excitation microscopy, they captured high-resolution images of prostate tissue, enabling detailed analysis of structural and biochemical features. Through automated image analysis and machine learning algorithms, they accurately classified prostate tissue samples based on cancer presence and severity. This study exemplifies how the integration of two-photon excitation microscopy with machine learning can provide rapid, objective, and precise assessment of cancer tissue, thereby enhancing diagnostic accuracy and guiding personalized treatment strategies in cancer biology research.

Stem Cell Research

TPM-ML integration has also found applications in stem cell research, facilitating the study of stem cell fate determination, lineage commitment, and tissue regeneration processes. TPM enables the visualization of dynamic cellular processes, such as cell migration, proliferation, and differentiation, providing insights into the mechanisms governing stem cell behaviour [46, 47]. ML techniques, including clustering algorithms and dimensionality reduction methods, have been employed to analyze large-scale imaging datasets and identify distinct cell populations based on their phenotypic characteristics [48]. These approaches enable researchers to decipher the complex regulatory networks underlying stem cell fate decisions and develop strategies for manipulating cellular behaviour for regenerative medicine applications. Kong et al. [49] investigated the differentiation status of mesenchymal stem cells (MSCs) using label-free microscopy and machine learning. They developed an innovative approach to assess MSC differentiation without exogenous labels, leveraging the inherent morphological changes associated with differentiation. By combining label-free microscopy with machine learning algorithms, the researchers achieved automated classification of MSCs into distinct differentiation states based on morphological features [49]. This approach offers a non-invasive, real-time assessment of MSC differentiation, enhancing our understanding of stem cell biology and facilitating the development of optimized differentiation protocols for therapeutic applications. Overall, the study demonstrates the potential of label-free microscopy and machine learning in stem cell research, highlighting their role in advancing stem cell characterization techniques for regenerative medicine (Fig. 8.6).

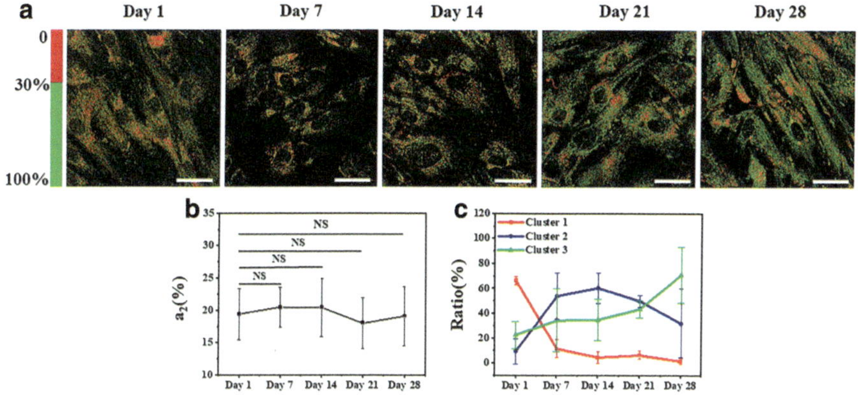

Fig. 8.6 Typical FLIM images and K-means ++ clustering results with three features input of induced adipogenic differentiation from Day 1 to Day 28. **a** FLIM a_2 images (percentage of protein-bound NAD(P)H). Scale bar: 50 μm. **b** The averaged peaks of the a_2 distribution curve at different times. NS: no significant difference. **c** The ratios of the cells in the three clusters. Figure reproduced from [49] with permission from MDPI

Further, Desa et al. [50] investigated the application of label-free optical imaging and sensing techniques for quality control in stem cell manufacturing. Their research focused on the development of non-invasive methods to assess stem cell quality throughout the manufacturing process without the need for exogenous labels or dyes [50]. By employing label-free optical imaging modalities researchers were able to visualize and characterize stem cell cultures in real-time. This approach enabled the monitoring of key parameters of stem cell quality, including cell morphology, viability, and functionality, thereby facilitating the optimization of manufacturing protocols and ensuring the reproducibility and safety of stem cell-based therapies. Overall, the study demonstrates the potential of label-free optical imaging and sensing techniques to enhance quality control in stem cell manufacturing, contributing to the advancement of regenerative medicine and cell therapy applications.

8.6 Conclusion

In conclusion, the integration of two-photon microscopy with machine learning stands as a pivotal advancement in biomedical research, particularly evident in cancer biology. Studies such as those conducted by Chen et al. [37] exemplify the power of this synergy, showcasing the precise characterization of tumour microenvironments through automated classification of breast cancer histologic grade. Similarly, the research by Desa et al. [50] underscores the transformative impact of label-free optical imaging for quality control in stem cell manufacturing, facilitating advancements in regenerative medicine. Furthermore, the investigation by Kong et al. [49] highlights the potential of label-free microscopy and machine learning in evaluating mesenchymal stem cell differentiation status, contributing to the development of optimized therapeutic protocols. These findings collectively underscore the interdisciplinary nature of modern biomedical research, where the convergence of cutting-edge technologies drives innovative solutions for understanding complex biological processes and improving healthcare outcomes. As we continue to explore the synergies between two-photon microscopy and machine learning, we pave the way for groundbreaking discoveries and transformative advancements in cancer biology and beyond.

Acknowledgements The authors thank the Indian Council of Medical Research (ICMR), (Project Number-ITR/Ad-hoc/43/2020-21, ID No. 2020-3286) Government of India, India for financial support. The authors extend their sincere thanks to the Science and Engineering Research Board (SERB), Department of Science and Technology (DST), Government of India, India (Project Number-SERB/MTR/2020/000058) for funding. The authors further thank the Department of Science and Technology, Government of India, India, (grant number: GST/DST/TWN/P-95/2021) for the financial support. The authors also thank Manipal School of Life Sciences, Manipal Academy of Higher Education, Manipal, Karnataka, India for providing the infrastructure and facilities.

References

1. P.T. So, C.Y. Dong, B.R. Masters, K.M. Berland, Two-photon excitation fluorescence microscopy. Annu. Rev. Biomed. Eng. **2**(1), 399–429 (2000)
2. M. Oheim, D.J. Michael, M. Geisbauer, D. Madsen, R.H. Chow, Principles of two-photon excitation fluorescence microscopy and other nonlinear imaging approaches. Adv. Drug Deliv. Rev. **58**(7), 788–808 (2006)
3. F. Helmchen, W. Denk, Deep tissue two-photon microscopy. Nat. Methods **2**(12), 932–940 (2005)
4. R.K. Benninger, D.W. Piston, Two-photon excitation microscopy for the study of living cells and tissues. Curr. Protoc. Cell Biol. **59**(1), 4–11 (2013)
5. L. Garziano, V. Macrì, R. Stassi, O. Di Stefano, F. Nori, S. Savasta, One photon can simultaneously excite two or more atoms. Phys. Rev. Lett. **117**(4), 043601 (2016)
6. F. Shi, J. Wang, J. Shi, Z. Wu, Q. Wang, Z. Tang, D. Shen, Review of artificial intelligence techniques in imaging data acquisition, segmentation, and diagnosis for COVID-19. IEEE Rev. Biomed. Eng. **14**, 4–15 (2020)
7. A.S. Panayides, A. Amini, N.D. Filipovic, A. Sharma, S.A. Tsaftaris, A. Young, C.S. Pattichis, AI in medical imaging informatics: current challenges and future directions. IEEE J. Biomed. Health Inform. **24**(7), 1837–1857 (2020)
8. K.K. Singh, A. Singh, A study of image segmentation algorithms for different types of images. Int. J. Comput. Sci. Issues (IJCSI) **7**(5), 414 (2010)
9. M.J. Iqbal, Z. Javed, H. Sadia, I.A. Qureshi, A. Irshad, R. Ahmed, J. Sharifi-Rad, Clinical applications of artificial intelligence and machine learning in cancer diagnosis: looking into the future. Cancer Cell Int. **21**(1), 270 (2021)
10. T. Hoeser, C. Kuenzer, Object detection and image segmentation with deep learning on earth observation data: A review-part i: Evolution and recent trends. Remote Sensing **12**(10), 1667 (2020)
11. M. Jogin, M.S. Madhulika, G.D. Divya, R.K. Meghana, S. Apoorva, Feature extraction using convolution neural networks (CNN) and deep learning. in *2018 3rd IEEE international conference on recent trends in electronics, information & communication technology (RTEICT)* (IEEE, 2018), pp. 2319–2323

12. W.L. Bi, A. Hosny, M.B. Schabath, M.L. Giger, N.J. Birkbak, A. Mehrtash,..., H.J. Aerts, Artificial intelligence in cancer imaging: clinical challenges and applications. CA: Cancer J. Clin., **69**(2), 127–157 (2019)
13. V. Kistenev Yu, D.A. Vrazhnov, E.E. Shneider, H. Zuhayri, Predictive models for COVID-19 detection using routine blood tests and machine learning. Heliyon. **8**(10), e11185 (2022). https://doi.org/10.1016/j.heliyon.2022.e11185
14. Z. Qin, C. Chen, S. He, Y. Wang, K.F. Tam, N.Y. Ip, J.Y. Qu, Adaptive optics two-photon endomicroscopy enables deep-brain imaging at synaptic resolution over large volumes. Sci. Adv. **6**, eabc6521 (2020)
15. W.Q. Malik, J. Schummers, M. Sur, E.N. Brown, Denoising two-photon calcium imaging data. PLoS ONE **6**(6), e20490 (2011). https://doi.org/10.1371/journal.pone.0020490
16. Bilalli, B, *Learning the impact of data pre-processing in data analysis*. Ph.D. thesis. Universitat Politècnica de Catalunya (Spain, 2018), p. 162
17. B. Bilalli, A. Abelló, T. Aluja-Banet, R. Wrembel, *Intelligent assistance for data pre-processing. Computer Standards and Interfaces* (2017), p. 57. Preprint https://doi.org/10.1016/j.csi.2017.05.004
18. M. Bevk, I. Kononenko, A statistical approach to texture description of medical images: a preliminary study. in *Proc 15th IEEE symp computer_based med systems*, vol. 64 (2002), pp. 239–244
19. D. Cousineau, S. Chartier, Outliers detection and treatment: a review. Int. J. Psychol. Res. **3**, 58–67 (2010). https://doi.org/10.21500/20112084.844
20. F.T. Liu, K.M. Ting, Z.-H. Zhou, Isolation forest. in *Proceedings of the 2008 Eighth IEEE international conference on data mining* (IEEE, Pisa, Italy, 2008), pp. 413–422
21. M. Celik, F. Dadaser-Celik, A.S. Dokuz, Anomaly detection in temperature data using DBSCAN algorithm. in *Proceedings of the 2011 international symposium on innovations in intelligent systems and applications* (IEEE, Istanbul, Turkey, 2011), pp. 91–95
22. Y.V. Kistenev, V.V. Nikolaev, O.S. Kurochkina, A.V. Borisov, D.A. Vrazhnov, E.A. Sandykova, Application of multiphoton imaging and machine learning to lymphedema tissue analysis. Biomed. Opt. Express. **10**(7), 3353–3368 (2019)
23. Y.V. Kistenev, D.A. Vrazhnov, V.V. Nikolaev, E.A. Sandykova, N.A. Krivova, Analysis of collagen spatial structure using multiphoton microscopy and machine learning methods. Biochem. Mosc. **84**, 108–123 (2019)
24. R.M. Haralick, S.R. Sternberg, X. Zhuang, Image analysis using mathematical morphology, IEEE Trans. Pattern Anal. Mach. Intell., **4**, 532–550 (1987)
25. A. Materka, M. Strzelecki, *Texture analysis methods—a review, Technical University of Lodz, Institute of Electronics, COST B11 report* (1998), pp. 9–11
26. J. Bigun, G.H. Granlund, J. Wiklund, Multidimensional orientation estimation with applications to texture analysis and optical flow. IEEE Pattern Anal. Mach. Intell. **8**, 775–790 (1991)
27. Q. Kemao, Two_dimensional windowed Fourier transform for fringe pattern analysis: principles, applications and implementations. Opt. Lasers Eng. **45**, 304–317 (2007)
28. S. Arivazhagan, L. Ganesan, Texture classification using wavelet transform. Patt. Recon. Lett. **24**, 1513–1521 (2003)
29. T. Chang, C.C.J. Kuo, Texture analysis and classification with tree-structured wavelet transform. IEEE Trans. Image Process. **2**, 429–441 (1993)
30. G.M. Farinella, M. Moltisanti, S. Battiato, Classifying food images represented as bag of textons. in *2014 IEEE Int Conf Image Processing (ICIP)* (2014), pp. 5212–5216
31. F.H.C. Tivive, A. Bouzerdoum, Texture classification using convolutional neural networks. in *2006 IEEERegion 10 Conf* (TENCON 2006) (2006), pp. 1–4
32. R.M. Haralick, K. Shanmugam, I. Dinstein, Textural features for image classification. IEEE Trans. Systems, Man, Cybernetics, SMC **3**, 610–621 (1973)
33. L. Sità, M. Brondi, P. Lagomarsino de Leon Roig, S. Curreli, M. Panniello, D. Vecchia, T. Fellin, A deep-learning approach for online cell identification and trace extraction in functional two-photon calcium imaging. Nat. Commun. **13**, 1529 (2022). https://doi.org/10.1038/s41467-022-29180-0

34. S. Soltanian-Zadeh, K. Sahingur, S. Blau, Y. Gong, S. Farsiu, Fast and robust active neuron segmentation in two-photon calcium imaging using spatiotemporal deep learning. PNAS, **116**(17), 8554–8563 (2019). www.pnas.org/cgi/doi/10.1073/pnas.1812995116
35. A. Song, J.L. Gauthier, J.W. Pillow, D.W. Tank, A.S. Charles, Neural anatomy and optical microscopy (NAOMi) simulation for evaluating calcium imaging methods. J. Neurosci. Methods **358**, 109173 (2021)
36. F. Sobhani, R. Robinson, A. Hamidinekoo, I. Roxanis, N. Somaiah, Y. Yuan, Artificial intelligence and digital pathology: Opportunities and implications for immuno-oncology. Biochim. Biophys. Acta (BBA)-Rev. Cancer, **1875**(2), 188520 (2021)
37. X. Chen, U. Leischner, N.L. Rochefort, I. Nelken, A. Konnerth, Functional mapping of single spines in cortical neurons in vivo. Nature **475**(7357), 501–505 (2011)
38. S. Xu, M. Momin, S. Ahmed, A. Hossain, L. Veeramuthu, A. Pandiyan, T. Zhou, Illuminating the brain: advances and perspectives in optoelectronics for neural activity monitoring and modulation. Adv. Mater. **35**(42), 2303267 (2023)
39. K.K. Ghosh, L.D. Burns, E.D. Cocker, A. Nimmerjahn, Y. Ziv, A.E. Gamal, M.J. Schnitzer, Miniaturized integration of a fluorescence microscope. Nat. Methods **8**(10), 871–878 (2011)
40. K. Runge, C. Cardoso, A. De Chevigny, Dendritic spine plasticity: function and mechanisms. Front. Synaptic Neurosci. **12**, 36 (2020)
41. M.J. Pittet, R. Weissleder, Intravital imaging. Cell **147**(5), 983–991 (2011)
42. M. Angelo, S.C. Bendall, R. Finck, M.B. Hale, C. Hitzman, A.D. Borowsky, G.P. Nolan, Multiplexed ion beam imaging of human breast tumors. Nat. Med. **20**(4), 436–442 (2014)
43. G. Xi, Q. Wang, H. Zhan, D. Kang, Y. Liu, T. Luo, S. Zhuo, Automated classification of breast cancer histologic grade using multiphoton microscopy and generative adversarial networks. J. Phys. D Appl. Phys. **56**(1), 015401 (2022)
44. E.F. Gomes, E. Paulino Junior, M.F. de Lima, L.A. Reis, G. Paranhos, M. Mamede, A.M. de Paula, Prostate cancer tissue classification by multiphoton imaging, automated image analysis and machine learning. J. Biophotonics **16**(6), e202200382 (2023)
45. K. Ramamurthy, A.R. Varikuti, B. Gupta, N. Aswani, A deep learning network for Gleason grading of prostate biopsies using EfficientNet. Biomed. Eng/Biomed. Tech., **68**(2), 187–198 (2023)
46. B.H. Zinselmeyer, J. Dempster, D.L. Wokosin, J.J. Cannon, R. Pless, I. Parker, M.J. Miller, Two-photon microscopy and multidimensional analysis of cell dynamics. Methods Enzymol. **461**, 349–378 (2009)
47. G. Raju, S. Nayak, N. Acharya, M. Sunder, Y. Kistenev, N. Mazumder, Exploring the future of regenerative medicine: Unveiling the potential of optical microscopy for structural and functional imaging of stem cells. J. Biophotonics **17**(1), e202300360 (2024)
48. D. Peralta, Y. Saeys, Robust unsupervised dimensionality reduction based on feature clustering for single-cell imaging data. Appl. Soft Comput. **93**, 106421 (2020)
49. Y. Kong, J. Ao, Q. Chen, W. Su, Y. Zhao, Y. Fei, L. Mi, Evaluating differentiation status of mesenchymal stem cells by label-free microscopy system and machine learning. Cells **12**(11), 1524 (2023)
50. D.E. Desa, T. Qian, M.C. Skala, Label-free optical imaging and sensing for quality control of stem cell manufacturing. Curr. Opin. Biomed. Eng. **25**, 100435 (2023)

Chapter 9
Deep Learning in Biomedical Applications of Raman Spectroscopy

Ajinkya Anjikar, Nidhi Prahlad Rao, Rajapandian Paneerselvam, Kolleboyina Jayaramulu, Chandrabhas Narayana, Tatsuyuki Yamamoto, and Hemanth Noothalapati

Abstract Chronic diseases, non-communicable diseases, age-related diseases, the burden of diseases, and pandemics are never-ending problems that humans have been going through since our evolution. The biomedical field is continuously upgrading and making required changes that can decrease the incidence and mortality rate associated with such diseases. One persistent problem still needs to overcome which is the diagnosis of these diseases at its early stages such as at the cellular or molecular level. Efforts in terms of cellular biology such as DNA sequencing, and chromosome testing have been made over the years to understand and resolve the problem but still, we are facing the problem with an exponential increase in the number of these kinds of diseases. An appropriate and concrete revolution is a need of time. Vibrational spectroscopy with recent technological advancements proven a favorable disease diagnostic tool by researchers and practitioners in recent years. Raman spectroscopy, a vibrational spectroscopic tool is a label-free, reliably non-invasive, rapid, easy-to-use, efficient, and sensitive to biomolecular changes in the human body. This book

A. Anjikar
The United Graduate School of Agricultural Sciences, Tottori University, Tottori, Japan

N. P. Rao · T. Yamamoto · H. Noothalapati (✉)
Faculty of Life and Environmental Sciences, Shimane University, Matsue, Japan
e-mail: nvhnag@life.shimane-u.ac.jp

R. Paneerselvam
Department of Chemistry, SRM University AP, Amaravati, Andhra Pradesh, India

K. Jayaramulu
Hybrid Porous Materials Laboratory, Department of Chemistry, Indian Institute of Technology Jammu, Jammu & Kashmir, India

C. Narayana
Chemistry and Physics of Materials Unit, International Centre of Materials Science and School of Advanced Materials, Jawaharlal Nehru Centre for Advanced Scientific Research, Bangalore, India

Rajiv Gandhi Centre for Biotechnology, Thycaud, Thiruvananthapuram, India

H. Noothalapati
Department of Chemical Engineering, Indian Institute of Technology Hyderabad, Hyderabad, India

A. Gogoi and N. Mazumder (eds.), *Biomedical Imaging*, Biological and Medical Physics, Biomedical Engineering, https://doi.org/10.1007/978-981-97-5345-1_9

chapter is a dedicated explanation of the Raman spectroscopic applications assisted by Artificial Intelligence methods which proved a never seen before advancement in disease diagnosis applications. We comprehensively studied all the variants of Raman spectroscopy such as Spontaneous, Surface Enhanced, Coherent Anti-Stokes, and Stimulated forms used in recent years, and their valuable contribution as compared to the standard medical tools. This chapter is intended for readers who want to learn about the deep-learning field of Artificial Intelligence and how it is contributing to the disease diagnosis process with the help of Raman spectroscopy. Through our efforts, we hope that we can trigger the right ruminating spots of the readers which by the end of this chapter can attempt to use and contribute to making the clinical transition of Raman spectroscopy a reality. Finally, we bid to make a revolution in the biomedical field by introducing Raman spectroscopy which can help to lessen the load of diseases to a great extent and make human life worth living as wholesome.

Abbreviations

AI	Artificial Intelligence
CARS	Coherent Anti-Stokes Raman Scattering
CNN	Convolutional Neural Network
DA	Discriminant Analysis
DL	Deep Learning
RS-DL System	Raman Spectroscopy-Deep Learning System
DT	Decision Tree
KNN	K-Nearest Neighbor
LDA	Linear Discriminant Analysis
ML	Machine Learning
NN	Neural Network
PCA	Principal Component Analysis
PLS	Partial Least Squares
ResNet	Residual Neural Network
RF	Random Forest
SERS	Surface Enhanced Raman Spectroscopy
SRS	Stimulated Raman Spectroscopy
CARS	Coherent Anti-stokes Raman Spectroscopy
SVM	Support Vector Machine

9.1 Introduction

At the dawn of Raman spectroscopy for biomedical applications, it was primarily analyzed, studied, and interpreted using statistical means or basic Chemometric methods that, to some extent, are still applicable. Although statistical studies with the help of computational resources were effective, they were not sufficient for the automated application of Raman spectroscopy, especially in the biomedical field. Therefore, there was a need to develop and apply novel analytical methods to make the best use of molecular information contained in Raman hyperspectral data. Based on an extremely strong mathematical foundation of statistical science, Machine Learning (ML) evolved, drawing a separate modelling horizon. With the help of ML, it was possible to create a program to perform analytically exhaustive work and create models that make predictions based on training or learning from a set of samples. Despite certain advances of what the ML paradigm offered for solving problems by analyzing Raman spectra to a marginal extent, it failed to solve an inherently humungous problem known as the "Curse of dimensionality." In general, the fingerprint region of Raman spectra contains > 1000 wavenumbers which contribute to that number of dimensions. In Biomedical applications, the normal range of wavenumbers is between ~600 and 1800 cm^{-1} with 1200 dimensions if each wavenumber is considered one dimension. In addition, ML models generally desiderate human intervention, such as removing unnecessary information from the available Raman spectra, feature extraction, dimensionality reduction, etc. Furthermore, Raman spectra are too complex, and it is recommended to measure and analyze them by following a particular protocol [1]. Thus, it was exigent in biomedical applications of Raman spectroscopy to reduce human-made errors to make accountable decisions in crucial situations such as disease diagnosis [2].

In recent years, due to the advent of computational architecture and technological development, it has become possible to handle high-dimensional data, also termed "Big data analysis," with the help of Deep Learning (DL) such as Neural Networks (NN). DL helps to overcome the problem of human intervention, is self-capable of feature extraction, can perform analysis with or without pre-processing of Raman spectra, and provides human comparative classification accuracy. Figure 9.1 shows the recent trend in the form of the number of publications between 2017–2022 in two major databases ('PubMed' and 'Web of Science'). The number of publications was obtained by using keywords such as biomedical, disease diagnosis, Raman spectroscopy, Surface Enhanced Raman spectroscopy, Coherent Anti-Stokes Raman spectroscopy, and Stimulated Raman spectroscopy by using appropriate combining functions (AND, OR). It is clear from Fig. 9.1 that applications of DL in Raman spectroscopy by researchers and practitioners are steadily increasing. Based on these trends, this chapter intends to discuss the fundamentals of DL, the reasons behind its popularity, and the recent revolutionizing work done by researchers in the field of biomedical applications using Raman spectroscopy in detail.

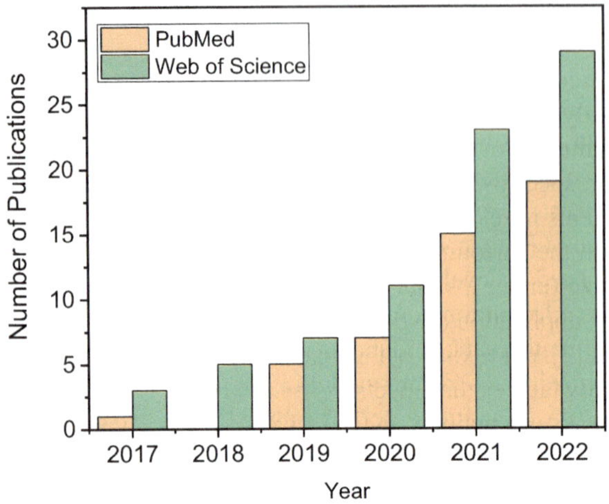

Fig. 9.1 Number of publications per year that employ Deep Learning for biomedical applications of Raman spectroscopy between 2017–2022 in 'PubMed' and 'Web of Science' databases (Keywords used include biomedical, disease diagnosis, Raman spectroscopy, Surface Enhanced Raman spectroscopy, Coherent Anti-Stokes Raman spectroscopy, and Stimulated Raman spectroscopy by using appropriate combining functions (AND, OR)

9.2 Fundamentals of Deep Learning

9.2.1 Building Blocks of Deep Learning

The sole factor due to which the Deep Learning field offers many advantages over traditional analytical methods is its building blocks, such as perceptron, which eventually enclose the Deep Learning field with its various variants and applications. It is also known by the term "Artificial neuron" as it mimics the functions and properties of biological neurons. In a biological neuron, the signals received by the body part such as touch or smell get translated to quantifiable inputs and give the impact in the form of output whether and how to react in response to those inputs [3]. The first attempt to simulate the activity of biological neurons was made in 1899 in terms of electrical cable behavior and electronic analogs [4]. The operation can be modelled in computational terms where artificial neurons try to gauge the intensity of those inputs concerning outputs and train a model in a supervised manner to predict unseen data. Figure 9.2a and b shows a representative Raman spectrum to understand the working of the building blocks of DL. The wavenumbers are shown in terms of inputs. x_1, x_2, ... x_n. The working of a single Perceptron is shown in Fig. 9.2c, where it takes the inputs, makes a weighted average based on randomly assumed weights/coefficients, and the neuron fires an output response y based on the intensified input features [5] by using an activated function. Activated functions using mathematical operations can be used in the form of sigmoidal functions, rectified linear unit functions, etc.

However, the Single Perceptron considers all features as independent inputs. Sometimes it is necessary to find the relative patterns in the Raman spectra, so a loop in the Perceptron Unit is desired, which can remember the previous input/state for the correct prediction. For such problems, the Recurrent unit is used, as shown in Fig. 9.2d. It performs similarly in operation to a Perceptron but has the additional quality of making a note of previous input by following the concept of "Memory." Although the Recurrent unit overcomes the problem of perceptron, it lacks credibility when the unit needs to remember more step size of inputs (for example: more than 5–10 wavenumbers region). Therefore, two more units, Long Short-Term Memory (LSTM) and Gated Recurrent Unit (GRU), can be used. In addition to Recurrent unit operation, these two units offer gates to retain the input information for sequential data and make pattern recognition operation effective. Figure 9.2e shows the GRU unit, and Fig. 9.2f shows the LSTM unit.

GRU has two gates, namely, the update gate and the reset gate. The update gate is responsible for deciding which information to retain for a long period and which information is irrelevant to the data fed to the unit. The reset gate decides the amount of past information to be retained and is useful for making the prediction. On the other hand, LSTM uses cell and gate concepts like RNN and GRU but in a more sophisticated manner. The cell state has a forward propagation to carry forward the information, and gates, however, are responsible for either adding or removing the information to ongoing input. This can be achieved in a manner as follows: The LSTM consists of three gates that are input, forget, and output. The forget gate determines

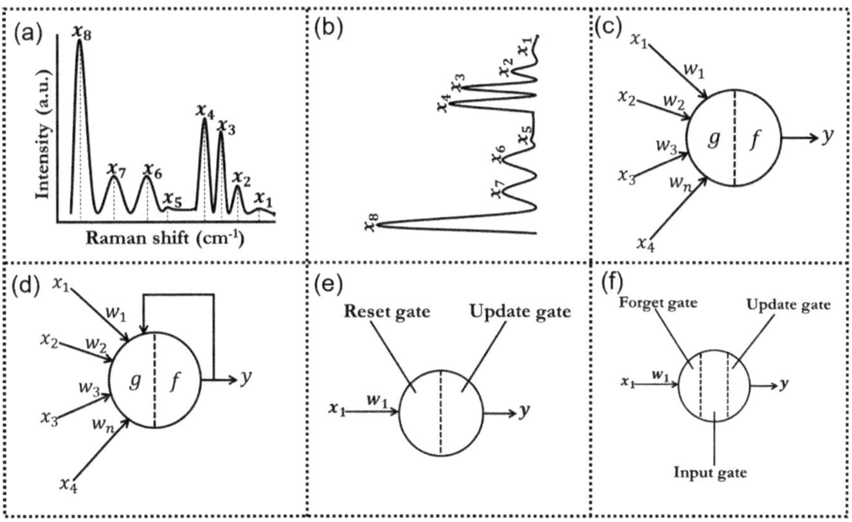

Fig. 9.2 Building blocks of deep learning. **a** Representative Raman spectrum, **b** Raman spectrum as an input with features in terms of wavenumbers fed to Neural Network model, **c** Perceptron or artificial neuron, **d** Recurrent unit, **e** Gated Recurrent Unit (GRU) and **f** Long Short-Term Memory (LSTM) unit

the information to add to the current input, the Input gate does the same to the current input, and the output gate updates the previous input storage for future inputs. GRU does not perform the Forget operation of LSTM separately and performs it in the Update gate, making less computation. Building blocks are fundamental units that can be used to form a network called a Neural Network to handle high-dimensional data.

9.2.2 How Does the Network Form?

The fundamental units form a network similar to the connectionism mechanism [6]. Terminology is used to make a network of any number of units. The vertical stack of the building blocks is called a layer. The layer which receives the input from the Raman spectral data is called the Input layer. Generally, it equals the number of features present in the Raman spectrum. The layer that provides a prediction is called as Output layer. The layer/layers in between the Input and Output layers is/are called Hidden layers/layers. Each unit in each layer performs the same operation but with updated forms of input and weights from the previous units. A schematic of a neural network is shown in Fig. 9.3. Neural Networks can be imagined as a pattern recognition tool to model or formulate nonlinear, complex relationships present within the data[7]. It is possible that the data can be fed to the Neural Network for as much time as possible via a technique called Epoch, as shown in Fig. 9.4.

In this way, we make sure that the model has enough time and computational feasibility to learn all the possible patterns present in the Raman spectra. At the end of one forward propagation, the model prediction gets cross-verified to the actual target labels using a loss function, and the task of achieving the global minima of the loss function is achieved by gradient across all layers in the backward propagated manner by using Gradient Descent based Optimizer. In a deep neural network where the number of hidden layers is more than one, a gradient can become zero, which halts the backpropagation operation. This is also known as the vanishing gradient problem. To avoid this problem, a constant in terms of Bias is always added to each of the neurons in addition to the input feature. The forward and backward propagation can be viewed with the help of Accuracy and Loss curves derived based on the Epochs, which can be used to judge the performance and quality of the model. Since all the neurons perform similar operations, the model tends to overfit the data, and thus, a regularization parameter can be used in the form of Dropout and Batch Normalization. In Dropout, selectively or randomly, some of the neurons from any of the layers are restricted to firing any output after the activation function. In Batch Normalization, the data gets normalized before entering any of the layers except the output layer so that the weights get the same distribution and also helps to train the network easily. The best Gradient Descent-based Optimizer, activation function, batch size, number of epochs, number of hidden layers, weight initialization method, dropout rate, and number of neurons in the hidden layer are all hyperparameters of

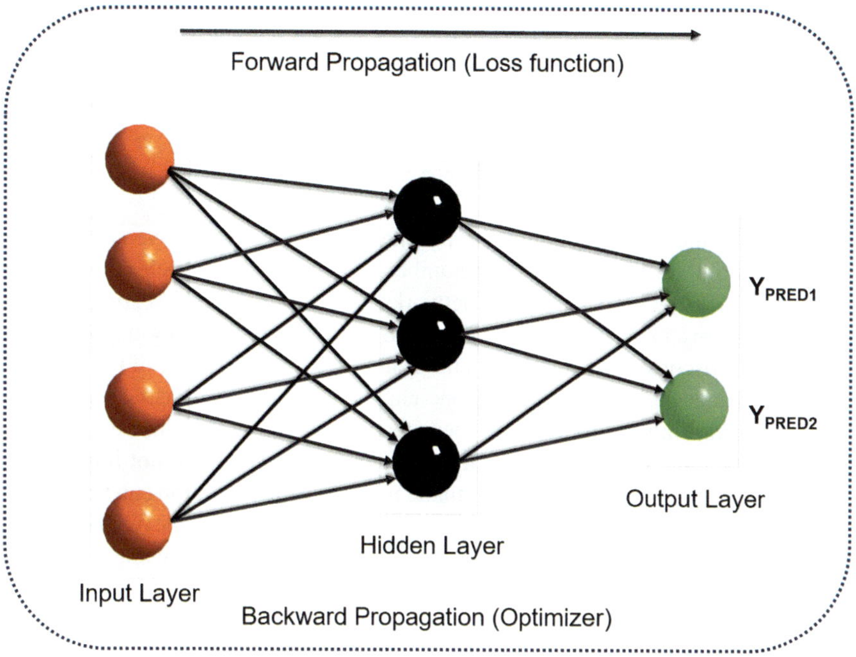

Fig. 9.3 Schematic of neural network with any of building blocks perceptron or artificial neuron, recurrent unit, GRU, or LSTM. The vertical stacking of such units is layer namely input, hidden, and output which receives, computes, and predicts the Raman spectra respectively

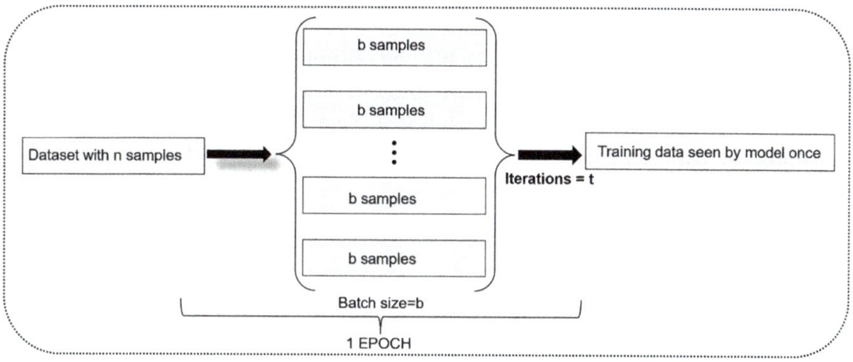

Fig. 9.4 Understanding of iterations, batch size, and epoch of DL model. The dataset with n samples or Raman spectra was divided into batches of arbitrary size b and fed to the neural network to avoid overfitting

the model, which may vary with the nature of Raman spectral dataset and can be tuned using any hyperparameter optimization techniques.

9.2.3 Is ANN Capable of Solving All Problems?

Theoretically, the Perceptron model with one hidden layer can solve any complex real-world problem with the desired number of neurons. Nevertheless, it also has some drawbacks. As we know, the networks given in the above section learn all the features present in the Raman spectral dataset. In the imaging application of Raman spectroscopy, we could have a thousand spectra at one point of tissue/organ/cell. Also, images are not necessarily one-channel data and might consist of several channels. Such Raman spectral data needs feature extraction to reduce the unnecessary computational cost for the features/wavenumbers, which are not important in making a decision/prediction. An automatic mathematical operation can be incorporated before the actual Perceptron model so that it can only handle the main important features. One such operation is Convolution by using a kernel. By using a Convolutional kernel of any desired shape, we can introduce sparse interactions and weight sharing [8], which cannot be possible to achieve in a simple neural network. One more advantage of the Convolution block is the Pooling layer which helps to reduce the dimension, which is uncommon to other ML methods, and thus, feature extraction can be performed without any human intervention. A flattening layer followed by a sub-sampling layer sends the important features to the simple neural network architecture to make the decision. In addition to the hyperparameters mentioned for the simple neural networks, the number of convolution filters, kernels, and convolution blocks can be optimized to obtain a stable and optimum model based on the Raman spectral dataset in consideration. There are a few variants of convolution blocks that overcome the problems in Convolutional Neural Networks (CNN), including vanishing and exploding gradients such as the Residual block and Inception module, which we will briefly discuss next.

9.2.4 The Reason Behind the Real Hype of Convolution

Convolutional Neural Network involves convolution operation. Let's assume a 2-dimensional image, and then Convolution means a matrix multiplication by a kernel or any size (generally 3×3 or 5×5). A general example of understanding the convolution operation is using filters to change the appearance of the photo on a smartphone or any other device. This operation can be done by multiplying each pixel value by a filter which can perform matrix multiplication and convert the original image to any other shade. This is possible because the pixel values of any image are fixed within a certain range and are universal. The kernels can be viewed as "receptor fields"; thus, we can detect the colors which were first claimed by Hubel and Wiesel in 1962 on the visual cortex cells of cats [9]. So, with the kernel of size, for example, 3 × 3 as shown in Fig. 9.5a with red boundaries moving through 9-pixel values at a time, hover over the image and produce a filtered image, and the operation is called Convolution or a dot product which results in sparse connections. It is obvious that when the input is one-dimensional, the Convolution and other parameters are also one-dimensional. Stride is known as a value by which the filter hovers over the image or pixel values. Figure 9.5b shows Convolution with stride equals 1. Padding is also sometimes used to keep the dimension of input the same as the original input. Figure 9.5c shows Convolution with zero padding. A pooling layer, sometimes called a sub-sampling layer, follows the convolution layer, which is the dimension or spatial size reduction process. There are two types of pooling that can happen, average pooling and max pooling, with another filter size not necessarily the same size as that convolution filter size. In average pooling, the average of pixel values presents in a filter size (say 3 × 3) is calculated and generates a new image size with reduced dimension, while in max-pooling, the maximum value is taken among those nine values and results in a new image dimension. A Batch normalization layer and an Activation layer generally followed the pooling layer. Thus, the process, including Convolution, pooling, batch normalization, and activation, can be termed a convolution block, as shown in Fig. 9.5d. Afterward, a feature map obtained from the sub-sampling layer is now utilized by the simple Artificial Neural Network as discussed before but this time only with valuable features. Finally, an output layer is present, giving an output predicted by the model [8]. By stacking these convolution operation blocks, some renowned architectures have been made, such as LeNet [10] AlexNet [11], and VGG [12], which provide better performance in image recognition tasks. Forward and backward propagation, Epochs, batch size, and other hyperparameters are the same as those discussed for Artificial Neural Networks. Interestingly, backpropagation operation is not limited to changing the weights or gradients within the single network; it can also be used to upgrade another network by passing the information. Generative Adversarial Networks (GANs) use the same trick, and two networks operate in an adversarial manner. Figure 9.5e shows a block diagram of GANs in which the network that gives the gradient information is called the Discriminator model, and the one that uses it is called the Generator model. The architectures of both models can be of any choice and depend on the requisite application. The Discriminator uses ground truth and data generated by the Generator to classify as a simple binary classification problem. The generator model, however, tries to fool the Discriminator by learning the weights from the trained Discriminator and attempting to make input to same as ground truth to a realistic extent [13]. For example, the Generator is provided by a simple Gaussian distribution noise and wants to make a Raman spectrum of silicone from that random noise. The Discriminator in the first encounter can easily distinguish the random noise and Silicone Raman spectrum, and thus, the weights at this point of the Discriminator get frozen and passed to the Generator, which trained according to the learned weights and eventually produces the Raman spectrum from random noise to the realistic similarity to actual silicone Raman spectrum. With the increasing popularity of Convolution, a steep increase in introducing deeper networks was obvious, and it became a custom to add a greater number of layers to form a deeper network and try to solve the problem. However, the approach came with a cost of computation load and a serious problem of vanishing gradients. As a result, Christian Szegedy et al. proposed the network, which is famous by the name "GoogLeNet," which uses Convolution of various scales (different filter sizes) and then aggregates the abstract features. The idea is to start with a simple convolution block/s depending on the requirement and data at hand, and then.

the output of one layer feed to a simple 1×1 independently, 1×1 followed by 3×3, pooling followed by 1×1 Convolution, and 1×1 followed by 5×5 Convolution forming nested convolutions and finally concatenating all filters output and forming an "Inception module" [14] as shown in Fig. 9.5f. It is also possible that the network is only comprised of convolution operations and without a dense layer that can generate the same size of images as that of input by "convolutionalization." Such networks are called Fully Convolutional Networks or also famous by the name "U-Net." The input image downsamples the same as normal convolution block operation to a basic spatially important size, and then it upsamples to the original size, but this time, only the spatially important region is highlighted because of convolution and deconvolution operations [15]. Applications such as semantic segmentation, spatial importance, and pixel-wise analysis are possible with such networks. In Raman spectroscopy, this can make significant importance as we always measure cellular data in the presence of media, which interferes with the analysis and classification. With the help of U-Net, we can get rid of background or less important information and work on only patches with meaningful biomolecular information. CNN is the most powerful neural network, but as we increase the number of layers or convolution blocks or stacked convolution operations, the network gets prone to vanishing gradient problems more severely, and the model also tends to have saturation in the performance, which leads to improper training of the deeper model thus, the residual block introduced by He et al. [16] can be used in such operations shown in Fig. 9.5g. The input from one convolution block is sent to two destinations, one is to the next convolution block for regular operation, and the other one is directed to the output of that layer and thus introduces a "skip connection." This results in fewer parameters than the plain CNN and better performance [17]. Lastly, one model fully trained and fully learned the weights of one dataset could be used on another dataset as it is with the concept of Transfer Learning [18], as shown in Fig. 9.5h. Since we discussed deep neural network architectures, it also needs loads of data, and in terms of Raman spectroscopy, getting a huge amount of data is a serious problem which is an obstacle to making such models with only Raman data all the time. Thus, Transfer Learning can be implemented to use an already trained model on some other dataset and can be used in Raman spectroscopic applications. This technique can also reduce the Bias since the dataset is not confined to any particular type but is a mixture of two or many types of populations. Transfer Learning seeks validation, especially in Raman spectroscopic study, and needs more modifications in order to become useful in actual Raman applications.

As we discussed various combinations possible due to convolution operation, we can see why such models are gaining popularity in solving the number of problems

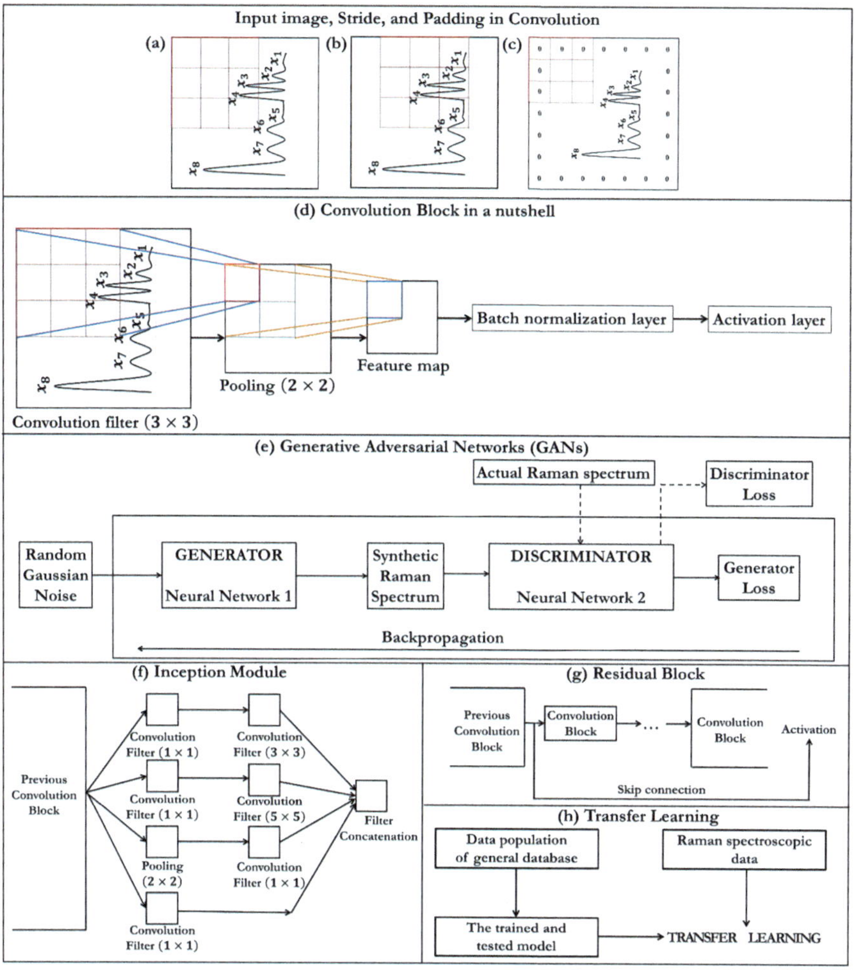

Fig. 9.5 Convolution operation and block **a** A representative 2-dimensional image of Raman spectrum with a highlighted kernel size of 3 × 3, **b** A representative 2-dimensional image of Raman spectrum with a highlighted kernel size of 3 × 3 with stride equal to 1 which hovers over the whole image, **c** A representative 2-dimensional image of Raman spectrum with a highlighted kernel size of 3 × 3 and zero padding which keeps the dimension intact to the original image size after the convolution operation, **d** Convolution block comprising of convolution operation, pooling, batch normalization, and activation, **e** Schematic of Generative Adversarial Networks (GANs), **f** Schematic of Inception module, **g** Schematic of Residual block, **h** Schematic of Transfer Learning concept

that were not possible or were difficult to solve before the DL revolution. Some interesting applications are discussed in the rest of the sections in all the possible variants of Raman spectroscopy one by one. The idea is to have a brief understanding of the research progress done in each Raman variant by using the DL methods so that a comprehensive idea would be gained in order to solve or handle a similar problem by the students, practitioners, and researchers in the relevant field.

9.3 Applications of Deep Learning: Spontaneous Raman Spectroscopy

Raman Spectroscopy (RS) has gained significant attention in both fundamental studies and advanced medical diagnostics for its ability to characterize biological samples such as single cells, tissues, and body fluids such as serum [19], urine [20] and saliva [21] at molecular level with high specificity and sensitivity. Multiple valuable processes such as cell differentiation [22, 23], tissue classification [24, 25], bodily fluid identification [26] and cellular imaging [27–29] have been made more robust with Raman Microspectroscopy. However, there are limitations while analysing data and to combat the possible downsides that RS could pose, Deep Learning has been utilized to improve efficiency and reliability of the technique.

9.3.1 Disease Diagnosis

Raman spectroscopy has been employed to make diagnostic decision for disease of various organs (Fig. 9.6). Especially, the application of deep learning for both non-communicable and communicable diseases are summarised in this section.

9.3.1.1 Non-Communicable Diseases: Cancers

Over the years, accurate cancer diagnosis has been an extremely important area of research using RS. Especially, Raman Microspectroscopy coupled with CNNs has proven to be useful for region-based cell discrimination in tongue squamous cell carcinoma (TSCC). Two convolutional kernels were used in the CNN model, which reduces noise generation and increases distinguishing ability. The model achieved 99.2% sensitivity and specificity, while also being the highest accurate model to discriminate TSCC. The spectral range of 1380–2250 cm^{-1} was observed to provide a large impact on the result of the study. This work is beneficial to guide marginal resection during surgery, as well as reduce the sampling problems and evaluation time during surgery. Furthermore, it can lessen the necessity for adjuvant therapy [30]. RS has various benefits as a standalone imaging method for disease diagnostics but using it in combination with other optical and imaging methods can combat the possible disadvantages of RS and improve results multi fold. One study has thrown light on basal cell carcinoma diagnosis where RS has been used in tandem with Mohs micrographic surgery. It has been observed that the specificity of RS is lacking due to high

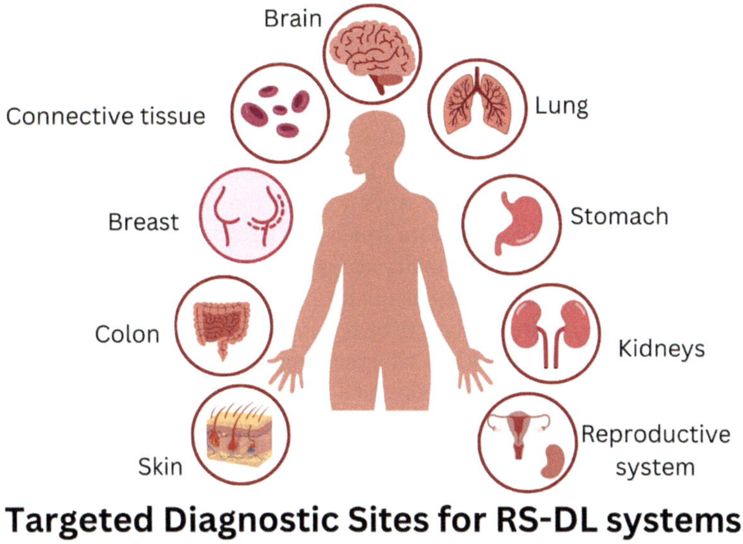

Fig. 9.6 Targeted diagnostic sites for RS-DL systems

spectral similarity between diseased tissue and normal tissue for basal cell carcinoma. To improve this specificity of RS, Reflectance Confocal Microscopy (RCM) has been used to identify morphological and cytological features of skin tissue such as details on the epidermis and hair follicles. A two-step classification model was created where firstly, a successful Raman based model in a previous work that was used to classify BCC tumours from normal tissue was employed. 191 RCM images were obtained from the same sites as Raman data and were implied for two ResNet50 networks. The networks identified the structures of hair and epidermal images positively corresponding to the predictions of the previously successful Raman based model. The specificity of this previous model was enhanced by shifting Raman spectra corresponding to selected images from false positives to true negatives. The new model removed 52% false positive predictions from the previous model while maintaining a 100% sensitivity. The specificity showed an improvement to 92.4% when integrating Raman spectra with RCM images compared to the previous 84.2%. This study has been acknowledged to be a promising tool for guiding tumour resection surgery [31].

Chondrogenic cancers are a rare type of cancer that plague the bones and cartilage of the body. These cancers are usually differentiated based on their appearance and are characterized in-vitro through biopsy techniques but need the expertise of a bone tumour pathologist for highly accurate diagnosis. Other technical limitations with laboratory methods might also be in the way of an efficient diagnosis. A study by Russo et al. has brought attention to enchondroma and chondrosarcoma diagnosis using a RS-DL system. The classification between enchondroma and chondrosarcoma is an essential step in prognosis and the following therapy prescriptions. The morphology of the tissue must be investigated thoroughly for classification. To achieve this, the authors have created a chondrogenic tumour classification through wavelet transform of Raman spectra (CLARA) by using CNN and ResNet18 to identify cancers of the bone accurately. This model was trained and tested on a dataset of 10 samples from patients with different chondrogenic tumour stages such as benign enchondroma as well as 3 malignant grades: First grade, second grade and third grade respectively. A hybrid 1D-2D classification has been implemented in this model on both raw Raman spectra as well as after wavelet transformation. The 1D model receives the pre-processed Raman data which acts as a classifier for all the samples. The samples that are identified as malignant are then transformed through a hybrid time frequency transformation, producing an RGB 2D image. This image is input into a pre-trained deep model to classify benign enchondroma and first grade cancers. CLARA has been noted to identify and grade tumours with 97% accuracy, and can differentiate between benign, stage one and a joint cluster of stage 2 and 3 malignant grade cancer. The authors note Raman signals for an increase in the DNA/RNA due to cellular proliferation along with progressive degradation of collagen at 728 cm^{-1}, 830 cm^{-1} and 1206 cm^{-1}, which characterizes progression in the stages of cancer. These results present to us the valuable time-associated changes that can be recorded in the progression of disease by implementing RS-DL in disease diagnosis [32].

In a work by Zhu et al., it is discussed that using deep learning to improve the accuracy of diagnosis comes with multiple drawbacks such as quality and quantity of samples, which are main factors governing the accuracy of the diagnosis. Along with this, the acquisition of samples is risky due to ethical constraints. A small sample size might lead to overfitting in the machine learning model and hence the authors have chosen to use Generative Adversarial Networks to create artificial Raman data that resembles training data. The concatenation of real data with the generated synthetic data was used to train classification modules, and this paper successfully discusses implying this knowledge of both creating synthetic data as well as data augmentation to improve cancer diagnosis [33]. Like other works reviewed in this book chapter, we can observe that medical diagnostics using Raman Spectroscopy-Deep Learning or RS-DL systems especially CNNs have shown stellar results.

The successful classification of glioma through serum samples has been established by using multiple DL processes to Raman spectra in one such study. Gliomas are brain tumours that have a poor prognosis rate and early diagnosis can aid in 38 glioma samples and 45 healthy samples were subjected to Raman spectroscopy and then partial least squares or PLS analysis for dimensionality reduction. The Raman peaks of healthy controls and patients were mainly observed at 568, 645, 724, 820, 950, 1004, 1156, 1190, 1339, 1448, 1515, and 1671 cm^{-1}. Data augmentation was done by adding different levels of noise to the reduced data and then classification was achieved by AlexNet, ResNet and GoogleNet fine tuning models. Fivefold cross validation was used to evaluate the performance of the models and the accuracy rates of the AlexNet, ResNet, GoogleNet fine tuning models were 98.50, 98.24 and 99.50% respectively. GoogLeNet clearly presents the best classification results, and its specificity and sensitivity are at 98.98 and 98.48% respectively. The area under the receiver operating characteristic (ROC) curve (AUC) of the diagnostic model was 0.9949. Since deep learning has a poor classification effect on small sample size of data, data augmentation can be a method used to generate more valuable data from existing data to increase the performance of the model. Gaussian noise has been rightfully employed in this study to do the same [34].

To inch a step further, a correlative study between serum samples of both lung cancer as well as glioma patients has been achieved. Since both cancers are common forms of malignancies in humans and there is a high probability that these similarities are reflected in the blood serum, the authors found it important to analyse the relationship if any between the two. The study has also focused on creating an instantaneous and low-cost method of screening to reduce misdiagnosis of both types of cancer. Raman peak similarities between both diseases have been found at 568, 645, 727, 822, 1004, 1157, 1190, 1448, 1517, 1520, and 1669 cm^{-1}. Peak differences were seen at 822, 1004, 1343, 1448, 1520–1528, and 1669 cm^{-1}. Raman spectra obtained from samples from lung cancer, glioma patients as well as control samples were used to extract features to create the CNN model. Layers of varying decibels of white Gaussian noise were added to train the data to enhance the data. This data was implied to multilayer perceptron (MLP), recursive neural network (RNN), convolutional neural network (CNN) as well as AlexNet to prepare models for diagnosis. The PLS-AlexNet model created was deemed the best as the accuracy of the model for parameters like classifying lung cancer and control, lung cancer and glioma as well as glioma and control which were 99, 95.2 and 100% respectively. The experimental accuracy for the tri-classification algorithm is above 85%. Multidirectional research such as this truly enhances the scope for cancer diagnoses [35].

Direct investigation at the site of tumour using RS is an effective diagnostic method but some conditions are much harder to diagnose due to difficulties in locating the tumour, possibility of inflicting trauma during invasive surgery or tumour size. The possibility of low penetration into tissues for in-vivo diagnosis using other spectroscopic techniques is also present. To combat such problems, a paper by Vrazhnov et al. has discussed discovering glioma tissues through biomarker detection in blood using an RS-DL system. The most common model of human glioma cell lines, U87 was transplanted intercranially into mouse brains. The separability of experimental and control groups for various stages of glioma after introducing glioma cells was identified in blood serum using deep learning methods. Informative Raman bands were also identified through this study. SVM, RF and XGBoost were selected for the study due to their generalization ability and allowance for selecting informative features. Spectra were divided into 4 groups based on weeks of the experiment as U87-1, U87-2, U87-3, and control. The groups' separability was calculated visually using PCA. The tumour and control groups showed both similarities and differences [36].

Although blood serum is a very popular and convenient sample used for cancer diagnosis, there have also been works involving the use of urine samples. Liu et al. have shown an interesting study on distinguishing early prostate cancer (PCa) and benign prostatic hyperplasia (BPH) since they are similar on pathological grounds through the prostate-specific antigen (PSA) test as well as radiological detection. To avoid traditional diagnostic methods such as biopsies, investigating urine samples through Raman Spectroscopy seems an ideal option. The differences between the intensities of certain characteristic peaks are 725, 812, 959, 1014, 1094, 1134, 1247, 1333, 1357, and 1467 cm^{-1}. All these peak differences were analysed for both diseases after being characterized into different macromolecule assignments such as lipids, proteins, and amino acids. The abnormal metabolic activity caused by PCa was identified and the data thus acquired was used to create a CNN model with cross validation whose results present mean diagnostic accuracy, sensitivity, and specificity for PCa as 74.95, 77.32 and 72.46% respectively [37].

Colorectal cancer can be caused by defective DNA mismatch repair. This condition can be identified by the resulting instability of microsatellites, used as a biomarker for colorectal cancer. This method of diagnosis is stated to not be not popular due to clinical guideline constraints and resource constraints. To develop a new method to detect it, an RS-DL system has been created by Thomas et al. The study focuses on discussing the potential of RS as well as a 1D CNN to categorize normal, microsatellite stable (MSS) and microsatellite unstable (MSI-H) adenocarcinoma in human colorectal samples. The CNN model that was created was also compared to a principal component analysis-linear discriminant analysis and a support vector machine model. Cross validation was applied to train a total of 30 samples, 10 from each condition, for 1490 spectra. The CNN model that was created showed a sensitivity of 82% and specificity of 51% while the comparative PCA-LDA model showed a sensitivity and specificity of 82 and 51% respectively. An occlusion study was done on the CNN to highlight particularly active areas of the Raman spectra and it was found that the region between 680 and 1020 cm^{-1} was significant to the disease condition. The authors of this study have also explicitly stated that the scale of this study is too small to be implemented clinically, but is a stepping stone to a more robust implementation of such RS-DL systems clinically in the future [38].

Scientists have been exploring possibilities of label free diagnosis especially with respect to breast cancer. An attempt on the same lines has been reported by Selvarani et al. who have created a RS-DL system that uses a hybrid deep learning neural network model to characterize breast cancer blood plasma from control samples and use the distinguishing spectral features obtained as biomarkers for breast cancer. 6 HDL models were implemented on 62,199 Raman spectra from 26 plasma samples, namely: 1D-CNN-GRU, CNN-BiLSTM-AT, 1D-CNN-LSTM, GRU-LSTM, RNN-LSTM, and OGRULSTM. The results of the study show that the 1D-CNN-GRU model outperformed the other models for the detection of breast cancer with a classification accuracy of 98.90%, Cohen-kappa score of 0.941, F1-score of 0.969 and lowest score of test loss at 0.102776 and MSE at 0.0230. This study has highlighted the improvement in using RS-DL models off late with respect to the use of probabilistic techniques along with the advantages of deep learning [39]. Kothari et al. have taken up the approach of using stochastic neural networks "for transparent, statistically rigorous, probabilistic classification of healthy and tumour tissue". They have rightfully observed that the shifts in signals that are usually considered by researchers for diagnosing breast cancer are possibly so due to losing lipid signals instead of detecting pathognomonic shifts in the composition of nucleic material or protein. In order to test this hypothesis, they have combined Raman spectroscopy with unsupervised k-means as well as stochastic non-linear neural networks. The NN were first trained on Raman spectra with human manual histopathology diagnostics as a "gold standard". The K-means prediction using spectral data was compared to histopathology and produced 93.2–94.6% accuracy, 89.8–91.8% sensitivity and 100% specificity. NNs trained on this k-means prediction showed probabilities of correctness for autonomous classification. The autonomous system displayed an extended dataset from the original 74 spectra, 5 patients to 203 spectra, 8 patients. This shows that the stochastic nature of NNs allowed the generation of multiple models of tissue classification. They have also observed an increase in DNA/RNA signal intensity in the fingerprint region of 600–1800 cm^{-1} and global loss of high wavenumber signals at 2800–3200 cm^{-1} to be especially sensitive Raman signs of tumour, which presents a promising outlook for RS to be used in a similar method for diagnosing other diseases as well [40].

Study by Yin et al. has implemented a polarized RS system to obtain richer structural information from breast cancer affected tissues. This study has shown that the orientation of collagen fibres in the breast tissue shifts from a parallel conformation to a vertical conformation due to cancerization and these changes are consistent with previous studies. As tumour progression is observed, the anisotropic conformation of collagen in the cancer weakens. To diagnose breast cancer more accurately, this study applies a polarized method of Raman Spectroscopy where Raman spectra are acquired at a set of linear polarization angles ranging from 0 to 360°. The resulting data is input as a matrix into a 2D CNN framework. This system has shown an average discrimination accuracy of 96.01% for test samples and is better than the results obtained from a KNN classifier and 1D CNN which are both based on non-polarized Raman spectra. This study experimentally demonstrates the potential of polarized RS for breast cancer research, since results from the polarized spectra showed a superior capacity of reliable diagnosis compared to non-polarized spectra. The 2D CNN extracted the characteristics of the spectra and learnt spectral changes under different polarization angles, which proves its superiority over 1D-CNN models [41].

9.3.1.2 Non-Communicable Diseases: Neurodegenerative Disorders

A select instance of identifying biomarkers through RS-Deep Learning is reported in a study for Alzheimer's Disease (AD) using Graphene-Assisted RS. Rapid and accurate identification of biomarkers is essential in AD diagnosis. Raman spectra of mouse brains with and without AD were classified using machine learning in this study. Monolayer graphene was contacted with brain slices to increase accuracy from 77 to 98% in machine learning classification. This increase in accuracy is due to noise reduction, Raman signal enhancement and lowered laser heating effect of graphene. Raman fingerprints for AD biomarkers were better captured in the cortex region with this graphene assistance, which shows that RS was more sensitive to AD related biomolecules in the cortex than the hippocampus and thalamus. The spectra obtained were subjected to different algorithms including linear SVM, random forest, XGBoost and CatBoost. Since Raman spectra for AD and non-AD brain slices are visually similar, machine learning methods can identify minute differences and classify the two with increased accuracy [43].

9.3.1.3 Communicable Diseases

Although RS-DL methods are popular for non-communicable disease diagnostics, we can confidently say that they have been applied even for communicable disease, as seen in this paper by Hassan et al. They have created an efficient ResNet deep learning model based on Raman spectra of human blood sera to diagnose dengue virus infection. This model was modified by using transfer learning on spectroscopic data. The proposed model was evaluated on 2000 Raman spectra of which 1200 were Dengue virus infected serum samples and 800 were healthy. An accuracy of 96% on testing data was observed for dengue infection diagnosis. A minimum improvement of 6% in terms of AUC and 7% in terms of Kappa index was recorded over other advanced methods. The success of this model is attributed to the better residual learning capacity and generalization ability compared to other DL based models [43]. Blood plasma based diagnostic approaches are quick, cost effective and a convenient way to apply Micro Raman spectroscopy. Some diseases like HPV are easily transmitted to caretakers and show false negatives in conventional tests due to various reasons. A novel method of diagnosis of Hepatitis B virus using ResNet101 CNN has been reported and is able to detect minute variations between infected and healthy samples. This model showed a high rate of detecting even minute differences between healthy and infected samples, while outperforming PCA-SVM and PCA-LDA models. It is revealed that the developed model achieved high classification accuracy, sensitivity, and AUC of 99.7, 100, and 98.7%, respectively. The authors attribute the success of the method to the better generalization and learning capability of the model and note that this method can be extended to other disease diagnosis as well [44].

Membranous nephropathy is another disease that has been explored in terms of RS-DL systems. The fatality rate in patients with membranous nephropathy is high since it progresses to end-stage renal diseases such as renal failure or uremia. Like other life-threatening diseases, the diagnosis of membranous nephropathy is usually done by identifying the symptoms that patients display, as well as pathological examinations like kidney biopsies that are costly/ time inefficient, and the accurate diagnosis depends on the expertise of pathologists diagnosing them. Hence, Lu et al. have devised a RS-DL system that employs Raman spectra obtained from urine as well as serum into ResNet, AlexNet and GoogleNet models to obtain the most ideal diagnostic model. The use of two types of bodily fluid samples provided a holistic and credible approach to this study wherein results of spectral analysis show correlations between changes in levels of specific biomarkers in urine and serum samples of both disease and healthy samples. The decrease in protein and guanine in serum samples and an increase of the same in urine samples was one such example, which is a characteristic finding in patients afflicted by membranous nephropathy. The spectral data acquired were first pre-processed by PLS baseline correction and gaussian noise was added to important spectral bands after which fivefold cross validation was done to increase reliability of the results. Among the three models adapted for this study, AlexNet showed the best results with a classification accuracy of 0.89 for urine samples and 1 for serum samples. This study implies the credibility of using multiple samples to obtain a more reliable result from RS-DL disease diagnostic systems [45].

The disastrous effects of the recent COVID-19 pandemic have urged spectroscopists to implement their knowledge to diagnose coronaviruses through RS-DL systems as well. A study by Zhao et al. has adopted Raman spectroscopy with a CNN to classify the spike proteins of various coronaviruses such as HCoV-229E, HCoV-OC43, HCoV-NL63, HCoV-HKU1, SARSCoV, and MERS-CoV that affect the upper respiratory tract of humans and cause a great deal of discomfort and sometimes even death. 10,000 Raman spectra of the spike proteins of these viruses has been collected through mapping the samples at various spots and it is reported that the average accuracy of the spectra obtained in a measurement time of 0.5 s exceeded 97%. The dataset was then pre-processed with convolution filtering and background correction and normalized between 0 and 1. The spectra were then split into a 6:2:2 ratio for training, validation and test sets and then input to a Multilayer Perceptron or MLP CNN where several hidden layers were added between the input and output layers. The input layer had a size of 895, which is the spectral length. Three fully connected hidden layers after the input layer were present and the number of neurons in each layer was 512, 256 and 32 respectively. The end of the model was reported to be a fully connected classification layer and the number of neurons was equal to the number of classes of coronaviruses. CNN and Resnet were experimented with but their performances were similar yet the complexity of MLP as well as the computational resources required to train the model were significantly smaller. Different wavenumber characteristics were identified in this study and assigned to the specific coronaviruses. (Insert Image) The authors have noted that this study has spectral limitations but as a preliminary approach can be considered for application in practical scenarios. They have also said "In clinical applications, surface-enhanced Raman scattering can be applied to reduce the measurement time and further accelerate the detection speed" [46].

Disease diagnosis using RS is commonly observed in ex-vivo samples such as blood serum, urine, or tissue samples but there are instances of real time in vivo diagnostics as seen in a study where a deep learning guided fibreoptic Raman diagnostic platform has been developed for nasopharyngeal carcinoma (NPC). A multilayer "Raman-Specified Convolutional Neural Network" (RS-CNN) was established to acquire fingerprint and high wavenumber spectra within sub seconds. A total of 15,354 in vivo Raman spectra (control: 1761; NPC: 4147; and post-treatment (PT): 9446) from 888 tissue sites of 418 subjects (healthy control: 85; NPC: 82; and PT: 251) were acquired during endoscopy. The RS-CNN model shows an overall diagnostic accuracy of 82.09% to identify NPC in control and post treatment patients. This is observed to be better than the performance using PLS-LDA which proves that CNN is a more sensitive and holistic method of DL for diagnosis. Along with this, the Raman biosignatures in collagens, lipids and nucleic acids for NPC are discussed, which further proves the capability of this model to act as an efficient diagnostic tool, as well as a post treatment option for monitoring patients- a unique door that has opened for clinicians [47]. To combat Raman imaging speed, DCNNs have been employed while using a line scan imaging method. The reduced imaging time is due to widening the slit and laser beam and scanning the sample with a large scan step. Single CaSki cells were used to test the fast line scan imaging method and the hyperspectral Raman images obtained with spectral ranges between 1408 and 1485 cm^{-1} as well as 2798~3015 cm^{-1} were compared as obtained in both high resolution, low resolution and then recovered by CNN. The imaging time was reduced to 1 min by this method and this speed is 5 times faster than the conventional line scan imaging method [48].

9.3.2 Microbial Cellular Imaging and Classification

RS has proven to be an extremely valuable tool for single cell imaging, and an example of combining this with deep learning has been shown in a paper by Fu et al., where microbes have been identified at the single cell level using a ConvNet model. This combination of Raman tweezers and CNN has allowed first the characterization of Raman features in a cell, after which the cells are classified into species. 14 microbial species were classified using this model with an average accuracy of 95.64%. This study also introduces occlusion based Raman Spectra feature extraction or ORSFE to visualise the Ramanomic profiles of the cells and the weights of spectral features to classify microbes [49]. Another example of the same concept is observed in a study where common pharmaceutical contaminant microbes were classified using a CNN. This work focused on classifying samples comprising only individual microbes as well as microbes mixed with Chinese Hamster Ovary cells (CHOs).Gram Positive and Gram Negative bacteria as well as fungal species that were used in this study were classified with an accuracy of 95–100%.This study displays to us the multifaceted capability of Raman Spectroscopy-Deep Learning [50]. Interestingly, RS-DL systems have been employed for classifying genetic phenotypes in antibiotic resistant bacteria as well. Antibiotic resistance has been an extremely jarring phenomenon in the clinical field in recent times. A work by Zhang et al. has implemented RS-DL on *Klebsiella pneumoniae* and its various antibiotic resistance strains. 71 different strains including the classical *K. pneumoniae* as well as strains with carbapenem resistance and/or hypervirulence phenotypes that have higher mortality rates were used in this study. The minimum inhibitory concentrations (MICs) of 15 common antimicrobial agents on K. pneumoniae strains were calculated. 7455 spectra obtained from confocal Raman Spectroscopy were used to develop a CNN to rapidly identify antibiotic resistant genes (ARGs) or hypervirulence encoding factors and resistant phenotypes from K. pneumoniae strains such as blaNDM, blaKPC, blaIMP, blaOXA, mcr-1, mcr-8 and virulence genes like rampA and rampA2. The CNN model that was created simplified the classification system and showed higher accuracy in identifying K. pneumoniae with high resistance and virulence when compared to other methods like Support Vector Machine and Logistic Regression Models. This RS-DL system showed "highly accurate and rationally designed treatment plans against bacterial infections within hours". This study is extremely important since it highlights the dangers of antibiotic resistance in species that form part of the normal biotic flora of humans which can disseminate to other areas and in turn become life threatening. Conventional methods of screening drugs for ARGs like Antimicrobial Susceptibility Test (AST), PCR, MALDI-TOF might fail in the clinical setting due to time constraints, necessity of extensive knowledge on strains or heterogeneity of analyte but RS-DL systems can overcome such problems [51].

9.3.3 Forensic Applications

A model using CNN has been created to distinguish Raman spectra of human blood and animal blood. Identifying species using their blood samples is a very important tool in forensic science and wildlife research. This method boasts the non-invasive advantage of Raman Spectroscopy, while identifying various Heme assignments, tryptophan as well as glucose Raman peak assignments. DL models using PLS-DA, SVM and CNN were considered, and the CNN model gave a prediction accuracy of 96.33% [52]. Another study has addressed the same concept with a focus on the effects of time on blood preservation, as well as the biochemical differences that could arise due to a variety of physiological states. A total of 3138 spectra from 20 different animal species including human blood were analysed by creating a 1D CNN model, and this model showed a recognition accuracy of around 97.33%. This study has outlined the benefits of Deep Learning in forensics and wildlife crime since it can differentiate not only between human and animal blood but also the species [53].

9.3.4 Biochemical Process Analysis

Process analysis technology for the manufacture of phytochemicals and botanical drugs are usually done by complex and time-consuming methods such as HPLC that consume large amounts of organic solvents as well as time. To combat this, a group in China has come out with a study pertaining to the rapid quantitative analysis of pharmacodynamic constituents of Guanxinning tablets, a phytopharmaceutical used to treat heart conditions. Danshensu, ferulic acid, rosmarinic acid and salvianolic acid B from the tablets were estimated through Raman spectra as well as high performance liquid chromatography HPLC-diode array detection and the soluble solids obtained from the tablets were determined through oven drying. A CNN calibration model was created wherein only spectral characteristic bands were screened by competitive adaptive reweighted sampling (CARS) algorithm. The performance of the CNN model was estimated by root mean square error of calibration (RMSEC), root mean square error of cross validation (RMSECV), root mean square error of prediction

(RMSEP) coefficient of determination of calibration (R2c and coefficient of determination of validation (R2c), coefficient of determination of cross-validation (R2 p). The R2cv), p values for soluble solid, salvianolic acid B, danshensu, ferulic acid, and rosmarinic acid are 0.9415, 0.9246, 0.8458, 0.8667, and 0.8491, respectively. This study evaluates the benefits of using RS-DL systems to evaluate drug quality in real time [54]. Another study involving the improvement of Process Analytical Technology has been reported by Nagy et al., where an artificial neural network was created to predict the dissolution profile of immediate release tablets using spectral data. 6 tablets with varying caffeine and disintegrant (potato starch) concentrations were used to obtain near IR (NIR) and Raman (RA) spectra in reflection (RE) and transmission modes (TR). After a principal component analysis, an ANN was trained at each hidden neuron number from 1 to 10 to identify the ideal number of neurons for the hidden layer. A small neuron number size showed the most optimal results. The average similarity values the four tested spectral methods NIR RE, NIR TR, RA RE and RA TR the Raman methods provided the most accurate predictions. The similarity values (f_2) of the ANN methods were above 59 for the validation tablets. Thus, this study shows that Raman spectroscopy is a more robust method for classifying pharmaceuticals according to concentration. This method can prove useful to improve quality control systems that exist in the pharma industry while also keeping in mind costs and time [55].

9.3.5 Environmental Studies

Off late, RS-DL systems have been a valuable addition to the world of medical diagnostics and imaging, but researchers have also ventured into using RS for environmental concerns that impact society and the scientific universe. An example of this can be seen through the work of Post et al., where the focus is on monitoring aquatic systems. This work has created a deep UV Raman Spectrometer to detect levels of nitrates and nitrites, pharmaceuticals such as acetaminophen, metformin hydrochloride etc. and microplastic polymers like polyethylene, polypropylene, polyester etc. in environmental aquatic systems. A convolutional neural network has been created to recognise RS data of individual pollutants which was previously heavily dependent on human intervention through mathematical filters and machine learning algorithms. Hence this study provides the ability to automatically detect chemicals in real time and allows reliable online measurements. Although this study gives a unique outlook for RS and Deep Learning in environmental systems, it comes with many challenges. The system was unable to identify Raman inactive pollutants, and selected pharmaceuticals in low concentrations such as in the μg/l or ng/l range were also not identified. The microplastics were detected and identified although the measurements were grim due to the heterogeneous distribution in the flow experiment [47].

Another example of environmental application has been observed on similar grounds where micropollutants in wastewater have been monitored in real-time using

Laser-Induced Raman and Fluorescence spectroscopy as well as artificial intelligence. Samples of wastewater from different treatment stages from a wastewater treatment plan were analysed though a Deep-UV laser-induced Raman and Fluorescence spectroscopy combined with CNN-based AI model. This was done to gain sensitive inputs from fluorescence measurements and specificity from Raman measurements which when aided by AI provides a robust piece of information. Since Raman Spectroscopy has detection limits of alert pharmaceuticals like carbamazepine and naproxen in the range of a few micromoles per litre, combining it with fluorescence methods is optimal. AI-Raman support provided a monitoring rate for nitrate levels after biological treatment of water which was over 95% [56].

9.4 Applications of Deep Learning: Surface Enhanced Raman Spectroscopy (SERS)

This section discusses researchers' revolutionizing use of DL methods in SERS applications in recent years. One of the most important things in modelling is the quality of the dataset, and good quality can be maintained by creating a database with standard Raman spectra. Shi et al. [57]. set up a database of SERS Raman spectra of deoxyribonucleic acid (DNA) due to their anti-interfering ability for AI applications. Afterward, they showed a pioneering sensing strategy with the help of the Convolutional Neural Network (CNN) model. To make sure that the database has prominent and abundant intrinsic fingerprint information and to present a proof-of-concept, the authors used three kinds of representative tumor suppressor genes, p16, p21, and p53 fragments from the database and showed that the AI-based model could perform a specific recognition of both single DNA targets and DNA mixtures with an accuracy rate of 90.28%. In particular, each Raman spectrum was divided into three regions, 600–800, 800–1200, and 1200–1800 cm^{-1}, and the number of peaks (pksN) and the corresponding number of heights (pksV) were defined. These extracted parameters of peak number and peak height can be used as the eigenvalue of the sample. Thus, each spectrum is divided into three sections, and each section contributes peak height and peak number, resulting in 6 independent parameters. Selective DNA targets were encoded with dependent vectors in 9-dimensional space to define the output. Furthermore, human plasma was employed as a real and hybrid sample to evaluate sensing performance, and $\sim$ 76% accuracy was obtained to discriminate 9 DNA targets spiked in human plasma. The sensing ability of the CNN model for DNA mixtures with and without different ratios of mixing was also verified with SERS mapping spectra. The results show a promising use of the CNN model in handling DNA sequential SERS spectra, which can provide biologists a novel way to collect and categorize DNA information.

On the other hand, Saifuzzaman et al. [58] demonstrated that the CNN model is a better choice for diagnosing Dengue Fever than the World Health Organization

(WHO) recommended standard clinical test known as Enzyme-Linked Immuno-Sorbent Assay (ELISA). SERS spectra enriched with Non-Structural protein 1 (NS1) were used in the study. NS1 is a recent antigen-based biomarker whose Raman fingerprint is identifiable and hence is valuable to use in this study. Principal Component Analysis (PCA) and Cattle Scree Test were used to use the optimum features, which were fed to CNN for better performance. To determine the optimal number of PCs based on the variance explained, a graphical heuristic method, also known as the Scree test or Elbow method used, which plots the number of PCs against the Eigenvalues. A total of 490 types of classifiers were tested by varying the number of hyperparameters, and the best Internet-enabling model reported an across-the-board 100% for all performance indicators considered in the study. As DL models showed better performance than the conventional WHO-recommended method for Dengue Fever diagnosis, they also ought to outperform Machine Learning methods. Hyunku Shin et al. [59]. hypothesized and evinced that the Raman signal of cancer cell exosomes can be detected from the plasma exosomes. Exosomes are extracellular vesicles found in blood, and a notable contributor for analysis through liquid biopsies. An accurate diagnosis of early-stage lung cancer using exosomes can be achieved by nano-plasmonic sensing technique with the features of cell exosomes through deep learning. ResNet was used for the classification of cell exosomes and lung cancer diagnosis. In the cell exosome classification task, the model provided an average accuracy of 94.8%, and for the lung cancer diagnosis task using human plasma-derived exosomes, the values of 10 neurons at the last Fully Connected layer was extracted and clustered through PCA. Mahalanobis distance criteria were used on the PCA clusters to check the relative statistical similarity between cancer cell exosomes and plasma exosomes. The mean value of the relative similarity was used as a diagnostic scale, and it was found that the patients showed about 1.5-fold higher similarity than the healthy controls with statistically significant differences. Moreover, the Jonckheere-Terpstra test was carried out, which yielded a statistically significant ascending trend by the cancer stages. PCA-LDA, PLS-DA, SVM, CNN, and VGG-16 models were also incorporated, and it was found that the ResNet performed better in terms of accuracy for the classification of cellular exosomes and patient prediction. Thus, the DL model not only showed the superiority for the diagnosis but also showed the potential to detect early-stage lung cancer (stage-I) efficiently without using any specific biomarkers. It is also worth mentioning that since the neuron output values showed a cancer progression which was confirmed by the statistical analysis, the DL-based model makes meaningful feature assignments during the training or learning period, which helps the model to achieve superior performance.

Along the same line of application, Cheng et al. [60] considered the antibody-free liver cancer screening based on 3D Nano-plasmonics Biosensing Chips (NBC) with the help of Deep Neural Network (DNN) modeling as a Point-of-Care and fast diagnostic tool through SERS detection. In this direct serological detection platform, healthy and Hepato Cellular Carcinoma (HCC) diagnosis was presented by screening and automatically identifying liver cancer within minutes and achieving 91% identification accuracy. Additionally, the DNN model was found to be better at tasks than LDA, and PCA-LDA, and the intelligent model, which is affordable and facile, is possible by combining DL applications to SERS spectral that can be used in rapid liver cancer diagnosis. Rapid liver cancer diagnosis is essential as it can improve the mortality rate in patients with early stage liver cancer. Chen et al. [37] also established the intelligent diagnostic model for the early, non-invasive diagnosis of Prostate cancer (PCa). Urinary Raman spectra of Benign Prostatic Hyperplasia (BPH) and PCa were obtained with differences in intensities of characteristic peaks of lipids, nucleic acids, and some amino acids, which also indicates the changes in metabolism due to PCa. The proposed model, as an alternative to conventional diagnostic methods such as prostate-specific antigen test (PSA) and radiological detection, consisted of CNN architecture similar to that of LeNet-5 and was used without any additional feature extraction algorithms. Thus, a non-invasive method for liquid prostate biopsy based on urine Raman spectroscopy which gives a mean diagnostic accuracy of 74.95%, was developed which can automatically extract the meaningful and abstract features which are responsible for the accurate diagnosis of PCa.

In chemical analyte detection, Prajapati et al. [61] used the new paradigm in Raman analysis to showcase 2D patterned hierarchical SERS substrate and imaging-analysis protocol for chemical identification of analytes by hyperspectral imaging. This engineered hierarchical substrate in which plasmonically active regions on a micron-scale were used with rhodamine and fluorescein and applied to detect glucose in biologically relevant concentrations with > 90% confidence especially abnormal levels of glucose, for monitoring type II diabetes. The topical protocol includes automated hexagonal pattern recognition in optical microscope images via a sequence of detection, K-means clustering, and analysis algorithms using Convolutional Neural Nets (CNN). A Haar cascade classifier, a supervised method based on OpenCV, was used, followed by a shallow CNN model to detect the hexagonal structures with a confidence interval higher than 0.8 and an accuracy of 95% to reject fragments and noisy patches. Afterward, the K-means clustering was used for position coordinate by minimizing the variance of position coordinates with respect to the cluster's centroid. Finally, hexagonal structures cropped out from each cluster, and a Deep-CNN model was used, which gave positive detection capability with 95% accuracy where positive images constitute images with complete hexagons and negative ones with resized images of background noise, corners, and edges, making way for machine-recognizable chemical identification tool a reality.

Ciloglu et al. [62]. focused on the bacterial identification process as the molecular similarity between bacterial species is high using Stacked Autoencoder (SAE)-based Deep Neural Network (DNN) to discriminate methicillin-resistant Staphylococcus aureus (MRSA) and methicillin-sensitive Staphylococcus aureus (MSSA) bacteria's SERS signals. The subtle spectral difference originates from the cell wall structures of sensitive and resistive bacteria is always present, and the SAE-DNN model characterizes and discriminates the two types with a mean accuracy of $97.66 \pm 0.26\%$ without hardcore pre-processing and feature extraction, as the model inherently performs the feature selection for the classification. The model performance was compared with state-of-the-art classifiers SVM, LDA, KNN, DT, and NN. The Mann–Whitney U Test with a significance level of 0.05 was used to show the statistical significance of the SAE-DNN model as compared to other models and with p-value as a measure. It was confirmed that the SAE-DNN outperformed other methods in consideration for the reliable identification of antibiotic-resistant and susceptible bacteria. It also provides a solution to efficiently handle a large amount of SERS spectral dataset, which can be made with minimum sample preparation procedure, label-free, easy use, rapid, and less incubation time method compared to conventional phenotypic Antimicrobial Susceptibility Testing (AST).

So far, it is evident that the DL boosts the diagnostic applications of SERS spectra in cellular, chemical analyte, and bacterial applications. Fang et al. [63]. presented the usage of DL in circulating tumor cells (CTCs) which shed from primary or metastatic lesions and enter peripheral blood, and are used as biomarkers to accurately identify tumor cells and blood cells. The ResNet model was used, and it was found that significant differences in nucleic acid-related characteristic peaks helped the residual network to identify 11 kinds of SERS spectra with 100% accuracy. The study further validates the model's performance by checking the effect of pre-processing, different residual blocks in the model, CNN, and PCA-KNN models, and the ResNet has better diagnostic capability in CTCs diagnostic application. To understand the working of ResNet, T-distributed stochastic neighbor embedding (t-SNE) was used. The t-SNE diagram of original spectra data and that of feature peaks obtained by manually extracting the characteristic peaks of 730 and 1330 cm^{-1} from the original data shows the same correlation. Therefore, it is preferable to say that the ResNet has extracted the essential features of various cells to perform the requisite task. The characteristic peaks mentioned correspond to nucleic acid, which is primarily different in tumours and blood cells.

In another bacterial study of Urinary Tract Infections (UTIs), Fu et al. [64] proposed an intelligent identification strategy for antibiotic resistance of pathogenic bacteria of 6 species, antibiotic sensitivity, and multidrug resistance (MDR) using a CNN without using complex pre-processed SERS Raman spectra. First, in the pathogenic bacteria identification task, SERS spectra of 6 types of bacteria were tested with an average recognition accuracy of $97.8 \pm 1.4\%$ by the CNN model and found that it was higher than the PCA-KNN model, and the other CNN model trained on pre-processed spectral SERS data. Second, independent CNN models were developed to identify SERS spectra of sensitive and resistive strains of E. coli, E. faecalis, and P. aeruginosa and recognize the drug susceptibility of pathogenic bacteria that causes UTIs. The detection accuracy was above 95%, which shows the recognition ability of the CNN model. Third, 5 E. coli strains, 5 P. aeruginosa strains, and 2 E. faecalis strains, with all strains showing different degrees of resistance to antibiotics, were analysed, and found 100% identification accuracy of E. faecalis, 99% for E. coli, and 97% for P. aeruginosa (As seen in Fig. 9.7). And finally, spectra of 18 different strains were used to train and evaluate the corresponding CNN model for each strain and achieved an average recognition accuracy of 96.6%. It proved that the CNN models or DL models are far more sensitive than typical multivariate statistical methods in recognizing subtle differences between species, antibiotics, tolerability, and multidrug resistance of pathogenic bacteria, which can also detect the harmful microbes in UTIs.

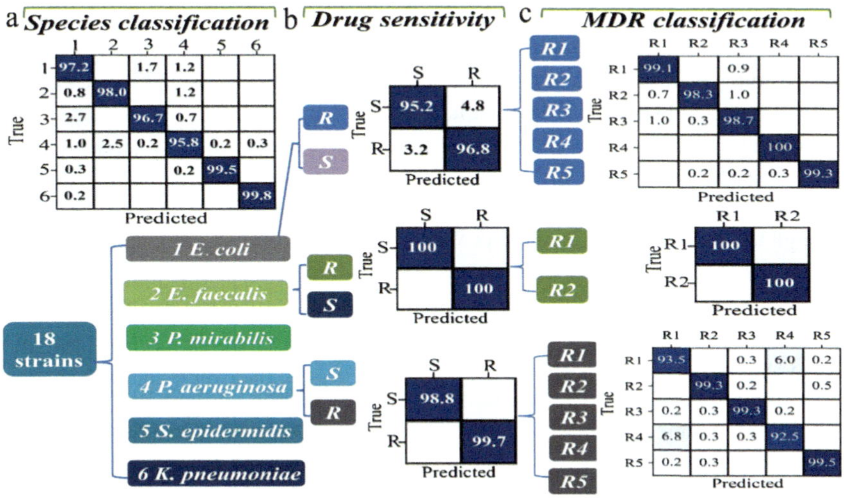

Fig. 9.7 Prediction results of independent test sets by 3-level CNN classification models to classify pathogenic bacteria in UTIs by Qiuyue Fu et al. where **a** Species classification. **b** Drug sensitivity reorganization. **c** MDR pathogen classification

Huang et al. [65] also established a database based on the spike protein of SARS-CoV-2 from experiments and theoretical calculations to provide standard references. S protein was chosen as the detection target, and spectra collected from the throat swabs of healthy people were used as negative samples and labelled as 0. Negative spectra superimposed with the intensity of randomly modulated normal Raman spectra of the S protein were used as positive training data and labelled as 1. A homemade computer program was used to perform the identification process. The number of positive predicted points based on the output scores from the ResNet model was used as a diagnostic scale. The mean number of positive points for COVID-19 patients was 13.2-fold higher than that for the healthy controls, and the difference was statistically significant. A portable Raman spectrometer was used for throat swabs or sputum samples, which can provide high manoeuvrability and make an on-site detection of SARS-CoV-2 within 20 min without considering the inactivation treatment with an identification accuracy of 87.7% by using ResNet. In drug design and discovery applications, the study of protein functionalization is of utmost importance and is generally studied by the fields such as Proteomics, epigenetics, and metabolomics. However, such processes are complex and require a secondary tag that might interfere with the study. Modified sites can be recognized via proteolytic cleavage followed by liquid chromatography-mass spectrometry but are impeded by the complexity of peptide mixtures and the elaborate synthetic design for tags. To overcome this problem, Peng et al. [66] employed a deep-learning-based SERS screening strategy (DLSS) to identify protein binding sites. The deep learning model was trained and could detect signal features of small molecules with > 99% accuracy without any secondary tags. By using an automated DLSS system, they identified the modification sites of fomepizole in alcohol dehydrogenase, which is coordinated with zinc along with three peptides. DLSS also works for the identification of amino-acid residues that covalently bond with ibrutinib in Bruton tyrosine kinase. Despite not capturing and targeting fomepizole-modified peptides using radioisotope, fluorophore, and biorthogonal tag, the DLSS method achieved high accuracy with ROC-AUC of 0.956 (95% CI: 0.918–0.994) and recognized ibrutinib with AUC of ROC 0.940 (95% CI: 0.890–0.991). The SERS signals of the fractions were predicted through the model to calculate the probability of each fraction containing the small molecule data to form a probability map. A probability chromatogram was obtained from the probability of Fomepizole-specific and ibrutinib-specific signal patterns against SERS signal patterns of HPLC fractions to confirm their presence in the SERS signals.

As the application of DL can be seen as more diverse in the scale of samples and field of applications, Mo et al. [67] introduced to solve the most commonly diagnosed type of cancer in women, i.e., breast cancer, as an alternative to Immuno Histo-Chemistry (IHC) method, which is considered a gold standard for detecting the expression of the Human Epidermal Growth Factor Receptor (HER2). The ratio of the number of HER2-positive spectra to the total number of spectra is used as a diagnostic scale. The average consistency between the proposed method and IHC can reach 95%, which shows that the proposed method can accurately identify the difference in HER2 content in different types of breast cancer tissues and make diagnosis results of breast cancer quantifiable. To check the responsible features for the classification, normalized intensities of each wavenumber and a region of 10 cm^{-1} near it (± 5 cm^{-1}) were set to zero and processed spectrum input to the deep learning model again for prediction. The variation in the influence of judgment of the network was studied in the form of saliency, and it was found that the differences between the Raman spectra of breast cancer and normal tissues originated from HER2. Yang et al. [68] showed that the recurrent neural Network (RNN)-based model with three LSTM layers can detect SARS-CoV-2 and can be used as a SERS sensor for RNA in human nasopharyngeal swab (HNS) specimens with 98.9% overall accuracy. The feasibility of the proposed study was checked by using blind test samples confirmed by RT-PCR test and tested by the model, and 97.2% accurate prediction for positive specimens and 100% for negative specimens were obtained. All the predictions were made in 25 min, and in comparison, to SVM, RF, BP, and CNN models, the RNN model gave the highest training and testing accuracy. The proposed method has comparatively higher accuracies compared to other COVID-19 diagnostic methods, such as 66% for rapid antigen test, 95% for nucleic acid test, 97.2% for dipstick detection, 91.2% for positive and 99.3% for negative specimens for breath detection. In addition, the feature extraction method was used to find out the main component responsible for the classification by using Occlusion. It was found that RNA hybridization was the determining factor. The highest contribution was made by features including 794 cm^{-1} and 1329 cm^{-1} corresponding to the ring stretching mode of adenine as well as ring-breathing modes of cytosine and thymine, and RNA hybridization were the determining factors.

To conclude, the SERS applications are favorable for rapid biomedical application, and DL-based methods help to achieve the task at hand with more ease, are less human interventive, and are robust result oriented. The prospects of the DL in the SERS application look more fruitful, and it could be the first Raman variant to use a real-time disease diagnostic tool in day-to-day life clinical applications due to its multi-fold advantages over other Raman variants.

9.5 Applications of Deep Learning: Stimulated Raman Spectroscopy

Imaging applications of Raman spectroscopy, especially stimulated Raman spectroscopy are gaining wide acceptance due to the possibility of rapid image analysis for real-time applications of cells and tissues. For example, living cell dynamics can be understood with or without the effects of drugs [69]. But imaging Raman spectral dataset suffers from many problems, such as pixel value blurring due to background, media interference, and/or instrumental noise. DL can help to reduce the problem of such interferences to study cellular Raman imaging. To show such an application, Manifold et al. [70] have solved the problem in SRS images by using the DL in the form of a U-Net model to denoise the images obtained at different zoom, imaging power, imaging depth, and imaging geometries to improve the SNR of SRS images. In the first application of the denoising task, HeLa cells images of low and high power were considered without any complex pre-processing. Pairs of corresponding low and high SNR images provided the signal and truth images for the model. To verify the capability of the DL model, another denoising method Variance Stabilization Transform (VST) Poisson denoising, was used. The metrics such as Peak Signal-to-Noise Ratio (PSNR), Room Mean Square Error (RMSE), and Pearson Correlation Coefficient (CC) values showed that the DL model denoising was a better option in terms of recovery of cellular features, including lipid droplets, nuclei. In the second application, SRS images of murine brain tissue are enriched in more heterogeneity and no true background, so with two-color images of low and high SNR images in transmission mode, the U-Net model was able to recover the axons presence, nuclei contrast, spatial resolution, and imaging fidelity. The depth concern in the tissue sample was solved by using the DL model. The SRS images of murine brain tissue at a depth of 175 μm, which was never done in an SRS imaging application, were used. In addition, to VST, the CANDLE and PURE-LET methods were also used for the comparison of denoising of SRS images. Once again, the denoised image by the DL model showed axons in the lipid channel and better nuclei resolution even if the images were recorded at high SNR only. Instead of low SNR, low-distance (non-zoomed) images were recorded, and instead of high SNR, high-distance (zoomed) and the DL model were able to recover the spatial resolution, which also shows the DL model superiority over instrumental limitations. Finally, to overcome the problem of epi-SRS geometry images due to low light recovery at the detector, the DL model was trained on transmissive mode and denoised the epi-SRS images. The DL model makes the epi-SRS images look like transmissive-SRS images, which opens many opportunities for SRS imaging application in in-vivo studies.

The epitomic performances of the Raman spectroscopy assisted by DL methods indicate its unprecedented disease diagnosis usefulness. At the same time, we discussed the applications related to samples at static conditions. The problem of diagnosis becomes multi-fold complex when it comes to measuring the changes in a dynamic state or flowing state. Suzuki et al. [71] have encountered such a complex problem of disease diagnosis in blood where the DL method was able to diagnose in a continuous flow of cells with a throughput of $\sim$ 140 cells/s. So, a chemical imaging flow cytometry by SRS and with the aid of DL can not only achieve the cell characterization of micro-algal cells but also able to provide marker-free cancer detection. VGG-16 with five convolution segments for discrimination of Euglena gracilis in the presence and absence of nitrogen showed an excellent accuracy of > 99% for all cultures. In cancer detection, Circulating Tumor Cells (CTCs) were measured by using an SRS imaging cytometer in addition to whole blood cells (Red Blood Cells), Peripheral blood mononuclear cells (PBMCs), human colon cancer (HT29 cells) as models for CTCs and Jurkat cells as a model to blood cancer cells. The CNN model was able to classify the whole blood cells with >98%, PBMCs with >98%, Jurkat cells with >94%, and HT29 cells with >93% classification accuracy. A fastest-to-date multi-color SRS microscopy is shown to perform a disease diagnosis for a flowing system that could be employed in real-life scenarios. It is also important to note that some forms of tumors look the same as normal to the naked eye, especially when the sample is fresh. Multi-color SRS microscopy can be used on unprocessed fresh samples to diagnose the disease. It is also customary that the AI-based models are not trustworthy and random and might change to give different predictions at different times, but Zhang et al. [72] diagnosed histologic images of human larynx tissues with the help of the ResNet34 model to classify normal and neoplasia SRS images with a validation accuracy of 95.9% with a stable standard deviation of 0.4%. Moreover, SRS images of laryngeal Squamous Cell Carcinoma (SCC) tissue on a fresh sample can be accurately predicted by the model with 100% accuracy.

On the contrary, Hollon et al. [73] showed that the SRH images could help to diagnose the intraoperative brain tumor of a patient at the bedside with the help of DL in the form of an Inception-ResNet-v2 model. Once again, the developed model surpasses the performance of conventional pathological techniques and proves the extraordinary skills the DL model possesses as a characteristic extraction for disease diagnosis. A two-arm, prospective, multicenter, noninferiority clinical trial was conducted to compare the diagnostic accuracy of SRH image classification by pathologists following conventional means and the DL model following the SRH approach resulting in 93.9% overall diagnostic accuracy by pathologists and 94.6% for the proposed approach claiming the diagnostic accuracy higher than the human intelligence. An activation maximization method was used to understand the representations learned by the DL model. An image can be generated that maximally activates a neuron among the layers of the neural network. Neuron-specific findings reveal that different neurons identify the lipid-rich axons in gray matter, high nuclear density, and lipid droplets, which are the high-grade glioma-specific features, large nuclei with nucleoli and cytoplasmic vesicles in metastatic tumor cells, and pyramidal neurons which confirms that the DL model identifies specific histomorphologic, cytologic, and nuclear features. Hill et al. [74] analyzed four different types of tissues from the brain, lung, kidney, and liver to show the effectiveness of SRS with the CNN denoising algorithm. Similar to Bryce Manifold et al. work, a low SNR and a high SNR SRS image were used to train the denoising U-Net algorithm with 40 fields of view, and Signal to Background (SBR) was used as a metric to check the capability of the DL model. A depth of 210 μm in ep-imaging of the brain for a 2 mm sample yields a 40 μm (24%) increase in achievable imaging depth, which sets a milestone in terms of depth of imaging by the SRS-cum-DL method for a biological sample.

Hollon et al. [75] continued the work on the intraoperative SRH and detected the diffuse glioma recurrence and pseudo progression using the Inception-ResNet-v2 model with a 95.8% diagnostic accuracy. Semantic segmentation was used on a pixel level to highlight the regions of the images considered for the classification by the DL model and ensure that the model used tumour recurrence-specific regions for accurate diagnosis. Cell counting can be thought of as a method to detect the disease at its early stages. Zhang et al. [76] set up a framework for automatic cell counting and density estimation using SRS images assisted by the U-Net model with medical semantic segmentation, which can perform tumour tissue characteristic analysis, cancer diagnosis, and surgery planning processes. A real human tumour brain specimen cell counting was achieved with > 98% of AUC for cell counting correlation between SRS and histological images with haematoxylin and eosin (H&E) staining. A comparison between segmentation results of SRS images followed by distance transform and watershed segmentation algorithm and H&E images followed by K-Means clustering showed that the cell segmentation gives pixel-level real-time segmentation and cell counting. Dice coefficient (DICE) and Percentage Error (PE) were also used as metrics to evaluate the performance, and the DL model accomplished the task with the best performance metric (Fig. 9.8).

Liu et al. [77] also worked on intraoperative Raman spectroscopic applications and diagnosed the gastroscopic biopsy with single-shot femtosecond SRH imaged in < 60 s via the Inception-ResNet-v2 model. The cancer class includes two cases, differentiated and non-differentiated cancer. The differentiated cancer class consists of moderately differentiated tubular and papillary adenocarcinoma, while the non-differentiated ones with poorly differentiated adenocarcinoma, mucinous adenocarcinoma, and signet ring cell carcinoma. The non-cancer class consists of normal adenoma, intestinal metaplasia, and gastritis. One neural network with binary outputs first classifies cancer and normal cases with an AUC of 98.8%, and the second neural network further classifies the cancer class into differentiated and non-differentiated ones with an AUC of 94.2%. Quantitative evaluation of SRS images via the DL model and rating of professional pathologists on H&E stains of corresponding specimens showed a high level of concordance ($\kappa = 0.899$). In the identification of non-differentiated and differentiated types, the proposed model achieved diagnostic accuracy higher than human intelligence, which once again proves the performance quality of the DL model. Semantic segmentation results of SRH images confirmed by

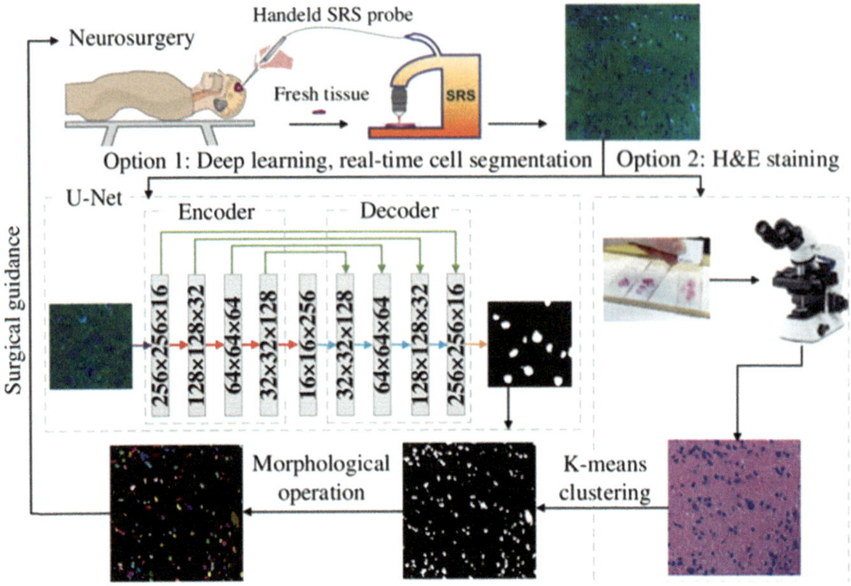

Fig. 9.8 An overview of the cell counting framework as reported by Qianqian Zhang et al.

the pathologists ensured the DL model used representative information and also very well known the boundaries of normal and tumor regions on a pixel level information.

Jiang et al. [78] presented OpenSRH, the first and only public dataset of clinical SRH images from the most common brain tumor diagnoses, full pathologic annotations, whole slide tumor segmentations, and raw and processed optical imaging data for end-to-end model development and validation. ResNet50 and vision transformers (ViT-S) were used for two tasks, one is histologic brain tumor classification, and another is contrastive representations learning. The model trained on the ImageNet dataset to encash the advantages of transfer learning. In the first task, patch-level predictions from the same whole slide or patient for slide-based or patient-based prediction, both the models performed decently, showing the utmost effect of transfer learning. In the second task, unsupervised contrastive learning (SimCLR) and supervised contrastive learning (SupCon) were used on both models. With SimCLR, two augmented positive pairs of the same image from other images were identified, and with SupCon, discrimination of one class image from the rest of the images was done. SimCLR predicted 85.6% patient-level accuracy for ResNet and 78.3% for ViT-S. SupCon performed better than transfer learning and SimCLR in patch-based tasks. Thus, the DL models can also leverage the transfer learning approach, which can solve the data scarcity problem, which is common in Raman spectroscopic study to efficiently train the model and gives optimum performance.

9.6 Applications of Deep Learning: Coherent Anti-Stokes Raman Spectroscopy (CARS)

CARS has been used in tandem with deep learning to investigate images of cells and tissues of medical importance. The benefit of this approach for cell imaging is that CARS provides a strong signal for the C–H stretching region, which shows valuable information especially pertaining to lipids and proteins.

A study by Weng et al. in 2017 has employed the advantages of CARS with respect to its ability to provide cellular-level images and pathologically resolve related features on lung tissues. Lung cancer is one of the most common cancers in the world, and reducing diagnosis time will ensure that patients receive timely therapeutic care. The study involves differentiating a fair variety of cancers afflicting the lung, namely adenocarcinoma, squamous cell carcinoma and small cell carcinoma cases using CARS imaging. Snap-frozen lung tissues that were previously histologically classified manually were used for the study. The cellular images obtained after CARS were taken in the form of 'z-stacks' or a series of images acquired at different image depths and at different locations of the lung tissue. A CNN model called Google Inception v3 Model was modified and retrained to fit the CARS images. 89.2% accuracy was obtained using this model, and it proves to be a steppingstone in the direction of instantaneous cancer diagnosis [79]. Similarly, another attempt at instantaneous non-invasive cancer diagnosis using CARS has been reported by Aljakouch et al. for cervical cancer. Liquid Pap smear samples with multiple types of cell conditions: negatives for intraepithelial lesions or malignancy, atypical squamous cells of undetermined significance, low-grade squamous intraepithelial lesions and high grade squamous intraepithelial lesions were screened using a combination of CARS with Second Harmonic Generation (SHG) Imaging and Two Photon Excited Autofluorescence (TPF) Imaging. Molecular Information obtained from the Raman spectra was analysed, alongside the CARS/SHG/TPF Images which were subjected to analysis by DCNNs. CARS microscopic imaging at 2935 cm^{-1} was observed to monitor protein in the cells. Raman micro spectroscopy was employed to confirm the CARS findings. For the different smear types, six wavenumbers were selected: 1011, 1375, 1657, 2845, 3042, and 3049 cm^{-1}. DCNNs were trained on the resulting images from CARS of normal, LSIL, and HSIL cells to identify these cells. The different pap smears were classified with 100% accuracy, and this method seems meritorious over other non-invasive tests such as the expensive HPV-DNA test or the less sensitive and less specific Papanicolaou test [80] (Fig. 9.9).

Tissue imaging using CARS has also been a successful venture, especially for in vivo endoscopic imaging as shown by Hashimoto et al. Their study has specifically pointed out the importance of CARS and deep learning in terms of endoscopically visualizing and identifying the exact location of nerve tissues in vivo for nerve-saving surgery. Myelinated nerves were visualized using U-net with a VGG16 encoder which was pre-trained with fluorescence images, since fluorescence images visualize lipid distribution similarly to CARS images. Nerve segmentation of myelinated nerves in rabbit prostates was also achieved with a mean accuracy of 0.962 and an F1 value

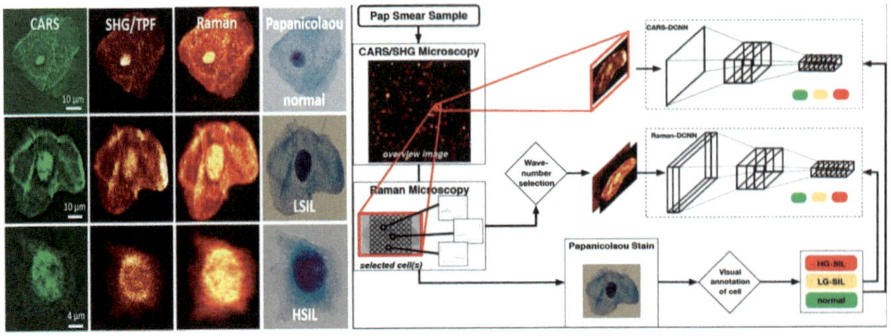

Fig. 9.9 A comparative image of various staining techniques for Cervical Cancer diagnosis as reported by Aljakouch et al. (Left) and a schematic diagram for the working of CARS-DL system (Right)

of 0.860 [81]. This study showed problems with a low imaging speed, and hence the authors proposed another method as a separate study where noise reduction with deep learning boosted the speed of nerve imaging. Fine tuning of the models involved pre-training with CARS microscopy images focused on the CH_2 symmetric stretching vibration of lipids at 2845 cm^{-1} found on myelin sheaths of nerves and re-training with CARS endoscopy images. The Equivalent Imaging Rate (EIR) was used as the evaluation metric to assess the imaging rate improvement using deep learning models. The EIR turned out to be 5 times better than the previous study mentioned above. These results prove the efficacy of combining CARS with deep learning to improve clinical research [82].

Since CARS conventionally requires multiple processes that are dependent on human intervention, these studies have demonstrated that using deep learning can reduce the burden on human analysis while increasing the efficiency of clinical imaging.

9.7 Conclusions

Raman spectroscopy is a label-free, rapid, molecular fingerprint guide which has the upper hand in providing advantages as a disease diagnostic tool with the help of Artificial Intelligence based methods. It would not be wrong to envision that it can be included in pathological diagnosis systems along with conventional techniques and create a revolution in clinical methods. It could be possible to have a human-friendly, easy-to-use tool that can change the arithmetic related to incidences and mortality related to any trivial and severe disease. From the review of recent progress in AI-based Raman spectroscopy, it is evident that the DL has revolutionized handling, processing, and analysing the Raman spectra with never seen before performance metrics. Given the progress, it is reasonable to expect that DL assisted Raman spectroscopy in biomedical field would solve the diagnostic problem and might help to understand new and accurate biomarkers, which could help in effective drug discovery/delivery.

References

1. O. Ryabchykov, S. Guo, T. Bocklitz, Analyzing Raman spectroscopic data. Phys. Sci. Rev. **4** (2019). https://doi.org/10.1515/psr-2017-0043
2. K. Eberhardt, C. Stiebing, C. Matthaus, M. Schmitt, J. Popp, J. Advantages and limitations of Raman spectroscopy for molecular diagnostics: an update. Expert Rev Mol Diagn **15**, 773–787 (2015). https://doi.org/10.1586/14737159.2015.1036744
3. S. Kohli, S. Miglani, R. Rapariya, Basics of artificial neural network. Int. J. Comput. Sci. Mob. Comput. **3**, 745–751 (2014)
4. L.D. Harmon, Studies with artificial neurons, I: Properties and functions of an artificial neuron. KYBERNETIK, (1961)
5. A. Duykuluoğlu, The significance of artificial neural networks in educational research: A summary of research and literature. Technium BioChemMed **2**, 2734–7990 (2021)
6. P.H. Sebastian Bader, A. Seda, in *Mathematical aspects of logic programming semantics*
7. M. Gniadecka et al. Melanoma diagnosis by Raman spectroscopy and neural networks: structure alterations in proteins and lipids in intact cancer tissue. J Invest Dermatol. **122**, 443–449 (2004). https://doi.org/10.1046/j.0022-202X.2004.22208.x
8. W. Wang, Y. Yang, Development of convolutional neural network and its application in image classification: a survey. Opt. Eng. **58**, (2019). https://doi.org/10.1117/1.Oe.58.4.040901
9. D.H. Hubel, T.N. Wiesel, Receptive fields, binocular interaction and functional architecture in the cat's visual cortex. J. Physiol. **160**, 106–154 (1962). https://doi.org/10.1113/jphysiol.1962.sp006837
10. Y. LeCun et al., Backpropagation applied to handwritten Zip code recognition. Neural Comput. **1**, 541–551 (1989)
11. A. Krizhevsky, I. Sutskever, G.E. Hinton, (ed.) *Neural information processing systems* (2012)
12. K. Simonyan, A. Zisserman, *Very Deep convolutional networks for large-scale image recognition*. arXiv (2014)
13. NIPS-2014-generative-adversarial-nets-Paper.pdf
14. C. Szegedy et al. *Going deeper with convolutions*. arXiv (2014)
15. J. Long, E. Shelhamer, T. Darrell, Fully convolutional networks for semantic segmentation. arXiv (2015)
16. K. He, X. Zhang, S. Ren, J. Sun, *Deep residual learning for image recognition*. arXiv (2015)
17. M. Shafiq, Z. Gu, Deep residual learning for image recognition: A survey. Appl. Sci. **12**, (2022). https://doi.org/10.3390/app12188972
18. A. Hosna et al., Transfer learning: a friendly introduction. J Big Data. **9**, 102 (2022). https://doi.org/10.1186/s40537-022-00652-w
19. P.A. Jadhav et al., Serum Raman spectroscopy: Evaluation of tumour load variations in experimental carcinogenesis. J Biophotonics, e202300424 (2024). https://doi.org/10.1002/jbio.202300424
20. P. Saha et al., Raman Spectroscopy of urine: An exploratory study on stratification of oral cancers and tobacco habitu s. Transl. Biophotonics: Diagn. Ther. Iii, **12627**, (2023). https://doi.org/Artn 126272x10.1117/12.2670121
21. A. Hole et al., Saliva Raman spectroscopy: understanding alterations in saliva of tobacco habitués and oral cancer subjects. Vib. Spectrosc., 103414 (2022). https://doi.org/10.1016/j.vibspec.2022.103414
22. K. Iwasaki et al., Identification of molecular basis for objective discrimination of breast cancer cells (MCF-7) from normal human mammary epithelial cells by Raman microspectroscopy and multivariate curve resolution analysis. Int. J. Mol. Sci. **22**, (2021). https://doi.org/10.3390/Ijms22020800

23. M. Rahul et al., DNA fingerprint analysis of Raman spectra captures global genomic alterations in Imatinib-resistant chronic Myeloid leukemia: A potential single assay for screening Imatinib resistance. Cells **10**, 2506 (2021). https://doi.org/10.3390/cells10102506
24. H. Noothalapati et al., Towards the development of a non-bioptic diagnostic technique for eosinophilic esophagitis using Raman spectroscopy. Vib. Spectrosc. **85**, 7–10 (2016). https://doi.org/10.1016/j.vibspec.2016.03.016
25. R. Zakaria et al., Development of in-situ Raman diagnosis technique of eosinophil esophagitis. Spectrochim Acta A Mol Biomol Spectrosc., **285**, 121804 (2022). https://doi.org/10.1016/j.saa.2022.121804
26. A. Anjikar et al., Removal of hemolysis interference in serum Raman spectroscopy by multivariate curve resolution analysis for accurate classification of oral cancers. Sens. & Diagn., **2**, 390–397 (2023). https://doi.org/10.1039/D2SD00137C
27. I.M. Hossain et al., Direct Visualization of structurally similar polysaccharides in single yeast cells in vivo by multivariate analysis assisted Raman microspectroscopy. J. Phys. Chem. B, (2023). https://doi.org/10.1021/acs.jpcb.3c00632
28. K. Iwasaki et al., Visualizing wax ester fermentation in single Euglena gracilis cells by Raman microspectroscopy and multivariate curve resolution analysis. Biotechnol. Biofuels **12**, (2019). https://doi.org/10.1186/S13068-019-1471-2
29. H. Noothalapati et al., Studying anti-oxidative properties of inclusion complexes of alpha-lipoic acid with gamma-cyclodextrin in single living fission yeast by confocal Raman microspectroscopy. Spectrochim Acta A Mol Biomol Spectrosc. **197**, 237–243 (2018). https://doi.org/10.1016/j.saa.2018.02.011
30. H. Yan et al., Diverse region-based CNN for tongue squamous cell carcinoma classification with Raman spectroscopy. IEEE Access **8**, 127313–127328 (2020). https://doi.org/10.1109/access.2020.3006567
31. M. Chen et al., Deep learning on reflectance confocal microscopy improves Raman spectral diagnosis of basal cell carcinoma. J Biomed Opt. **27**, (2022). https://doi.org/10.1117/1.JBO.27.6.065004
32. P. Manganelli Conforti, M. D'Acunto, P. Russo, Deep learning for chondrogenic tumor classification through wavelet transform of Raman spectra. Sensors (Basel). **22**, (2022). https://doi.org/10.3390/s22197492
33. M. Wu et al., Deep learning data augmentation for Raman spectroscopy cancer tissue classification. Sci Rep. **11**, 23842 (2021). https://doi.org/10.1038/s41598-021-02687-0
34. X.C. Tian et al., Application of Raman spectroscopy technology based on deep learning algorithm in the rapid diagnosis of glioma. J. Raman Spectrosc. **53**, 735–745 (2022). https://doi.org/10.1002/jrs.6302
35. C. Chen et al. Rapid diagnosis of lung cancer and glioma based on serum Raman spectroscopy combined with deep learning. J. Raman Spectrosc. **52**, 1798–1809 (2021). https://doi.org/10.1002/jrs.6224
36. D. Vrazhnov et al., Discovering glioma tissue through its biomarkers' detection in blood by Raman spectroscopy and machine learning. Pharm. **15**, (2023). https://doi.org:ARTN20310.3390/pharmaceutics15010203
37. S. Chen et al., Raman spectroscopy reveals abnormal changes in the urine composition of prostate cancer: an application of an intelligent diagnostic model with a deep learning algorithm. Adv Intell Syst-Ger. **3**, (2021). https://doi.org/10.1002/aisy.202000090
38. N. Blake, et al., Deep learning applied to Raman spectroscopy for the detection of microsatellite instability/MMR deficient colorectal cancer. Cancers **15**, (2023). https://doi.org/10.3390/cancers15061720
39. R.V. Selvarani, P.S.H. Jose, A label-free marker based breast cancer detection using hybrid deep learning models and Raman spectroscopy. Trends Sci., **20**, (2023). https://doi.org/10.48048/tis.2023.6299
40. R. Kothari et al., Raman spectroscopy and artificial intelligence to predict the Bayesian probability of breast cancer. Sci Rep-Uk. **11**, (2021). https://doi.org:ARTN648210.1038/s41598-021-85758-6

41. L. Shang et al., Polarized micro-raman spectroscopy and 2D convolutional neural network applied to structural analysis and discrimination of breast cancer. Biosens. (Basel) **13**, (2022). https://doi.org/10.3390/bios13010065
42. *Rapid biomarker screening of Alzheimer's disease by interpretable machine learning and graphene-assisted Raman spectroscopy*
43. M. Hassan et al. Diagnosis of dengue virus infection using spectroscopic images and deep learning. PeerJ Comput. Sci. **8**, e985 (2022). https://doi.org/10.7717/peerj-cs.985
44. S. Ali, M. Hassan, M. Saleem, S.F. Tahir, Deep transfer learning based hepatitis B virus diagnosis using spectroscopic images. Int. J. Imaging Syst. Technol. **31**, 94–105 (2020). https://doi.org/10.1002/ima.22462
45. X. Zhang et al., Rapid diagnosis of membranous nephropathy based on serum and urine Raman spectroscopy combined with deep learning methods. Sci Rep **13**, 3418 (2023). https://doi.org/10.1038/s41598-022-22204-1
46. W. Mo et al., Classification of coronavirus spike proteins by deep-learning-based raman spectroscopy and its interpretative analysis. J Appl Spectrosc+. **89**, 1203–1211 (2023). https://doi.org/10.1007/s10812-023-01487-w
47. Post et al., Application of laser-induced, deep UV Raman spectroscopy and artificial intelligence in real-time environmental monitoring—solutions and first results. Sensors, (2021)
48. H. He et al., Speeding up the line-scan Raman imaging of living cells by deep convolutional neural network. Anal Chem **91**, 7070–7077 (2019). https://doi.org/10.1021/acs.analchem.8b05962
49. W. Lu, X. Chen, L. Wang, H. Li, Y.V. Fu, Combination of an artificial intelligence approach and laser tweezers Raman spectroscopy for microbial identification. Anal Chem. **92**, 6288–6296 (2020). https://doi.org/10.1021/acs.analchem.9b04946
50. M.K. Maruthamuthu, A.H. Raffiee, D.M. De Oliveira, A.M. Ardekani, M.S. Verma, Raman spectra-based deep learning: A tool to identify microbial contamination. Microbiol. **9**, e1122 (2020). https://doi.org/10.1002/mbo3.1122
51. J.Y. Lu et al. Identification of antibiotic resistance and virulence-encoding factors in Klebsiella pneumoniae by Raman spectroscopy and deep learning. Microb Biotechnol **15**, 1270–1280 (2022). https://doi.org/10.1111/1751-7915.13960
52. J. Dong, M. Hong, Y. Xu, X. Zheng, A practical convolutional neural network model for discriminating Raman spectra of human and animal blood. J Chemometr. **33**, (2019). https://doi.org/10.1002/cem.3184
53. S. Huang et al., Blood species identification based on deep learning analysis of Raman spectra. Biomed Opt Express. **10**, 6129–6144 (2019). https://doi.org/10.1364/BOE.10.006129
54. Y. Tao, J.Q. Bao, Q. Liu, L. Liu, J.Q. Zhu, Application of deep-learning algorithm driven intelligent Raman spectroscopy methodology to quality control in the manufacturing process of guanxinning tablets. Molecules **27**, (2022). https://doi.org:ARTN696910.3390/molecules27206969
55. O. Peterfi, Z.K. Nagy, E. Sipos, D.L. Galata, Artificial intelligence-based prediction of in vitro dissolution profile of immediate release tablets with near-infrared and Raman spectroscopy. Period Polytech-Chem, (2023). https://doi.org/10.3311/PPch.20755
56. C. Post et al., Possibilities of real time monitoring of micropollutants in wastewater using Laser-Induced Raman & Fluorescence Spectroscopy (LIRFS) and Artificial Intelligence (AI). Sensors **22**, (2022). https://doi.org:ARTN466810.3390/s22134668
57. H. Shi et al., Setting up a surface-enhanced Raman scattering database for artificial-intelligence-based label-free discrimination of tumor suppressor genes. Anal Chem **90**, 14216–14221 (2018). https://doi.org/10.1021/acs.analchem.8b03080
58. T.A. Saifuzzaman, K.Y.L., A.R.M. Radzol, P.S. Wong, I. Looi, Optimal scree-CNN for detecting NS1 molecular fingerprint from salivary SERS spectra. IEEE Xplore, (2020)
59. H. Shin et al. Early-stage lung cancer diagnosis by deep learning-based spectroscopic analysis of circulating exosomes. ACS Nano **14**, 5435–5444 (2020). https://doi.org/10.1021/acsnano.9b09119

60. N. Cheng, J. Fu, D. Chen, S. Chen, H. Wang, An antibody-free liver cancer screening approach based on nanoplasmonics biosensing chips via spectrum-based deep learning. NanoImpact **21**, 100296 (2021). https://doi.org/10.1016/j.impact.2021.100296
61. K.N. Prajapati, A.A. Nair., S. Ravi, P. Silva, J. Mitra, *Hyperspectral imaging with Raman scattered photons: A new paradigm in Raman analysis*. arXiv (2021). https://doi.org/10.48550/arXiv.2106.02876
62. F.U. Ciloglu et al. Drug-resistant staphylococcus aureus bacteria detection with the combination of surface enhanced Raman spectroscopy and deep learning techniques. Sci Rep-Uk **11**, (2021). https://doi.org/10.21203/rs.3.rs-536102/v1
63. X. Fang, et al., Fast discrimination of tumor and blood cells by label-free surface-enhanced Raman scattering spectra and deep learning. J. Appl. Phys. **129**, (2021). https://doi.org/10.1063/5.0042662
64. Q. Fu et al., Rapid identification of the resistance of urinary tract pathogenic bacteria using deep learning-based spectroscopic analysis. Anal Bioanal Chem **413**, 7401–7410 (2021). https://doi.org/10.1007/s00216-021-03691-z
65. J. Huang et al., On-site detection of SARS-CoV-2 antigen by deep learning-based surface-enhanced Raman spectroscopy and its biochemical foundations. Anal Chem **93**, 9174–9182 (2021). https://doi.org/10.1021/acs.analchem.1c01061
66. M. Peng et al., Deep learning-based label-free surface-enhanced raman scattering screening and recognition of small-molecule binding sites in proteins. Anal Chem **94**, 11483–11491 (2022). https://doi.org/10.1021/acs.analchem.2c01158
67. W. Mo et al. Combined morphological and spectroscopic diagnostic of HER2 expression in breast cancer tissues based on label-free surface-enhanced Raman scattering. Anal Chem **95**, 3019–3027 (2023). https://doi.org/10.1021/acs.analchem.2c05067
68. Y. J. Yang et al., Rapid detection of SARS-CoV-2 RNA in human nasopharyngeal specimens using surface-enhanced Raman spectroscopy and deep learning algorithms. Acs Sensors, (2022). https://doi.org/10.1021/acssensors.2c02194
69. Y. Zhang, H. Hong, W. Cai, Imaging with Raman spectroscopy. Curr. Pharm. Biotechnol. **11**, 654–666 (2010)
70. B. Manifold, E. Thomas, A.T. Francis, A.H. Hill, D. Fu, Denoising of stimulated Raman scattering microscopy images via deep learning. Biomed Opt Express **10**, 3860–3874 (2019). https://doi.org/10.1364/BOE.10.003860
71. Y. Suzuki et al., Label-free chemical imaging flow cytometry by high-speed multicolor stimulated Raman scattering. Proc Natl Acad Sci USA **116**, 15842–15848 (2019). https://doi.org/10.1073/pnas.1902322116
72. L. Zhang et al., Rapid histology of laryngeal squamous cell carcinoma with deep-learning based stimulated Raman scattering microscopy. Theranostics **9**, 2541–2554 (2019). https://doi.org/10.7150/thno.32655
73. T.C. Hollon et al., Near real-time intraoperative brain tumor diagnosis using stimulated Raman histology and deep neural networks. Nat Med **26**, 52–58 (2020). https://doi.org/10.1038/s41591-019-0715-9
74. A.H. Hill, B. Manifold, D. Fu, Tissue imaging depth limit of stimulated Raman scattering microscopy. Biomed Opt Express **11**, 762–774 (2020). https://doi.org/10.1364/BOE.382396
75. T.C. Hollon et al., Rapid, label-free detection of diffuse glioma recurrence using intraoperative stimulated Raman histology and deep neural networks. Neuro Oncol **23**, 144–155 (2021). https://doi.org/10.1093/neuonc/noaa162
76. Q. Zhang et al., Automatic cell counting from stimulated Raman imaging using deep learning. PLoS One **16**, e0254586 (2021). https://doi.org/10.1371/journal.pone.0254586
77. Z. Liu et al., Instant diagnosis of gastroscopic biopsy via deep-learned single-shot femtosecond stimulated Raman histology. Nat Commun **13**, 4050 (2022). https://doi.org/10.1038/s41467-022-31339-8
78. C. Jiang et al., *OpenSRH: optimizing brain tumor surgery using intraoperative stimulated Raman histology*. arXiv (2022)

79. S. Weng, X. Xu, J. Li, S.T.C. Wong, Combining deep learning and coherent anti-Stokes Raman scattering imaging for automated differential diagnosis of lung cancer. J Biomed Opt **22**, 1–10 (2017). https://doi.org/10.1117/1.JBO.22.10.106017
80. K. Aljakouch et al., Fast and noninvasive diagnosis of cervical cancer by coherent anti-stokes Raman scattering. Anal Chem **91**, 13900–13906 (2019). https://doi.org/10.1021/acs.analchem.9b03395
81. N. Yamato, M. Matsuya, H. Niioka, J. Miyake, M. Hashimoto, Nerve segmentation with deep learning from label-free endoscopic images obtained using coherent anti-stokes raman scattering. Biomol. **10**, (2020). https://doi.org/10.3390/biom10071012
82. N. Yamato, H. Niioka, J. Miyake, M. Hashimoto, *Improvement of nerve imaging speed with coherent anti-stokes Raman scattering rigid endoscope using deep-learning noise reduction* (2020)

Chapter 10
Covid-19 and Pneumonia Detection from Chest X-Ray Images by Deep Learning Model

Santanu Roy

Abstract This chapter presents an overview of automatic Covid-19 and Pneumonia detection from Chest X-Ray (CXR) Images by using Deep Learning models. More specifically, in this chapter, a deep analysis of existing Convolution Neural Network (CNN) models (for Covid-19 detection), is presented from the perspective of data imbalance. Covid-19 disease had a disastrous effect on human life, since 2020. One of the effective ways of diagnosing Covid-19 disease is to detect it from Chest X-Ray (CXR) images, because it is observed that Covid-19 may damage the human respiratory system (or, lung region). Hence, the impact changes due to Covid are very prominent on the Chest X-Ray (CXR) images. This chapter is further divided into two types of Covid-19 detection categories: (I) Covid-19 and Pneumonia detection by pre-trained CNN models, (II) A framework for alleviating class imbalance problem from the CXR dataset: Here, the architecture of the standard CNN model is not changed, but modified (or, weighted) loss function or, Ensemble models or, Re-sampling techniques are deployed in order to overcome the class imbalance problem. Furthermore, some of the existing CNN model's results are implemented and analyzed on an imbalanced (and readily available) CXR dataset in which there are four classes: (a) Covid, (b) Normal, (c) Viral Pneumonia (VP), (d) Lung Opacity (LO). Experimental results reveal that, Xception model has outperformed other existing pre-trained models. Furthermore, experimental results suggest that, Mobile Freeze-Net with Attention-based loss function, not only provides the highest accuracy on this CXR dataset, but also, has improved its generalization ability considerably.

S. Roy (✉)
Computer Science and Engineering, Pandit Deendayal Energy University (PDEU), Gandhinagar, India
e-mail: Santanu.Roy@sot.pdpu.ac.in

NIIT University, Rajasthan, India

A. Gogoi and N. Mazumder (eds.), *Biomedical Imaging*, Biological and Medical Physics, Biomedical Engineering, https://doi.org/10.1007/978-981-97-5345-1_10

10.1 Introduction

Covid-19 disease has been affecting on mortality and health of the global population considerably [1], for the last three years. This disease is caused by Severe Acute Respiratory Syndrome-2 (SARS-2) [2], which is a very contagious virus and it spreads more by the physical contact of human beings. Another important feature of SARS-2 is that it directly affects human immune system, moreover, it can damage our respiratory system severely [3]. COVID-19 patients generally have symptoms like cough, fever, breathing trouble, tastelessness, fatigue, headache, etc. However, recent studies showed that, it can be spread without necessarily developing Covid symptoms [4], which makes Covid-19 more dangerous and unpredictable. The mortality rate due to COVID-19 disease had increased by 10–12% (average from 2020–21) in USA than the normal mortality rate. In North USA it was 25–27% [5]. These numbers were significantly higher in 2020–21 and thus, deadly Corona virus has been the most concerning disease, in the history of mankind. WHO already declared Covid-19 as a pandemic in March, 2020. However, in recent months (from April 2022 to January 2023), the influence of Covid-19 disease has slightly decayed in terms of death. Total, there were four different Covid waves so far, at different times during 2020–2022. 3rd wave and 4th wave had comparatively lesser confirmed cases and lesser deaths [6] than that of wave 1 and wave 2. This was possible due to the heavy consumption of several vaccines for Covid-19, which were invented by numerous scientists in UK [7], in 2021. However, different kinds of Covid variants use to strike on different occasions and intervened in human life considerably in recent times. For example, the variant BF.7 had struck in China [8] in the end of 2022. Thus, it seems like the Covid disease will influence the human life for quite long time. Besides that, distinguishing Covid-19 affected patients from other variants of patients like Pneumonia, Viral Pneumonia (VP), Tuberculosis (TB) etc. from Chest X-Ray images dataset, is also a very challenging task for doctors and researchers. Diagnosing appropriate disease (whether it is VP or Covid) is very much necessary for further treatment of the patients. Therefore, an automatic diagnosis system for Covid-19 and Pneumonia disease from CXR images, is still a hot research topic, in recent advent.

Several techniques have been employed so far by researchers in order to diagnose Covid-19 disease effectively. The various sources for Covid-19 detection are Radiology images (e.g. Chest X-Ray images [9], Computational Tomography (CT) images [10]), Reverse Transcriptase-Polymerase Chain Reaction (RTPCR) etc. [11]. However, the most frequently employed RTPCR technique is a long and manual process and it takes generally 1–2 days for the final Covid diagnosis. As a result, RTPCR has been fraught with the problem of less accuracy and less sensitivity for Covid-19 detection [12]. According to numerous researchers, Deep learning models have been able to extract significant features from CT images [10, 13] in order to detect Covid disease efficiently. However, generating such CT images in hospitals is a very costly method than producing CXR images. That means, the availability of CT images' dataset is very limited [14]. Therefore, Covid-19 disease detection from Chest X-Ray (CXR) images by machine learning or deep learning technique has been the automatic choice so far, in the recent advent. Moreover, SARS-2 directly attacks the human respiratory system and consequently, the effect of that is very much prominent in human lung system, thus, choosing Chest X-Ray images for detecting Covid-19 disease is very much feasible [14]. Furthermore, developing such an automatic detection system (of disease detection) from CXR images, is not only useful for the diagnosis of Covid-19 disease, but also, it is helpful for differentiating diseases like Pneumonia, Viral Pneumonia, Lung Opacity, Tuberculosis etc. Because all these diseases have direct effects in human lung system as well and sometimes it is very difficult task for a doctor to separate amongst several variants of Pneumonia diseases. Hence, automatic Covid-19 and Pneumonia detection from CXR images is very essential step that would reduce the workload of doctors and furthermore, this would leverage the doctors to further conduct the treatment in right direction.

10.2 Existing CNN Models for Covid-19 and Pneumonia Detection from CXR Dataset

Numerous researchers have employed pre-trained CNN models, in order to detect Covid-19 disease and Pneumonia from Chest X-Ray (CXR) images. Moreover, some of the researchers [12, 15–17] have employed novel CNN architecture specially dedicated to Covid detection. However, conventional CNN models are fraught with class imbalance problem [18], since most of the medical dataset created in hospitals are indeed highly imbalanced. Therefore, numerous researchers [18–20] have employed novel frameworks in order to diagnose Covid disease from imbalanced CXR datasets. Here, they did not change the CNN model architecture, rather they changed the framework either by using several over-sampling or under-sampling techniques [21], or they incorporated Weighted loss function [19] or, Ensemble technique [20] in order to alleviate the class imbalance problem from the CXR dataset. In this research, first, some of the pre-trained CNN models (trained from the ImageNet dataset) are explored. Thereafter, some of these novel frameworks are explained in depth which works in the direction of alleviating class imbalance problem from CXR Images.

10.2.1 Basic Concept of CNN Model

The idea of a Convolutional Neural Network (CNN) was introduced in the early 2010s, especially for computer vision task. A CNN model [22] consists of three different parts: (I) Convolution Layers, (II) Pooling Layers, (III) Fully Connected (FC) Layers followed by an output layer, shown in Fig. 10.1. Convolution layers automatically extract the features in 2-Dimension, by utilizing Convolution operation. The significance of convolution operation is that it is translation invariant [23]. That means, after applying some translation (or, linear transformation), convolution operation can extract the same (invariant) features from the images. Thus, Convolution operation is very popular not only for Neural network, but also in conventional image processing. In the convolution layer of CNN model, the values of all (convolution) filter coefficients are chosen randomly. All these filter co-efficient values are trainable in CNN model, which means these values are automatically updated in each epoch, during training of the model by back-propagation algorithm. Generally, several variants of Gradient Descent (GD) algorithm, for example, Stochastic GD, Adagrad [24], RMS Prop, and Adams optimizers [25] are widely utilized in this back-propagation algorithm for training CNN model. On the other hand, in Pooling layer, no parameter is trained. Pooling layers [22] are utilized specially to reduce the size of the image (to half) after every convolution layer so that overall number of hyper-parameters is controllable. Next, there is flatten layer in which all 2-D features (coming from the last convolution layer) are converted into 1-D features. This flatten layer is the input of Fully Connected (FC) Layers. The FC layers are equivalent to hidden layers used in ANN model. Due to reducing the image size in each convolution layer (by pooling layer), the number of input feature vectors in flatten layer reduces significantly. Thus, Pooling layer helps to mitigate the overfitting problem in CNN model to a certain extent. Hidden layers or FC layers in neural network incorporates non-linearity, thus, more complicated task can be accomplished by the CNN model. However, unlike ANN model, CNN model does not need higher number of FC layers (or, hidden layers). Because most of the significant features are automatically extracted in the convolution layer of the CNN model. Another important property of CNN model is that 2-D spatial dimension is reduced as we go deeper into the CNN model, however, the spectral dimension increases in the deeper layers. In the final output layer, generally Softmax [22] activation function is utilized for multi-class classification task and Sigmoid activation function is employed for binary classification task. Wherein, for other layers, ReLU activation function is generally deployed. Moreover, batch normalization layer [26] is frequently employed after every convolution layer. This batch normalization layer enables the deep learning network, to normalize or scale all the feature values around zero value after every convolution block. This enables the neural network to converge faster and reduces the overfitting problem a little bit. Moreover, this batch-normalization layer may also reduce the exploding ReLU problem [27], which happens due to the positive bias of ReLU. A block diagram of a typical CNN model is shown the Fig. 10.1.

The main advantage of any CNN model over ANN model for computer vision task is that, unlike ANN it can preserve all the 2D neighbor pixel information while doing feature extraction. Secondly, CNN model utilizes Pooling layer after every convolution block which helps to reduce the number of hyper-parameters (to train), especially in the flattened layer. As a consequence, the overall number of hyper-parameters can be controlled in CNN model and therefore, it may reduce the overfitting problem in the CNN model considerably.

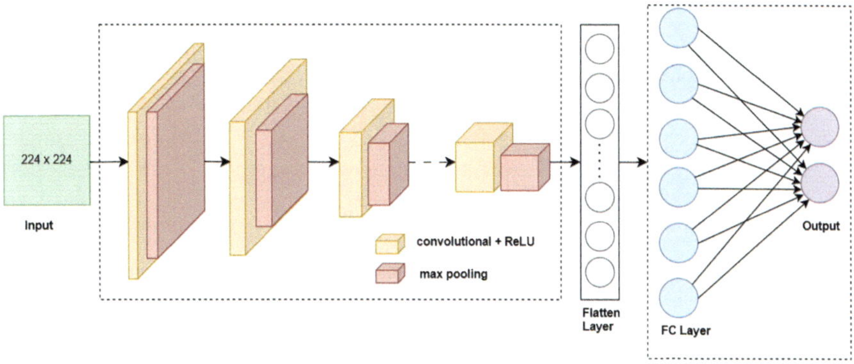

Fig. 10.1 Block diagram of CNN model

10.2.2 Pre-Trained CNN Models

There are some standard CNN models available in Keras library, like Alex-Net [28], VGG-16 [29], VGG-19, Inception-V2 [30], Xception [31], MobileNet-V3 [32], ResNet-50 [33], ResNet-101 etc. Numerous researchers already performed these models on a very large dataset ImageNet [34] which has more than 3,00,000 number of images and 1000 classes. All the pre-weights of these models have been saved in Keras library. Hence, while training a standard CNN model, one can do transfer learning from ImageNet dataset to a small dataset. As a result, CNN model can converge faster. More specifically, from the very first epoch, CNN model will learn in the right direction. Thus, a pre-trained model may give higher accuracy, precision, recall and F1 score etc. even on a very small dataset and furthermore, it can avoid the overfitting problem.

VGG-16 and Alex-Net Model: The VGG-16 model, the modified version of Alex-Net [28, has the simplest CNN model architecture so far, proposed by Visual Graphics Group in 2015 29]. Block diagram of VGG-16 model is further shown in Fig. 10.2. Unlike Alex-Net, VGG-16 model incorporates non-overlapping Max-Pooling layer which has the pooling size 2×2 and stride $= 2$. Whereas, Alex-Net model employs overlapping pooling with pool size is 3×3 and stride $= 2$. Over-lapping pooling may preserve more spatial information, however, it is more computationally costly than the non-overlapping Max-Pooling, with pooling size 2 $\times$ 2 and stride $= 2$. Moreover, unlike Alex-Net architecture, VGG-16 takes all the convolution filter size same, i.e., 3×3. On the other hand, Alex-Net architecture uses 11×11 kernel size at the 1st convolution layer and 5×5 kernel size at the 2nd convolution layer. Therefore, the overall computational complexity of VGG-16 model has become considerably lesser than that of Alex-Net model. In Alex-Net, total 8 number of trainable layers (including convolution layer and FC layers) are utilized. Whereas, in VGG-16 total 16 number of trainable layers are deployed. Despite of that, VGG-16 model has been able to reduce overfitting and complexity of the model, due to its very simple architecture. There is another version of VGG

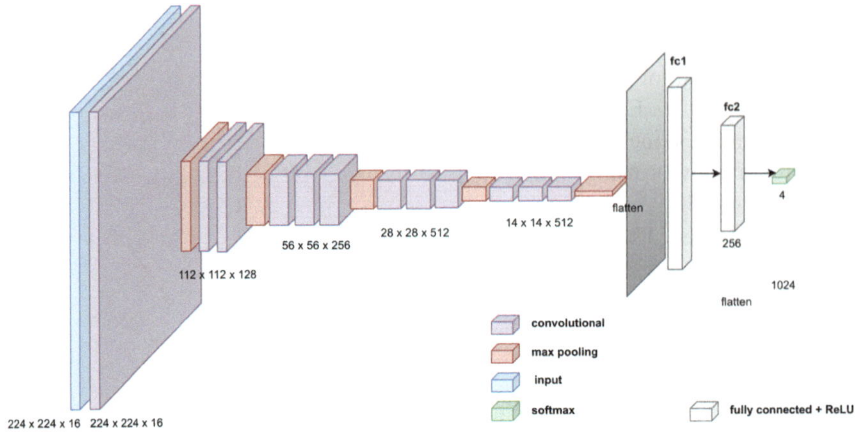

Fig. 10.2 Block diagram of VGG-16 model

model, which is known as VGG-19. Here, total 19 number of layers are deployed (extra 3 convolutional layers utilized). VGG-16 model is further implemented on an imbalanced CXR dataset. The results are further analyzed in Sect. 10.4.

Inception-V3: Inception-V3 (also known as Google Net) is one of the most popular inventions by Google [30], in the domain of CNN. Inception model is inspired by the concept of multi-scale (or, multi-resolution) feature extraction. In many images, the objects (or, foregrounds) may not always appear on the same scale. Thus, multi-scale 3×3, 5×5 filter sizes are deployed in the inception model. One unit of Inception block is shown in Fig. 10.3. Here, the purpose of deploying 1×1 filter prior to 3×3 or 5×5 kernel is to reduce (or, to control) the spectral dimension in CNN model [30] so that the overall number of hyper-parameters will be less. This 1×1 filter does not extract any features, rather it scales the features across different channels in the CNN model and thus, it also enables the model to learn features in the spectral dimension as well. A parallel connection of Max-pooling layer with padding size same and stride 1, is also deployed in the basic unit block of Inception-V3 which is shown in Fig. 10.3. The significance of this connection is that, Max-pooling with stride 1 does not reduce the size of the image, however, it reduces the redundant features inside images. Thus, a Max-pooling with stride $= 1$ followed by 1×1 filter, can produce images whose statistics differ a lot from the original images. Hence, this connection works as a data-augmenter (inside the model), to the best of my knowledge. This unit block, shown in Fig. 10.3, is repeated in the Inception-V3 model total 11 number of times, which is then followed by a flatten layer, 3 FC layers and an output layer. The most important thing about Inception model is that it utilizes less number of hyper-parameters (or controllable no. of hyper-parameters) than other previously mentioned CNN models, despite its complex architecture.

Xception Net: Xception Net [31] first time introduced the concept of Depth-wise separable convolution layers. Instead of conventional way of computing convolution,

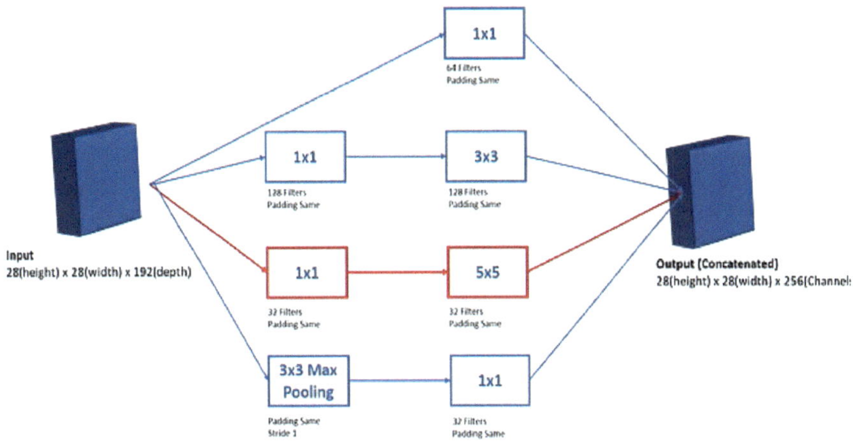

Fig. 10.3 One basic unit of Inception-V3 model

it factorizes the convolution layer into two separate layers, one is Depth-Wise convolution layer where it performs lightweight filtering by employing single convolutional filter per input channel. Another is 1×1 convolution layer, which is point-wise convolution layer. Here, the Point-wise multiplication further reduces the computational complexity considerably, compared to standard convolution layer. Employing Depth-wise separable convolution layers in the CNN model, it may lose very little information, however, this is bearable if it reduces the number of hyper-parameters 10 to 100 times. Many of the modern CNN architectures [35] are inspired from this Xception model, in which they have incorporated both traditional convolution layers and Depth-wise separable convolution layers for reducing overfitting in the model. The basic Xception model is however inspired from the basic Inception-V3 model. In other words, in the Xception Net, convolution layers of Inception-V3 are replaced by the Depth-wise separable convolution layers. Overall, the other components of the model architecture of Xception and Inception-V2 are very much the same.

ResNet-50: He et al. [33] first time observed that increasing the number of layers in the CNN model further increases both the training error and test error. They have found that this is not the case of over-fitting, rather this happens due to the vanishing gradient problem. In a deeper CNN model (having higher number of layers), the Gradient values diminish during the back-propagation algorithm, even after utilizing the ReLU activation function. In order to resolve this problem, authors have deployed small skip connections which is a direct connection from the input of one convolution layer to the output of the next convolution layer. A basic unit of a Res-Net model is further shown Fig. 10.4. This skip connection actually increases the strength of information flow through the deep learning model. Therefore, it is likely that Gradient values will not be reduced to 0, even if it is a heavy model. As a consequence, the training and testing error will not further increase simultaneously. Rather it can improve the accuracy considerably than that of a lighter CNN model (having lower number of layers). Hence, ResNet-50 outperforms all other CNN models like VGG-16, VGG-19, Inception-V2, etc. The ResNet model is also called as Residual model, due to this skip connection. A CNN architecture having this kind of skip connection is called Residual structure. This skip connection allows the Neural network model to incorporate more number of layers, without having large overfitting and Vanishing Gradient problem. However, the performance of the Residual model highly depends on the size of dataset. ResNet-50 model has total 50 number of layers, whereas, another ResNet-101 and ResNet-152 models have 101 and 152 number of layers respectively.

Mobile Net-V2: Another standard MobileNet-V2 model [32] is frequently employed by numerous researchers, on the Covid-19 CXR dataset. The architecture of Mobile-Net V2 can be further observed in detail from the Fig. 10.5. Mobile Net-V2 resembles all the features of original Mobile Net-V1 model [36], with the exception that it uses inverted residual blocks between two bottleneck layers. The total number of layers in Mobile Net-V2 is 154. Despite having higher number of layers (154), MobileNet-V2 uses residual connections which makes the complicated network simpler and it reduces the vanishing gradient problem from the Deep learning model. Moreover, due to utilizing Depth-wise Separable convolution layers, this model reduces its computational cost and the number of hyper-parameters significantly. Moreover, the Mobile-Net-V2 model employed Global Average Pooling (GAP) [37] instead of flattened layer. This GAP layer further takes global average of the image in spatial dimension and therefore, it reduces the overfitting problem

Fig. 10.4 Basic unit of ResNet-50

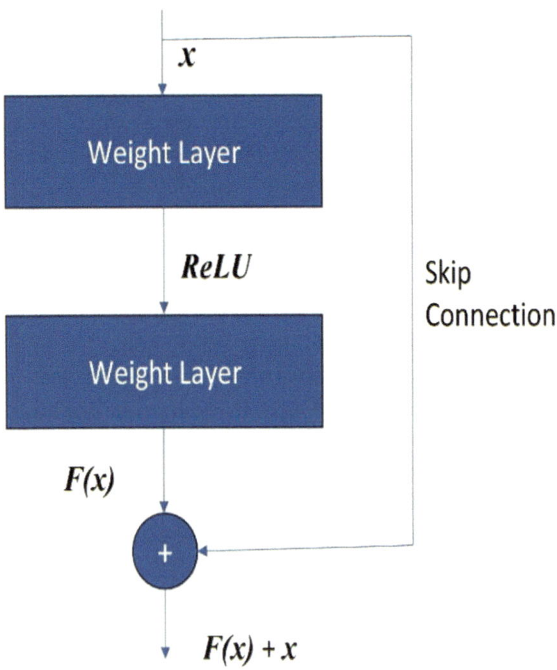

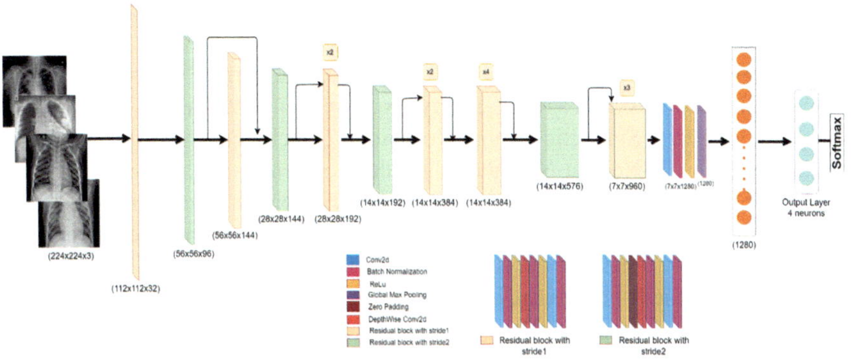

Fig. 10.5 Block diagram of mobile-Net V2

significantly from the CNN model. For example, if the output tensor size (of image) from the last convolution layer is $7 \times 7 \times 1280$, the number of neurons in the input of FC layer (which is GAP layer in this case), will be only 1280 instead of 62,720. Thus, it reduces the number of hyper-parameters by a significant margin at the first FC layer. Hence, MobileNet-V2 model has the ability to accomplish a complicated task, without having a great effect of over-fitting. We have employed this MobileNet-V2 (pre-trained) on an imbalanced CXR dataset which is further explored in the Results and Analysis Section.

10.2.3 CNN Frameworks for Alleviating Class Imbalance Problems from CXR Datasets

Class imbalance is a very common problem in computer vision and medical image diagnosis. Most of the existing datasets in medical field are considerably imbalanced, which means the number of images in one class is significantly higher than that of other classes. Due to class imbalance problem in the dataset, conventional CNN model may miss-classify some of the images of minor classes into major classes. Many researchers have proposed several frameworks in order to tackle this class imbalance problem from the CXR dataset. One of the ways to resolve the class imbalance problem is to employ re-sampling methods, that is, discarding images from major classes (under-sampling) randomly and generating synthetic images in minor classes (over-sampling) so that number of images per class will be comparable or similar. However, under-sampling is not very feasible method for a highly class imbalance problem. Excluding large number of images from a major class can reduce the challenges in a dataset, moreover, it can remove some images which had some important features. One of the very common methods of over-sampling is to deploy data-augmentation in minor class (by rotation, horizontal and vertical flipping etc.). These linear transformations can generate some synthetic images in minor classes and thus, can reduce the class imbalance problem from the dataset. However, the limitation of any over-sampling technique (addressed by Roy et al. [18]) is that, it may produce some synthetic images which are feature variant with respect to the final classification task. As a result, it may give less accuracy. Another popular over-sampling method is Synthetic Minor Oversampling Technique (SMOTE) [38]. This method employs K-Nearest Neighbor algorithm for generating similar images. For image dataset, SMOTE is not so feasible method, because large memory space is required to store all the images in a dataset. One of the most popular over-sampling methods in recent times is Generative Adversarial Network (GAN) [39]. By using GAN, many synthetic images can be generated in the minor class, hence, this method as a pre-processing automatically resolves the class imbalance problem. However, employing GAN as a pre-processing method is very costly and large number of images are required for utilizing GAN oversampling technique.

An entirely different way of resolving this class imbalance problem, is to incorporate Weighted Categorical Cross Entropy (WCCE) [18] on the standard CNN model. The conventional WCCE is presented by the following equation.

$$L_{WCCE}(y, p) = -\frac{1}{N}\sum_{i=1}^{N_c}\sum_{c=1}^{K}\beta_c \cdot \log(p_{i,c}) \tag{10.1}$$

Here, in Eq. (10.1), $p_{i,c}$ is the probability that the ith sample belongs to class c, β_c is the attention weight for a particular class c. In the conventional CCE, this β_c is equal for every classes i.e., $\beta_c = 1$. However, in WCCE this β_c is not same for every class, rather it is chosen inversely proportional to the number of instances in that class c. For example, there are two classes: 80 number of Apples, 20 number of Oranges. In order to resolve this class imbalance problem, Neural network should give little bit more attention to the "Orange" class than "Apple" class. Here, this weight β_c should be chosen 1:4 for Apple: Orange class, because the ratio of number of instances in Apple: Orange = 4:1. Hence, changing this attention weight, neural network will automatically give more attention towards minor class than the major class. As a result, CNN model will automatically resolve the problem of 'class imbalance'. This second method of incorporating WCCE on CNN model, is more efficient than employing oversampling or under-sampling. Because in case of oversampling by image processing technique or GAN, it may be effective in one dataset. However, there is no guarantee whether the same method will work efficiently in a different dataset or not. Nevertheless, employing only WCCE may not improve the accuracy of CNN model considerably. Hence, numerous researchers [18, 19] have deployed the combination of both over-sampling and WCCE loss function, in order to resolve class imbalance problem from the imbalanced CXR images.

Numerous researchers have employed various methodology in order to alleviate class imbalance problem for Covid-19 disease detection from CXR images. This is further explained below:

10.2.3.1 Ensemble of Deep Learning

Win et al. [20] have employed an Ensemble method of five CNN models in order to detect Covid and Pneumonia from a highly imbalanced CXR dataset. First, they employed pre-processing method (image contrast enhancement) along with Lung segmentation by Deep Lab V3 model. Furthermore, to alleviate class imbalance problem from the dataset, they incorporated five different approaches (I) Weighted Categorical Cross Entropy (WCCE) loss function in CNN model. Moreover, they employed (II) over-sampling by image processing, (III) Random under-sampling, (IV) Image data-augmentation by Image Data generator, (V) Hybrid sampling, that is, employing both under-sampling and over-sampling on the dataset. Thereafter, they combined all these results in an Ensemble classifier in which the final decision is made by majority voting of those five models. Despite their method seems like quite obvious and valid direction to follow, employing ensemble model of five different techniques increases the number of hyper-parameters five times. Moreover, their method is computationally very costly, since five different re-sampling methods they have employed.

10.2.3.2 Fuzzy Color and Stacking Based Data Augmentation

Togacar et al. [40] proposed a novel framework in which first they have chosen 458 images from different set of databases in which they have included three classes: Covid, Normal and Viral Pneumonia. Thereafter, they have created two different augmented datasets, by using fuzzy color-based image processing method and original dataset stacking technique. The main purpose of utilizing fuzzy color based pre-processing technique was to convert original images into such images which are having lesser noise. Moreover, a new dataset is made by stacking technique which is incorporated in order to fuse the original images with those processed images and this further provides better clarity and contrast. For experimentation, they have employed two very lightweight CNN models, they are Mobile-Net V2 and Squeeze-Net for feature extraction. Furthermore, an SVM classifier is utilized for final classification task. The reason for choosing those light weight models, is that the number of images in their dataset was very less. However, from our perspective, their method is not exactly automatic method because they manually excluded many images from the dataset in order to remove class imbalance.

10.2.3.3 Attention-Based VGG-16

An attention based VGG-16 model for Covid-19 detection from CXR images, was recently proposed by Sitaula et al. [15]. They have employed a parallel connection at the 4th pooling layer of VGG-16 model. This parallel connection consists of a Combination of both Max-Pooling and Average Pooling which is followed by a convolutional layer, of having only 1 filter, kernel size 7×7 and this is followed by a

Sigmoid activation function. Authors claimed that this parallel connection provides more attention towards spatial relationships of pixels inside CXR images, which was inspired from Se-Net model. Consequently, at the last convolution module, VGG-16 extracts more significant features which are relevant for the final classification task. Despite the fact that this model produces slightly better results than other standard CNN models like Inception-V3, ResNet-50, DenseNet-121, and Mobile-Net, the authors did not compare their results with the conventional VGG-16 model as it is. Thus, exactly how much improvement is achieved just because of imposing the attention block is not clear. Overall, they have experimented with three different classes, having 3–5 different classes. (I) Covid, (II) Normal, (III) Pneumonia Bacteria, (IV) Viral Pneumonia, (V) No findings. In this study, we have implemented their model as mentioned in their paper on a highly imbalanced CXR dataset.

10.2.3.4 SMOTE and Weighted CCE

Chamseddine et al. [19] have employed pre-trained models DenseNet-201 [41], VGG-19, Xception, ResNet-152, Mobile Net V2 etc. on three different CXR datasets. Transfer learning of all these models is done from ImageNet dataset to CXR dataset. The authors first proposed SMOTE oversampling technique [38] in order to resolve the class imbalance problem from CXR dataset. Thereafter, they have also proposed Weighted CCE (WCCE) [18] in which class weights are assigned inversely proportional to the number of images. Eventually they have proposed DenseNet-201 and VGG-19 model and on top of that, they have employed WCCE and SMOTE. However, the first dataset they employed was not very challenging. Thus, without using WCCE and SMOTE, DenseNet-201 model already achieved 97% accuracy. On top of it, SMOTE has enabled their model to improve its performance by 1% only, further deploying WCCE on the model it had actually decreased the performance by 1%. On the third dataset, WCCE has improved its performance by 3%. Further, this framework (i.e. WCCE on DenseNet-201) has been implemented on our employed CXR dataset which is explored in the Results and Analysis section.

10.2.3.5 SVD-CLAHE Boosting and Balanced Loss Function

Roy et al. [18] have recently proposed a novel data-augmentation technique called ‘SVD-CLAHE Boosting’ based on combining Over-sampling and Under-sampling methods. First, they employed Random Under-sampling (RUS) from an imbalanced CXR dataset, in order to randomly exclude few images from the major classes. Thereafter, they performed a novel SVD-based image processing technique and CLAHE 0.5, CLAHE 1.0, in order to generate some synthetic images, especially in minor classes. They had identified that conventional data-augmentation method like flipping (horizontal and vertical), rotation, zooming and cropping had the limitation that their method is not exactly feature invariant for CXR images. The authors claimed that their proposed data-augmentation method ‘SVD-CLAHE Boosting’

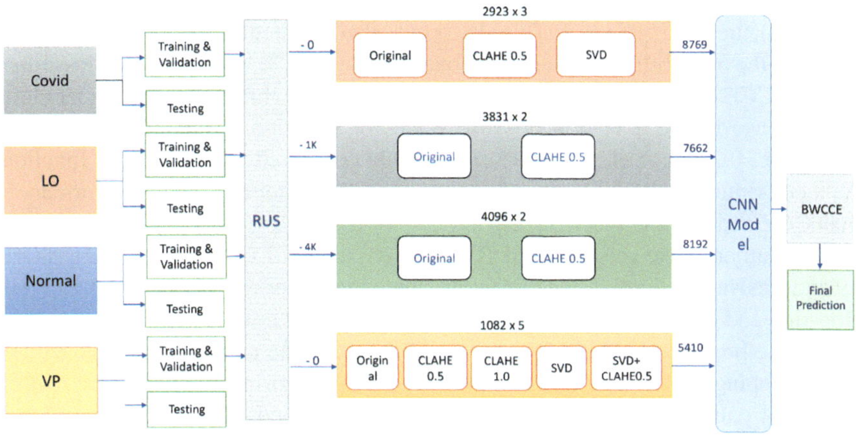

Fig. 10.6 Block diagram of SVD-CLAHE boosting method

neither produced images whose statistics are too much different from the original images, nor produced exactly the same replica of images. Rather, their method produced images, having slightly different statistics in terms of contrast and luminance. Thus, their proposed data-augmentation method overcomes the limitations of conventional data-augmentation methods. The entire scheme of their proposed Data-augmentation method for resolving class imbalance is presented in Fig. 10.6.

Furthermore, the authors have identified the limitation of conventional WCCE. They have identified that if one of the minor classes is given higher attention, then the performance in that class is improved considerably. However, in other major classes there is fluctuation of performances due to giving them very less attention. To overcome this problem, the authors have proposed a novel loss function, called as Balanced Weighted Categorical Cross Entropy (BWCCE) in which the attention weight for every class is assigned based on the probabilistic notion. The BWCCE loss function is further presented in the following two equations.

$$L_{BWCCE}(y, p) = -\frac{1}{N}\sum_{i=1}^{N_c}\sum_{c=1}^{K}\Upsilon_c \cdot log\left(p_{i,c}\right) \tag{10.2}$$

where Υ_c is the attention weight for class c and it is assigned by the following equation.

$$\Upsilon_c = \frac{1}{K-1} * \left(1 - \frac{n_c}{\sum_c n_c}\right) \tag{10.3}$$

where, 'K' is the number of classes and $K > 1$, n_c is the total number of images in class c, $\sum_c n_c$ is the total number of images in the entire dataset. Moreover, authors have proved one *Lemma* that if the attention weight Υ_c is chosen by the formula in

Eq. (10.3), then $\sum_{c=1}^{K} \Upsilon_c = 1$. This also proves that the attention weights are assigned based on the probabilistic notion. Moreover, unlike WCCE, their proposed loss function BWCCE will not give very higher attention (or, weightage) to a particular class. In other words, the deviation of these values of attention weights from the mean value (i.e., 0.25 for 4 class), decreased notably compared to WCCE loss function. As a consequence, BWCCE loss function provides better and stable performances than CCE and WCCE.

The authors have incorporated BWCCE on standard CNN models like VGG-19 and ResNet-50. Experimental results shown that, after employing SVD-CLAHE Boosting and BWCCE loss function on CNN models, it not only alleviates the class imbalance problem, but also, it converges faster than the previous frameworks. An early stopping criterion is also utilized based on monitoring validation loss with 5 epochs of patience. The authors claimed that their proposed framework produced the best results in terms of accuracy, precision, recall, F1 score and AUC, on the imbalanced CXR dataset so far. Further results are shown in the Results and Analysis Section.

10.2.3.6 Custom Weighted Balanced CCE

Tyagi et al. [43] recently have proposed a Custom Weighted Balanced Categorical Cross Entropy (CWBCCE), incorporated on ResNet-18 model, in order to overcome the class imbalance problem from the CXR dataset. Unlike the other recent works, they did not employ any over-sampling or data-augmentation technique for alleviating class imbalance problem. First, they observed that the employed CXR dataset not only has class imbalance problem, but also, there was huge intra-class variance in a particular class Covid. Thus, they had chosen attention weights of the proposed loss function not only inversely proportional to the number of images in that class, but also, it is chosen proportional to some statistical inference (intra-class variance), computed from the dataset. The equation of their loss function is given in the following equation.

$$L_{CWBCCE}(y, p) = -\frac{1}{N} \sum_{i=1}^{N_c} \sum_{c=1}^{K} \Upsilon_c \cdot log\left(p_{i,c}\right) \tag{10.4}$$

where Υ_c is the attention weight given in the class 'c', this is further shown in the Eq. (10.5).

$$\Upsilon_c = \frac{1}{K}\left[1 + \left(\frac{1}{K}\sum_{c=1}^{K} \mu(r_c) - \mu(r_c)\right) + \frac{\left(\frac{\sum_c n_c}{K} - n_c\right)^{\alpha}}{\sum_c n_c}\right] \tag{10.5}$$

Here in Eq. (10.5), $\mu(r_c)$ is the mean of correlation coefficient [44] values between any two images in that class. In other words, $\mu(r_c)$ is the measurement of intra-class similarity in the class c. This intra-class similarity is inversely proportional to the intra-class variance. Therefore, the final attention weight Υ_c should be inversely proportional to this $\mu(r_c)$. The factor $\frac{\left(\frac{\sum_c n_c}{K} - n_c\right)^{\alpha}}{\sum_c n_c}$ is class imbalance factor and this should also be inversely proportional to the number of images n_c in a class c. In order to give slightly less attention to this class imbalance factor, the authors have deployed "to the power α" to this factor, shown in this Eq. (10.5). Additionally, the authors have proved three Lemmas in order to support the theory behind their loss function.

Moreover, they have trained ResNet-18 model from scratch on a large ChexPert dataset (consisting of CXR dataset) and thereafter, they do the transfer learning from ChexPert datset to the employed Covid-19 CXR dataset. Subsequently, they have incorporated the proposed loss function CWBCCE on this ResNet-18 model. Their method of employing CWBCCE loss function has improved 3–4% (for 4 class problem) from the original ResNet-18 model with CCE loss function. Furthermore, they have shown one Grad-Cam diagram in which proposed CWBCCE loss function on ResNet-18 model gives more attention to the lung regions or suspected regions for disease than other regions. Whereas ResNet-18 model on CCE could not give more attention to the Region of Interest (RoI). This also validates their proposed loss function. Further results are shown in the Results and Analysis part.

10.2.3.7 Mobile Freeze-Net with Attention-Based Loss Function

Roy et al. [45] recently proposed a Mobile Freeze-Net model along with Attention-based Loss Function, for mitigating class imbalance problem from an imbalanced CXR dataset. First, the authors observed that only fine-tuning 50% of the model (and freezing the other 50% layers) may considerably reduce the number of trainable hyper-parameters. Therefore, the model will have enough time to study the challenges of the CXR dataset and thus, it resolves the class imbalance problem (a bit) automatically. Hence, they have proposed a Mobile Freeze-Net, in which first 50% layers of the Mobile NetV2 model [32] are frozen empirically and then train only the last 50% layers of the model. Thereafter, they observed that many of the images from "Normal" class were miss-classified into "Lung Opacity" class and vice versa by this Mobile Freeze-Net model and this happened due to large inter-class similarity present between the class "Normal" and "Lung Opacity". Subsequently, they proposed a novel attention-based loss function that provides more attention to the class, having higher inter-class similarity. For this, they had to compute the attention weights separately for each class from the statistical inference of the dataset itself, by employing a Monte-Carlo method [46] (Fig. 10.7).

The expression of the Attention-based CCE Loss function is expressed in the following Eq. (10.6).

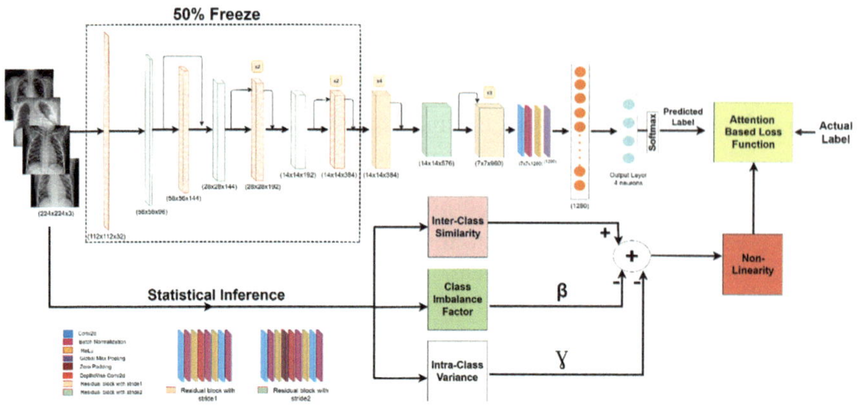

Fig. 10.7 Block diagram of Mobile-freeze net with attention-based loss function

$$L_{AttCCE}(y,p) = -\frac{1}{N}\sum_{i=1}^{N_c}\sum_{c=1}^{K}\alpha_c.log(p_{i,c}) \tag{10.6}$$

$$\alpha_c = r_{inter}(c_1, c_2) + \beta.class_{im}(c) + \Upsilon.\sigma_{intra}(c) \tag{10.7}$$

Here, α_c is the attention weight (for a particular class c), which can be further decomposed by three components shown in Eq. (10.7). $r_{inter}(c_1, c_2)$ is the inter-class similarity [44] between two classes c_1 and c_2, which must be proportional to α_c; this inter-class similarity factor has been computed by measuring correlation co-efficient between any two images from class "Normal" and class "Lung Opacity". Moreover, this "inter-class similarity" component has been given more weightage than the other two statistical inferences i.e., class imbalance factor $class_{im}(c)$ and intra-class variance factor $\sigma_{intra}(c)$. Thus, other two factors are multiplied by $\beta < 1$ and $\Upsilon < 1$ respectively, whereas, $r_{inter}(c_1, c_2)$ is not multiplied with any real constant.

The authors also claimed that, they have got the maximum accuracy of 95% from a challenging imbalanced CXR dataset, by incorporating this attention-based loss function on Mobile Freeze-Net. The main significance of their research is that, they have achieved a maximal performance from an imbalanced CXR dataset, without even deploying any re-sampling or, image processing method. Any image processing or pre-processing technique might work efficiently on a particular dataset, but there would not be any guarantee whether it would work for any other dataset or not. Therefore, their proposed framework not only provides the best result from this CXR dataset, but also, generalizes better than other existing CNN models. Their proposed framework is further implemented on the CXR dataset, the results are shown in Sect. 10.3.

10.3 Results and Analysis

This section is comprised of four parts. (I) The training specifications of all the implemented CNN models are presented in Sect. 3.1. (II) That is followed by a brief description of several quality metrics deployed in this study, mentioned in Sect. 3.2. (III) A brief explanation of the employed imbalanced CXR dataset is provided in Sect. 3.3. (IV) Eventually, in Sect. 3.4, several existing CNN models have been implemented on this imbalanced CXR dataset and thereby, their performances are compared and analyzed.

10.3.1 Training Specifications

All the existing CNN models have been built using Keras sequential API. Google Colab Pro platform is utilized which has availability of Tesla P100 GPU. The following training specifications are followed for overall all the existing CNN models:

1. Original dataset is divided with 60–20–20% ratio for training, testing and validation respectively. This splitting is done in a stratified way, which is feasible for class imbalance problem.
2. Adams-optimizer [25] is preferred for training all the models.
3. A very small batch size of 16 is employed while training all the CNN models, so that it can avoid the overfitting problem.
4. All the images are resized into 224×224 before feeding them into all the CNN models. For employing CNN model, this is a very standard input size followed by several researchers.
5. For training the standard pre-trained CNN models (i.e., transfer learning from ImageNet dataset), a fixed learning rate of 1e^{–4} is chosen, with early stopping which is monitored on the validation loss for 5 epochs patience.
6. Whereas, for training other existing CNN models (novel framework), exactly same learning rate is chosen as mentioned in their paper.
7. Two CNN models, Attention-based VGG-16 and Covid-Net, are only trained from the scratch. Here no early stopping is deployed, since it was noticed that for a train-from-scratch model early stopping stops the training too early. Instead, they are trained for a fixed number of epochs that is 25.
8. For pre-trained CNN models, Categorical Cross Entropy is utilized as a loss function which is usual for multi-class classification task.
9. For a spacial framework (like Custom based Balanced Loss function), Weighted Categorical Cross Entropy (WCCE) is employed as mentioned in their paper.
10. For training pre-trained CNN model, no dense layer or FC layer is employed to avoid overfitting problem.
11. Softmax activation function is deployed at the output layer, because it is multi-class classification task.

10.3.2 Quality Metrics

Standard 'accuracy' metric is used as an evaluation metric to check how accurately the model is working. Besides that, precision, recall, as well as F1 Score, are also evaluated in order to make sure that the model is not affected by any class imbalance problem. The mathematical formulae for the aforementioned metrics are given:

$$\boldsymbol{Precision} = \frac{\boldsymbol{TP}}{\boldsymbol{TP}+\boldsymbol{FP}} \tag{10.8}$$

$$\boldsymbol{Recall} = \frac{\boldsymbol{TP}}{\boldsymbol{TP}+\boldsymbol{FN}} \tag{10.9}$$

$$\boldsymbol{Accuracy} = \frac{\boldsymbol{TP}+\boldsymbol{TN}}{\boldsymbol{TP}+\boldsymbol{FN}+\boldsymbol{TN}+\boldsymbol{FP}} \tag{10.10}$$

$$F1\boldsymbol{csore} = \frac{2 * \boldsymbol{Precision}.\boldsymbol{Recall}}{\boldsymbol{Precision}+\boldsymbol{Recall}} \tag{10.11}$$

Here, in Equation (10.8)–(10.11), '*TP*' stands for True Positives, i.e., instances which are actually labeled as positive and they are classified correctly as positive instances, '*TN*' stands for True negatives which represent the instances where the actual label is negative, and it is predicted likewise. '*FP*' stands for False positives where the instance actually belongs to the negative label but is predicted to be positive. '*FN*' stands for False negatives, where instances are actually positive but are predicted as negative.

10.3.3 Imbalanced CXR Dataset and Its Challenges

We have implemented several existing CNN models on a publicly available Chest X-Ray (CXR) dataset in which there were four classes: (I) Normal, (II) Covid, (III) Lung Opacity (LO) and (IV) Viral Pneumonia (VP). This dataset is publicly available online on Kaggle website [47]. This is a highly imbalanced dataset, since the number of images per class differs considerably. This dataset has 10,192, 3616, 6012, and 1345 number of images for Normal, Covid, LO and VP class respectively. Researchers from Qatar University, and the University of Dhaka, Bangladesh had created this dataset. In this dataset, Covid class was made in three different updates which were taken from different hospitals. Thus, Covid class has a large intra-class variability inside this dataset. Moreover, some researchers noticed that, there were huge inter-class similarity between LO and Normal class. That means, many of the images from 'Normal' class look exactly same as the images from 'LO' class and vice versa. Due to these aforementioned challenges, conventional CNN models may not work effectively on this CXR dataset.

10.3.4 Comparison and Analysis

Here, two kinds of analysis are done, (I) First, the performances of several pre-trained CNN models like VGG-16, ResNet-50, Inception-V3, Xception, DenseNet-121, and MobileNet-V2 are compared in Table 10.1. In all these models, 100\% fine-tuning is employed (i.e., transfer learning is done from ImageNet dataset to CXR dataset). (II) The performances of several existing CNN models, which are deployed only to alleviate class imbalance problem, are also compared and analyzed in Table 10.2. The training specifications of all these models are already mentioned in Sect. 3.1.

From Table 10.1, this can be observed that the Xception model has outperformed all other pre-trained CNN models on this imbalanced CXR dataset. In my knowledge, Xception (pre-trained model) has many depth-wise convolution layers in its model architecture and therefore, it is a very lightweight CNN model. Therefore, despite having challenges in the CXR dataset (of class imbalance), their model was not induced by huge overfitting. Inception-V3 has the next best result on this CXR dataset in Table 10.1. Due to the fact that Xception is just a modification of the Inception model, overall, the Inception-v3 model also works in a very similar way, and as a consequence, it provides higher accuracy. On the other hand, the models like ResNet-50, DenseNet-121, and Mobile Net-V2 are affected by the overfitting problem due to using higher number of layers in their model architecture.

This is further prominent from Fig. 10.8 in which loss function vs number of epochs graph is shown for several pre-trained models. By looking at the Fig. 10.8 it is evident that VGG-16, Mobile-Net and ResNet-50 models' validation loss function is not so stable. On the other hand, the validation loss function of Xception model has gradually decreasing and there is smaller gap between validation and training loss. This indicates that Xception model performance does not have any effect by overfitting.

Table 10.2 presents another kind of model performance comparison of the numerous existing CNN models which are especially dedicated to Covid-19 and Pneumonia detection. The first two models "Attention-based VGG-16" and "Covid-Net" are trained from scratch and in the other 4 remaining frameworks, pre-trained models are utilized as mentioned in their paper. This is evident from the Table-2

Table 10.1 Comparisons of testing results for several pre-trained CNN models on an imbalanced CXR dataset

Models	Accuracy	Precision	Recall	F1 score
VGG-16	0.92	0.92	0.92	0.92
ResNet-50	0.85	0.86	0.85	0.85
Inception-V3	0.93	0.93	0.93	0.93
Xception	0.94	0.95	0.93	0.94
DenseNet-121	0.88	0.89	0.88	0.88
MobileNet-V2	0.88	0.88	0.89	0.88

Table 10.2 Comparisons of testing results for several existing CNN models proposed for Covid-19 detection from CXR dataset (Macro average)

Models	Accuracy	Precision	Recall	F1 score
Attention based VGG-16	0.92	0.92	0.92	0.92
Covid-Net	0.89	0.89	0.89	0.89
Covid-Net + Oversampling by HS	0.90	0.91	0.91	0.91
ResNet-50 + SVD-CLAHE boosting	0.93	0.96	0.93	0.94
ResNet-50 + SVD-CLAHE Boosting + BWCCE loss function	0.94	0.96	0.95	0.95
ResNet-18 + CWBCCE loss function	0.93	0.94	0.92	0.93
Mobile FreezeNet + Attention-based loss function	0.95	0.96	0.95	0.95

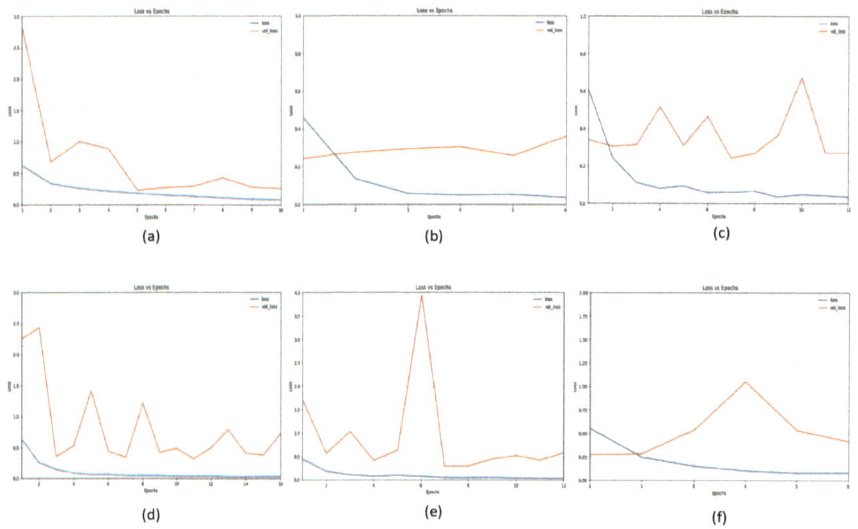

Fig. 10.8 Loss versus epochs for several pre-trained models **a** Xception **b** ResNet-50, **c** Mobile Net-V2, **d** Inception-V3, **e** VGG-16, **f** DenseNet-121

that the models trained from scratch could not perform well in this dataset. In fact, we have observed that the minutes per epoch for training Covid-Net model was approximately 6 min which is quite long. Due to training these models from scratch the (also number of trainable hyper-parameters was quite high for Covid-Net) often these models are prone to overfitting problem. On the other hand, pre-trained models like ResNet-50, ResNet-18 and Mobile Freeze-Net on their proposed framework have converged to minimum loss function too early compared to other experiments. Indeed, in these pre-trained models, from the very first epochs, the model was trained in the right direction. Therefore, these pre-trained models had enough time to study the challenges of this CXR dataset and thus, it resolves the class imbalance problem (a bit) automatically from the dataset, to the best of my knowledge. From Table 10.2

this is also evident that "Mobile Freeze-Net along with attention-based loss function", has outperformed other frameworks on this imbalanced CXR dataset. This framework has achieved the (macro) accuracy, precision, recall and F1 score of 0.95, 0.96, 0.95, and 0.95 which is still now the best result on this CXR dataset. Due to freezing extra 50% of layers in the Mobile Freeze-Net model, the number of hyper-parameters in the model has considerably reduced. As a consequence, the "Mobile Freeze-Net with attention-based loss function" framework converged (to minimum loss) much faster and it automatically alleviated the class imbalance problem in the CXR dataset. Another significance of their proposed framework is that, they had achieved this best accuracy without even using any pre-processing (or, re-sampling) method. Thus, it is likely that their framework would work in a more generalized way than other models.

10.4 Concluding Remarks

In this chapter, a brief description of numerous pre-trained CNN models and many existing CNN frameworks for Covid-19 detection from CXR images were explored. First, all these models or frameworks were explained along with its architecture. Thereafter, all these frameworks were implemented on a highly class-imbalanced and challenging CXR dataset, which was readily available on the Kaggle website. Experimental results indicated that pre-trained models have worked more effectively than novel CNN models trained from scratch. Furthermore, these results also revealed that "Mobile Freeze-Net with Attention-based loss function" had outperformed all other existing CNN frameworks on this CXR dataset and mitigated all the problems from the CXR dataset.

References

1. A.E. Gorbalenya et al., The species severe acute respiratory syndrome- related coronavirus: classifying 2019-nCoV and naming it SARS-CoV-2. Nat. Microbiol. **5**(4), 536–544 (2020)
2. Y. Feng et al., COVID-19 with different severities: a multicenter study of clinical features. Am. J. Respir. Crit. Care Med. **201**(11), 1380–1388 (2020)
3. Y. Xiang et al., COVID-19 epidemic prediction and the impact of public health interventions: A review of COVID-19 epidemic models. Infect. Dis. Model. **6**, 324–342 (2021)
4. V.J. Hall et al., SARS-CoV-2 infection rates of antibody-positive com- pared with antibody-negative health-care workers in England: a large, multicentre, prospective cohort study (SIREN). The Lancet **397**(10283), 1459–1469 (2021)
5. H. Wang et al., Estimating excess mortality due to the COVID-19 pan- demic: a systematic analysis of COVID-19-related mortality, 2020–21. The Lancet **399** (10334), 1513–1536 (2022)
6. H.M. Zawbaa et al., Effect of mutation and vaccination on spread, severity, and mortality of COVID-19 disease. J. Med. Virol. **94** (1), 197–204 (2022)
7. N.G. Davies et al., Estimated transmissibility and impact of SARS-CoV-2 lineage B. 1.1. 7 in England. Science **372** (6538), eabg3055 (2021)

8. L.J. Beesley et al., SARS-CoV-2 variant transition dynamics are associated with vaccination rates, number of co-circulating variants, and convalescent immunity. EBioMedicine **91**, (2023)
9. A.M. Ismael, A. S¸engu¨r,Deep learning approaches for COVID-19 detection based on chest X-ray images. Expert. Syst. Appl. (164), 114054 (2021)
10. K. Purohit et al., Covid-19 detection on chest x-ray and ct scan images using multi-image augmented deep learning model. in Proceedings of the seventh international conference on mathematics and computing (Springer Singapore, 2022)
11. M. Teymouri et al., Recent advances and challenges of RT-PCR tests for the diagnosis of COVID-19. Pathol.-Res. Pract. (221), 153443 (2021)
12. L. Wang, Z.Q. Lin, A. Wong, Covid-net: A tailored deep convolutional neural network design for detection of covid-19 cases from chest x-ray images. Sci. Rep. **10**(1), 19549 (2020)
13. Y. Song et al., Deep learning enables accurate diagnosis of novel coronavirus (COVID-19) with CT images. IEEE/ACM Trans. Comput. Biol. Bioinform. **18**(6), 2775–2780 (2021)
14. A. Narin, C. Kaya, Z. Pamuk, Automatic detection of coronavirus disease (covid-19) using x-ray images and deep convolutional neural networks. Pattern Anal. Appl. (24), 1207–1220 (2021)
15. C. Sitaula, M.B. Hossain, Attention-based vgg-16 model for covid-19 chest x-ray image classification. Appl. Intell. **51**(5), 2850–2863 (2021)
16. D. Das, K.C. Santosh, U. Pal., Truncated inception net: COVID-19 outbreak screening using chest X-rays. Phys. Eng. Sci. Med. **43**, 915–925 (2020)
17. A. Kumar, A.R. Tripathi, S.C. Satapathy, Y.D. Zhang, Sars-net: Covid-19 detec- tion from chest x-rays by combining graph convolutional network and convolutional neural network. Pattern Recogn. **122**, 108255 (2022)
18. S. Roy et al., SVD-CLAHE boosting and balanced loss function for COVID- 19 detection from an imbalanced Chest X-Ray dataset. Comput. Biol. Med. **150**, 106092 (2022)
19. Chamseddine, E et al., Handling class imbalance in COVID-19 chest X-ray images classification: Using SMOTE and weighted loss. Appl. Soft Comput. **129**, 109588 (2022)
20. K.Y. Win et al., Ensemble deep learning for the detection of Covid-19 in unbalanced chest X-ray dataset. Appl. Sci. **11** (22), 10528 (2021)
21. C. Seiffert, T.M. Khoshgoftaar, J. Van Hulse, A. Napolitano : RUSBoost: A hybrid approach to alleviating class imbalance, IEEE Trans. Syst. Man Cybern. 40 (1), 185–197 (2009)
22. I. Goodfellow, Y. Bengio, A. Courville, *Deep learning* (MIT press, USA, 2016)
23. Y. Xu et al., *Scale-invariant convolutional neural networks*. arXiv preprint arXiv. 1411 (6369), (2014)
24. R. Ward, X. Wu, L. Bottou, Adagrad stepsizes: Sharp convergence over nonconvex landscapes. J. Mach. Learn. Res. **21**(1), 9047–9076 (2020)
25. A. D´efossez et al., *A simple convergence proof of adam and adagrad*. arXiv preprint arXiv. 2003.02395 (2020)
26. N. Bjorck et al., Understanding batch normalization. Adv. Neural Inf. Process. Syst. **31**, (2018)
27. B. Hanin, Which neural net architectures give rise to exploding and vanishing gradients? Adv. Neural Inf. Process. Syst. **31**, (2018)
28. A. Krizhevsky, I. Sutskever, G.E. Hinton, Imagenet classificationwith deep convolutional neural networks. Adv. Neural Inf. Process. Syst. **25**, (2012)
29. K. Simonyan,A. Zisserman, *Very deep convolutional networks for large-scale image recognition*. arXiv preprint arXiv:1409.1556 (2014)
30. C. Szegedy et al., Going deeper with convolutions. in *Proceedings of the IEEE conference on computer vision and pattern recognition* (2015)
31. F. Chollet,Xception: Deep learning with depthwise separable convolutions. in *Proceedings of the IEEE conference on computer vision and pattern recognition* (2017)
32. M. Sandler etal., Mobilenetv2: Inverted residuals and linear bottlenecks. in *Proceedings of the IEEE conference on computer vision and pattern recognition* (2018)
33. K. He et al., Deep residual learning for image recognition. in *Proceedings of the IEEE conference on computer vision and pattern recognition* (2016)

34. J. Denget al., Imagenet: A large-scale hierarchical image database. in *2009 IEEE conference on computer vision and pattern recognition* (IEEE, 2009)
35. S. Roy, K. Bhalla, R. Patel, Mathematical analysis of histogram equalization techniques for medical image enhancement: a tutorial from the perspective of data loss. Multimed. Tools Appl., 1–30 (2023)
36. M. Siddhartha,A. Santra, *COVIDLite: A depth-wise separable deep neural network with white balance and CLAHE for detection of COVID-19*. arXiv preprint arXiv 2006.13873 (2020)
37. A.G. Howardet al., *Mobilenets: Efficient convolutional neural networks for mobile vision applications*. arXiv preprint arXiv. 1704.04861 (2017)
38. M. Lin, Q. Chen, S. Yan, *Network in network*. arXiv preprint arXiv: 1312.4400 (2013)
39. V.C. Nitesh et al., SMOTE: synthetic minority over-sampling technique. J. Artif. Intell. Res. **16**, 321–357 (2002)
40. A. Waheed et al., Covidgan: data augmentation using auxiliary classifier gan for improved covid-19 detection. IEEE Access **8**, 91916–91923 (2020)
41. M. Togacar,B. Ergen, Z. Comert, COVID-19 detection using deep learning models to exploit Social Mimic Optimization and structured chest X-ray images using fuzzy color and stacking approaches. Comput. Biol. Med. **121**, 103805 (2020)
42. S.H. Wang, Y.D. Zhang,DenseNet-201-based deep neural network with composite learning factor and precomputation for multiple sclerosis classification. ACM Trans. Multimed. Comput., Commun., Appl. (TOMM) **16** (2s), 1–19 (2020)
43. M. Tyagi,R. Santanu, V. Bansal, Custom weighted balanced loss function for covid 19 detection from an imbalanced CXR dataset. in *2022 26th International Conference on Pattern Recognition (ICPR)* (IEEE, 2022)
44. S. Roy, S. Lal, J.R. Kini, Novel color normalization method for Hematoxylin and Eosin stained histopathology images. IEEE Access **7**, 28982–28998 (2019)
45. S. Roy,R. Khurana, Mobile freeze-net with attention-based loss function for Covid-19 detection from an imbalanced CXR dataset. in *Proceedings of the 38th ACM/SIGAPP symposium on applied computing* (2023)
46. S. Raychaudhuri,Introduction to monte carlo simulation. in *2008 Winter simulation conference* (IEEE, 2008)
47. M.E.H. Chowdhury et al., Can AI help in screening viral and COVID-19 pneumonia? IEEE Access **8**, 132665–132676 (2020)

Chapter 11
Deep Learning Integrated Multiphoton Microscopy

Manikanth Karnati, Sanika Kulkarni, and Nirmal Mazumder

Abstract Deep learning (DL) has emerged as one of the most popular machine learning (ML) techniques for a variety of inference tasks in recent years. It has created transformative opportunities for image reconstruction and enhancement in optical microscopy. DL is being used to solve various inverse problems based on microscopy image data, from image transformations between microscopic imaging systems to adding new capabilities to existing imaging techniques. DL is assisting us in moving toward data-driven instrument designs that combine microscopy and computing techniques. This article gives an overview of recent work that has used ML techniques to various types of advanced microscopy and sensing systems and their biomedical applications.

11.1 Introduction

Optical microscopy (OM), also known as light microscopy, enlarges a sample by passing visible light through or reflected from it through single or multiple lenses. The image can be seen with the naked eye, captured on a photographic plate, or captured digitally. A light microscope is made up of a single lens and its attachments, or a system of lenses and imaging equipment, as well as the necessary lighting, sample stage, and support. The digital microscope, which uses a CCD camera to focus on the exhibit of interest, is the most recent advancement. Eyepieces are not required because the image is displayed on a computer screen. Fluorescence microscopes are any microscope that employs fluorescence to generate an image. They use optical sectioning to improve fluorescence image resolution [1]. The fluorophores are illuminated with a wavelength of light that absorbs and causes them to emit longer wavelengths of light (i.e., of a different color than the absorbed light). To distinguish the illumination, light from the much weaker emitted fluorescence, a spectral emission filter is used. The filters and diachronic beam splitter are chosen to match the

M. Karnati · S. Kulkarni · N. Mazumder (✉)
Department of Biophysics, Manipal School of Life Sciences, Manipal Academy of Higher Education, Manipal, Karnataka 576104, India
e-mail: nirmal.mazumder@manipal.edu

A. Gogoi and N. Mazumder (eds.), *Biomedical Imaging*, Biological and Medical Physics, Biomedical Engineering, https://doi.org/10.1007/978-981-97-5345-1_11

fluorophore which is used to label the specimen's spectral excitation and emission characteristics. In this method, a single fluorophore (color) distribution is imaged at a time. To create multi-color images of various fluorophores, multiple single-color images must be combined. The fluorescence imaging technique of two-photon microscopy (TPM) can image living tissue with a thickness of up to one millimetre (mm). Two-photon excitation, in contrast to standard fluorescence microscopy, necessitates the simultaneous excitation of two photons with wavelengths longer than the emitted light. In two-photon excitation microscopy, near-infrared (NIR) excited light is commonly used, which can also excite fluorescent dyes [1]. Multiphoton microscopy (MPM) works in the same way as TPM, but with more photons. MPM has numerous advantages. It improves depth imaging, with 1.5 mm deep inside the tissue and also improves image contrast at the longer wavelengths used in two-photon excitation The scattering coefficient is lower than for single-photon excitation wavelengths since excitation can only take place in the focal plane, and out-of-focus bleaching is avoided. Due to the longer wavelengths used, there is less phototoxicity [2]. MPM, as a result, allows for longer imaging times. MPM, as a result, allows for longer imaging times. Various ML algorithms can be used to visualize the images produced by this microscope [3].

For analysis, ML algorithms employ a variety of statistical, probabilistic, and optimization techniques. These algorithm's applications include automated text categorization, network intrusion detection, junk e-mail filtering, credit card fraud detection, customer purchase behavior detection, manufacturing process optimization, and disease modeling, to name a few [4]. As new data is fed into these algorithms, they tend to make more accurate predictions. Despite some differences in how ML algorithms are classified, they can be divided into three broad categories based on their purposes and how the underlying machine is trained. The three categories are supervised, unsupervised, and semi-supervised. In the supervised variant, a prediction model is built by learning a dataset with known labels and then predicting the outcome of unlabeled examples. In supervised ML algorithms, a labeled training dataset is used to train the underlying algorithm. The unlabeled test dataset is then fed into the trained algorithm, which divides them into groups based on their similarities. Classification and regression problems are two types of problems that benefit from supervised learning algorithms. In classification problems, the underlying output variable is discrete. The Support Vector Machine (SVM) algorithm can classify both linear and non-linear data. Each data point is first mapped into an n-dimensional feature space, where n represents the number of features. Figure 11.1 presents, the hyperplane that divides the data into two classes is then found, with the marginal distance between the two classes maximized and classification errors minimized [5]. A class's marginal distance is the distance between the decision hyperplane and the class's nearest instance. Each data point is first plotted as a point in an n-dimensional space (where n is the number of features), with the value of each feature being the coordinate value.

To complete the classification, we must first locate the hyperplane that separates the two classes by the greatest margin. The decision tree (DT) is one of the earliest and most well-known machine learning algorithms. The decision logic for classifying

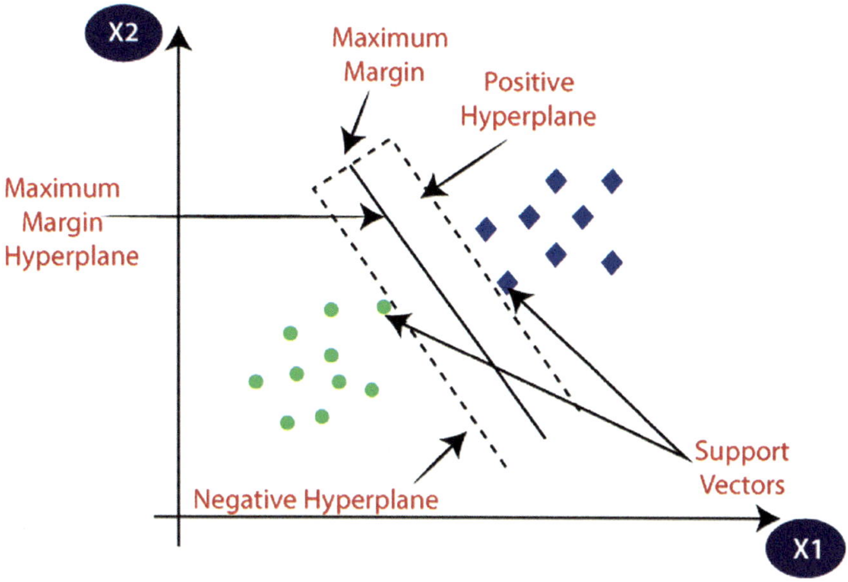

Fig. 11.1 A simple depiction of how the support vector machine operates. The SVM found a hyperplane (really a line) that maximizes the separation between the 'Diamond' and 'Circle' classes. The figures are reproduced with the kind permission from [6]

data items into a tree-like structure, i.e., tests and outcomes, are represented by a DT. The root node of a DT is usually the first or top-most node, with multiple levels. Input variables or attribute tests are represented by all internal nodes (those with at least one child). Based on the test result, the classification algorithm branches toward the appropriate child node, and the process of testing and branching continues until it reaches the leaf node. The leaf or terminal nodes represent the decision outcomes. One of the simplest and earliest classification algorithms is the K-nearest neighbor (KNN) algorithm. It's like a simplified version of an NB classifier. The KNN algorithm, unlike the NB method, does not require the use of probability values. The number of nearest neighbors considered to take a 'vote' in the KNN algorithm is the 'K'. For the same sample object, different values for 'K' can result in different classification results. Artificial neural networks (ANNs) are a class of ML algorithms inspired by the way the human brain's neural networks work. McCulloch and Pitts proposed them first, and they became popular in the 1980s [5].

Neurons in the biological brain are connected via multiple axon junctions, which form a graph-like architecture. These connections can be rewired (for example, through neuroplasticity), which aids in the adaptation, processing, and storage of information. ANN algorithms can also be represented as a network of interconnected nodes. According to the interconnection, one node's output becomes the input for another node's subsequent processing (as shown in Fig. 11.2). Depending on the transformation they perform, nodes are usually grouped into a layer matrix.

An ANN framework can have one or more hidden layers in addition to the input and output layers. Weights are assigned to nodes and edges to allow for the adjustment of communication signal strengths, which can be amplified or weakened through repeated training. ANNs can predict the test data based on the training and subsequent adaptation of the matrices, node, and edge weights. Convolutional Neural Network (CNN) is a DL algorithm that can take an input image, assign importance (learnable weights and biases) to various aspects/objects in the image, and distinguish between them. When compared to other classification algorithms, the amount of pre-processing required by a CNN is significantly less [7]. Multiple layers of artificial neurons make up CNN (as shown in Fig. 11.3). Artificial neurons are mathematical functions that calculate the weighted sum of multiple inputs and outputs of an activation value, similar to their biological counterparts. Each layer generates several activation functions that are passed on to the next layer when you input an image into a CNN. Basic features such as horizontal or diagonal edges are usually extracted by the first layer. This information is passed on to the next layer, which is responsible for detecting more complex features like corners and combinational edges. By reducing the dimensions, the computational power required to process the data is reduced. After that, the output layer will be visible. The majority of these applications were developed with supervised ML algorithms rather than unsupervised ML algorithms.

ML is a DL technique that helps to simulate how humans acquire knowledge. DL is an important component of data science, which includes statistics and predictive modeling. DL is a set of ML techniques that use multilayered neural networks to

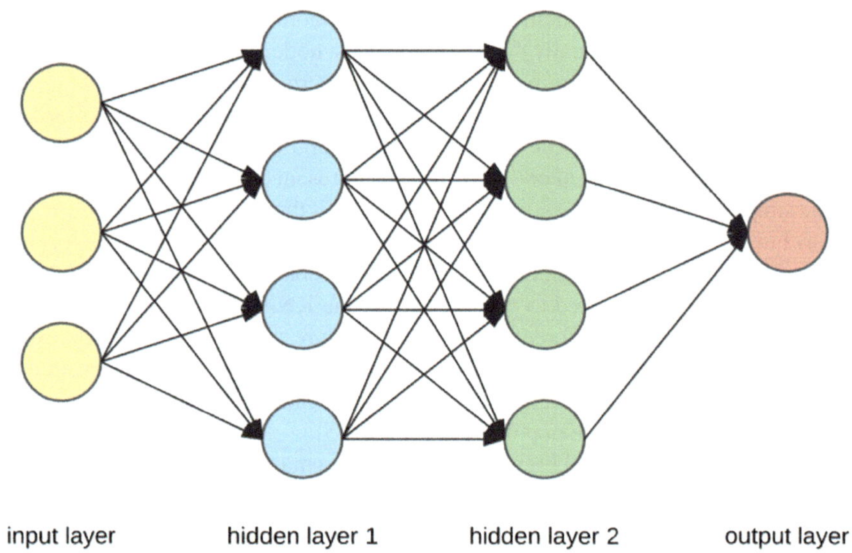

Fig. 11.2 An illustration of an artificial neural network with two hidden layers. The arrows connect nodes' outputs from one layer to their inputs from another one. The figures are reproduced with the kind permission from [8]

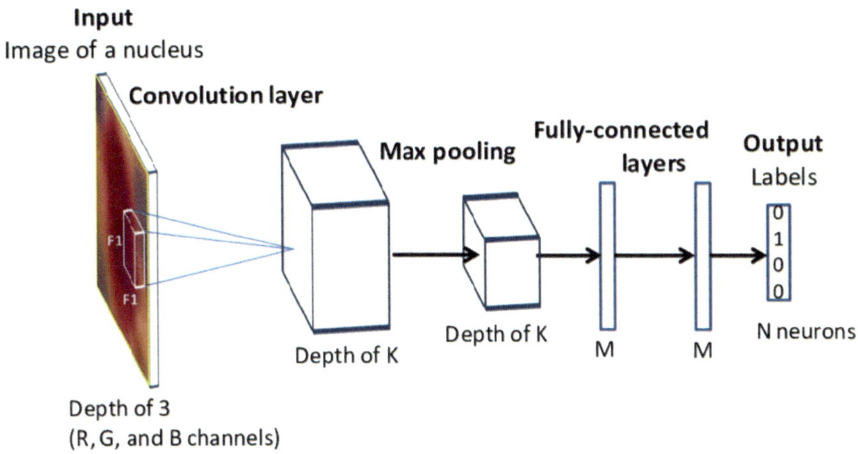

Fig. 11.3 A schematic illustration of the N-Net's architecture. It starts with a convolution layer, then a max-pooling layer, and finally two completely connected layers. The final layer is the output layer. This network receives an RGB image of a nucleus as input and produces a label as output. The figures are reproduced with the kind permission from [9]

automatically analyze signals or image data [10]. Artificial neurons are arranged in several layers in deep neural networks, each of which is typically. Deep networks are extremely effective for a variety of tasks, from image classification to video playback, and neural networks such as long short-term memory (LSTM) and convolutional neural networks (CNNs) have been developed. Because of their shift-invariance property, CNNs have been particularly effective at processing, transforming, and classifying images. Beyond classification, deep learning opens up a slew of new possibilities for solving traditional imaging inverse problems like deblurring, super-resolution, denoising, and pixel super-resolution, such as HR (high resolution) and LR (low resolution) imaging [11].

11.2 Application of Two-Photon Fluorescence Microscopy in Image Analysis

Since the invention of laser scanning confocal microscopy in the 1980s, which allowed three-dimensional (3D) imaging of both fixed and living biological tissue, HR fluorescence imaging has revolutionized biological research [1]. In recent years, DL-based image processing techniques have been successfully used in fluorescence microscopy imaging. In the absence of a precise physical model of the imaging process, the image super-resolution technique uses a deep neuronal network to describe the complicated connections between the registered pair of LR and HR images. As a result, both the radiative and non-radiative lifetimes have an impact on the effective measured lifetime [12]. The significance of this was not immediately apparent, but subsequent experiments showed that imaging samples at depths greater than 10–20 microns improved image quality significantly. The fluorescence lifetime is the average amount of time a fluorophore is excited before it relaxes, and it is usually measured in nanoseconds. FLIM (fluorescence lifetime imaging microscopy) can distinguish fluorophores with overlapping emission spectra that cannot be distinguished by traditional intensity-based imaging methods as long as the two fluorophores have a measurable lifetime difference because lifetime is an intrinsic property of fluorescence molecules. The fluorescence lifetime is the average time a fluorophore is excited before relaxing, and it is usually measured in nanoseconds.

The study shows how fluorescence lifetime in the NADH (nicotinamide Adenine Dinucleotide + Hydrogen) changes with different cellular metabolic states, allowing one to extract changes in the cellular metabolic state and accurately characterize normal and pathological states by measuring fluorescence lifetime [13]. On the right side of the semi-circle is the short lifetime, and on the left is the long lifetime. The ANN (Artificial neural network) method estimates the lifetime image in 0.9 s, which is 180 times faster than the curve-fitting technique based on the least-squares method, which takes 166 s. A weighted average of short and long lifetime images yields an average lifetime image. The ANN method outperformed curve-fitting tools in terms of lifetime estimation accuracy, with a success rate of 99.93% compared to 95.93% for the LSM (Least Square Mean) method, indicating that the ANN method outperformed curve-fitting tools in terms of lifetime estimation accuracy [13]. The histogram of TCSPC (Time correlated single photon counting) data is fed into EM (Expectation–Maximization), which produces lifetime and IRF (Instrument Response function) as outputs. The fluorescence lifetime imaging (FLI) network's architecture, which includes a common branch for temporal feature extraction and split branches for lifetime reconstruction and fractional amplitude for a short lifetime based on time-resolved CS data of Alexa Fluor dye, which is pH-insensitive over a wide molar range, the estimated intensity and lifetime for stable signals in imaging with six different concentrations ranging from 1000 nm to 31.25 nm [13]. A healthy and unhealthy embryo Blast cyst image using ML's lifetime and phasor approach, because different fluorophore compositions or excited state reactions can result in the same average lifetime measurements, the measured lifetime at a pixel cannot resolve the heterogeneity of fluorophores at that pixel.

They first show how to use ML to remove the noise FLI. Second, they demonstrated how to calculate FLI using fluorescence intensity images. Figure 11.4 shows how they looked at a network for FLI using compressive sensing data and a 3D CNN, FLI network for lifetime extraction, both of which are ML-based FLIM applications [13].

Three convolutional neural networks make up the 3D-dual HRGAN: an HR image generator, an image degrading generator, an HR image discriminator, and an LR image discriminator. The (discriminator) HR D distinguishes between the outputs of (generator) HR G and the real HR images in the primal GAN, which is made up of an HR image generator and a discriminator (as shown in Fig. 11.5) [14]. The HR image generator converts LR images into HR images that are predicted. 3D-dualHRGAN

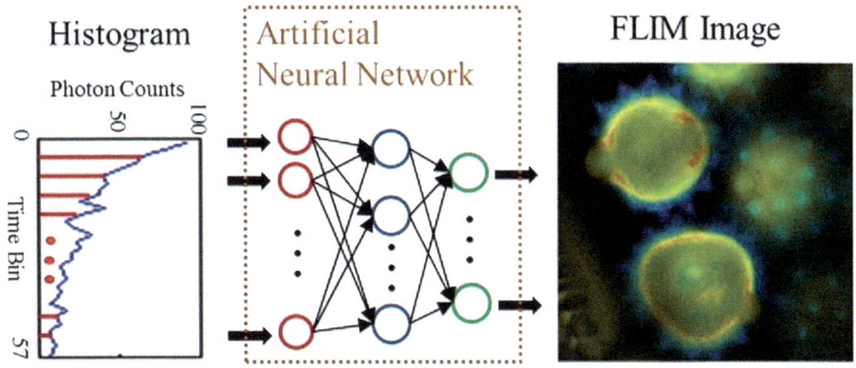

Fig. 11.4 Schematic representation of ANN in fluorescence lifetime imaging microscopy. The figures are reproduced with the kind permission from [4]

transforms 5 × volume images into 20 × volume images Peak widths of image intensity curves in the predicted image were smaller than LR images and similar to HR images. To map an HR volume image to an LR volume image, the image degrading generator is used. Blue, red, and green curves represented the LR, HR, and predicted HR images, respectively. Green curves represent up-sampled images, predicted images, and high-resolution images. They conclude that the predicted HR images are not only strictly derived from the LR images, but that the LR images' resolution has also been significantly improved, based on the findings [14].

TPF microscopy has made it possible to image Ca^2 in vivo in neurons located up to 500 μm below the cortical surface [6]. The advantage of Ca^2 imaging is that it allows for real-time analysis of individual cells and sub-cellular compartments. In vivo, high-resolution Ca^2 imaging has been limited to a single neuron at a time and has not been used to directly monitor neuronal networks. When the skull was

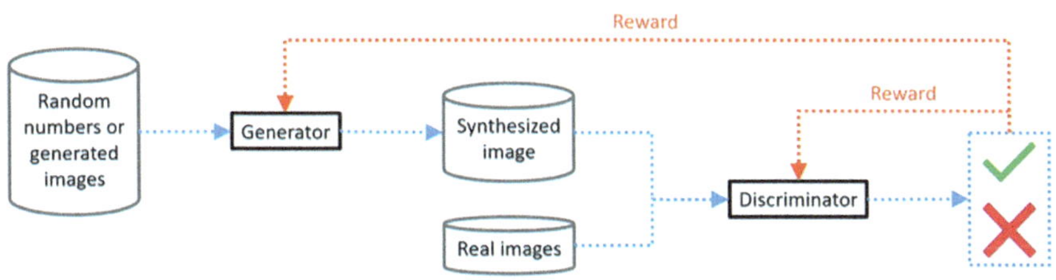

Fig. 11.5 Principles for generative adversarial networks (GANs). Cylinders represent data, and black rectangles represent neural networks. The primary line is shown in blue, while the feedback loops are in orange. The generator gets input data that enables it to create an image. The discriminator is fed either a genuine image or a synthesized image as input. It must then assess whether the example is real or created. The generator is awarded if it successfully deceives the discriminator, whereas the discriminator is rewarded if it correctly distinguishes between the true and produced images. The figures are reproduced with the kind permission from [14]

removed from the imaging field, depth resolution improved even more, allowing individual cells to be detected up to 300 μm beneath the cortical surface. No "Hot spots," indicating dye molecules captured within organelles, were visible in HR images of the thus unloaded cells. First, when imaging cell dendrites of microelectrode loaded cells, their recordings had a depth limit of 200 μm with the skull and 300 μm without it, compared to 500 μm when imaging cell dendrites without it [6]. Longitudinal imaging appeals because biological processes are dynamic, and time-lapse imaging is the best way to capture cellular transitions. Two-photon imaging in vivo was performed using a Prairie Ultima microscope and PrarieView software. For the contrast adjustment, ImageJ was used [15]. Although cell bodies can be imaged in L5, there are far too many brightly labeled cells to allow for clear imaging. In protocols and reviews, the benefits and drawbacks of the cranial window and thinned skull approaches to longitudinal TPF imaging have been discussed. The only way to image structures in deeper cortical layers like L5 cell bodies and L5 spines is to use the cranial window method.

In a small number of cases, imaging below layer 2/3 has been reported. To their knowledge, this is the first time that cellular details in layer 5 have been imaged repeatedly using longitudinal imaging of a biogenic H-line/familial Alzheimer's disease mouse [15]. Deep tissue imaging could only be done through a cranial window because the remaining bone from skull thinning prevents such details from being imaged. The energy exiting the lens was reduced by 17% when imaging a dendrite in layer 2/3, but the misalignment caused a dramatic reduction in signal and image quality. That even though daily alignment is not required to obtain signals on a TPM, the time spent on such procedures is more than compensated by optimized two-photon images. The ability to detect is the most basic rule of all imaging. In contrast to fine-scale imaging, the same lab tracked the appearance and disappearance of cells in the olfactory bulb over 9 months and found a net increase of around 13% [16]. It was only a matter of time after the first reports of longitudinal two-photon imaging established a baseline for dendrite stability that the technique was applied to the study of how neurons change over time in disease mouse models. These mice lines have been used in acute in vivo calcium imaging experiments to identify interneurons, but no longitudinal imaging applications have been discovered. Using 2D segmentation of mean images, it's difficult to distinguish between overlapping neurons that fire independently in time. Overall, $n = 40 \pm 23.6$ masks from the initial ABO marking were removed because they were not located on the soma of active neurons, accounting for $13.9 \pm 5.7\%$ of the initial neurons, and $n = 72.7 \pm 20.9$ neurons were added, accounting for $24.2 \pm 5.9\%$ of the final GT (gold-standard ground truth) neurons [16]. Alternative approaches currently available are even less suitable for generating GT. Simultaneous dual channel imaging of activity-independent nuclear tagged neurons, for example, yields consistent GT markings across all neurons. Alternative methods for generating GT that are currently available are even less suitable. For example, simultaneous dual channel imaging of activity-independent nuclear tagged neurons produces consistent GT markings across all neurons. The background-corrected video, the mean image, and the F/F trace of the average pixel intensities within the neuron's mask were all displayed in three panels, one for each segmented neuron. DL-based segmentation methods currently available are either unable to generalize to data from a variety of imaging systems or are computationally infeasible for large-scale angiograms. To handle the processing demands of large-scale angiograms, a vascular segmentation method that is not only generalizable to various TPM imaging setups but also fast and computationally efficient is required [16]. Traditional 3D OM methods based on linear one-photon absorption processes, either epi-illumination or plane illumination modes, are limited to imaging the tissue surface because highly anisotropic scattering blurs the images as depth increases [17]. Aside from the direct acquisition of a scattering-suppressed image, several hardware-based computational methods, such as adaptive optics and speckle correlation techniques, have been proposed to extract useful information from blurred signals to recover scattering images. For network application 2, they imaged another region of beads to obtain the scattered image and corresponding ground truth. For network application 2, ScatNet successfully restored a low-quality scattering image with significantly improved spatial resolution and high structure similarity [17]. Multi-view LSFM imaging was used to generate a set of LSFM stacks, which were then registered and fused to create a 3D image with complete sample signals, as previously mentioned. They used the three groups of scattering images and the ground-truth images to create a training system with hierarchical scattering models and compared the results to raw images in their ScatNet implementation, a deep network's structure. The goal of the restoration of scattered images is to create a high-quality output image from a degraded input image.

11.3 Application of Multiphoton Microscopy (MPM) Using Deep Learning

Daily, more medical images are created as medical imaging technology improves. Recent advances in digital staining, in which images captured with other modalities are converted into virtual H&E images (hematoxylin and eosin) make it easier for histopathologists to interpret MPM images, which we expect to dramatically increase the use of these techniques in clinical practice soon [3]. An upright Leica TCS-SP confocal laser scanning microscope modified for nonlinear optical imaging (NLO) was used to perform combined SHG and TPF imaging. A study used a total of 36 image samples from stage II lymphedema patients and 42 image samples from healthy volunteers [18]. SHG and AF images were recorded on a 512 pixel matrix after the first appearance of the stratum corneal was chosen. The image measured 75 $\times$ 75 m. Data processing involves image quality improvement, image segmentation and ROI selection, feature extraction, and the creation of predictive models that can be used to classify new data are all common image processing tasks. The SHG tissue image feature extraction was done using the histogram of oriented gradients (HOG) method. For image classification, a combination of SVM and a "majority vote" method was used. Random splitting of image patches into training and testing sets yielded a sensitivity of 0.79 ± 0.11 and a specificity of 0.77 ± 0.10 when averaged over 50 times [18].

CNNs have been used in image analysis of the vasculature, for example, in conjunction with MPM images. There has been no other study that has used DL methods to analyze MPM images of HCCs in the literature. They demonstrate how to use CNN to classify MPM images of unstained HCCs. Using the WHO grading system, an experienced pathologist determined the HCC grades from the H&E images, as well as identifying the tumor area on the section [19]. Figure 11.6 represents all of the HCC images that were horizontally flipped to create a new set of images. Both the new horizontally flipped images and the original HCC images were flipped vertically to double the number of images in the set. Multiphoton microscopy measurements yielded 217 HCC images, with 70 images graded G1, 68 images graded G2, and 79 images graded G3 [19]. The features of the G2 and G3 classes were difficult to distinguish from those of the G1, resulting in a lower true positive rate and AUC for their classifications. The classification results could be influenced by similar class distributions in the original dataset. It is primarily used in image precision evaluation to compare classification results to actual measured values.

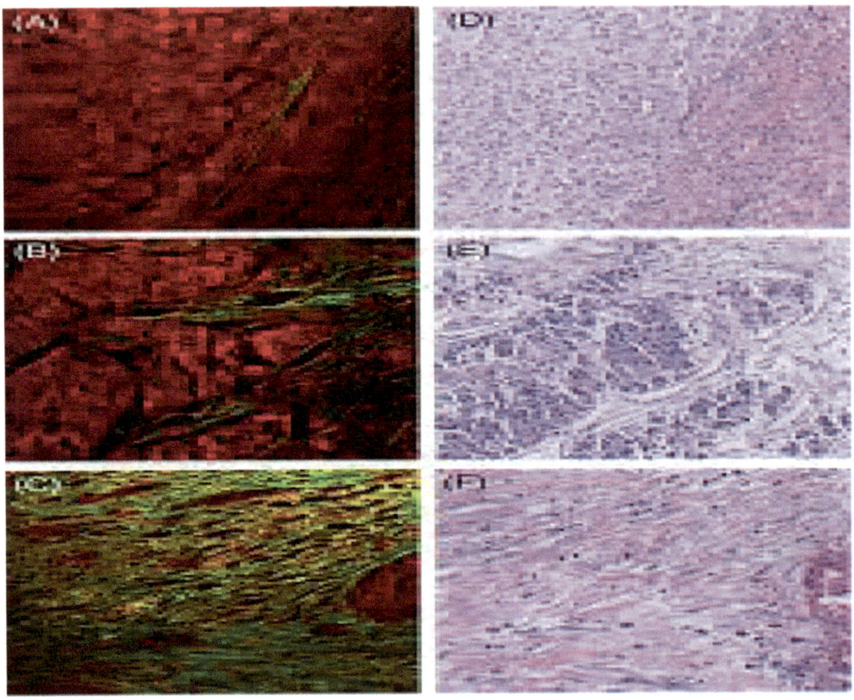

Fig. 11.6 Left: Representative multiphoton microscopy pictures (**a**–**c**), with second-harmonic generation signals in green and two-photon excitation fluorescence signals in red. Right: **d**–**f** matching hematoxylin and eosin stained pictures. The figures are reproduced with the kind permission from [1]

At first, CNNs were mostly used to analyze white light images. To obtain abundant image information and determine the detailed differences between MPM images of different HCCs, they decided to use multiple convolution layers that allow the extraction of higher-level features to obtain abundant image information and determine the detailed differences between MPM images of different HCCs. Because of the development of DL techniques for high-accuracy classification that require fewer training images, the application of CNNs to MPM image datasets has become more viable [19]. Figure 11.7 represents that they were able to accurately classify high, medium, and low differentiation grades of liver cancer in a small sample with nearly 90% accuracy. Because they do not change the information and structure in the image of HCC tissue, nor do they distort the image noise and background, rotation and flipping have become common methods for increasing sample sizes.

Supervised learning, which involves learning the task using labeled training images, is a common method for image classification. In general, labeled images are used to learn a better representation of the image data, making it easier to cluster and classify the images into distinct sets [20]. An 8-s pixel dwell time was used to improve the signal-to-noise ratio, and each image pixel was averaged 16 times, resulting in an imaging speed of 82 s per image. Bright-field images from adjacent stained sections, as well as multiphoton images of healthy and cancerous reproductive tissues. Due to the reduced field view, some of the image patches were found to be very dark, with only minor image features. Using label-free MPM images, automated and reliable ovarian tissue classification may be possible.

The speed with which these features are imaged has a significant impact on them, and imaging in vivo experiments is frequently slow, causing these features to stand

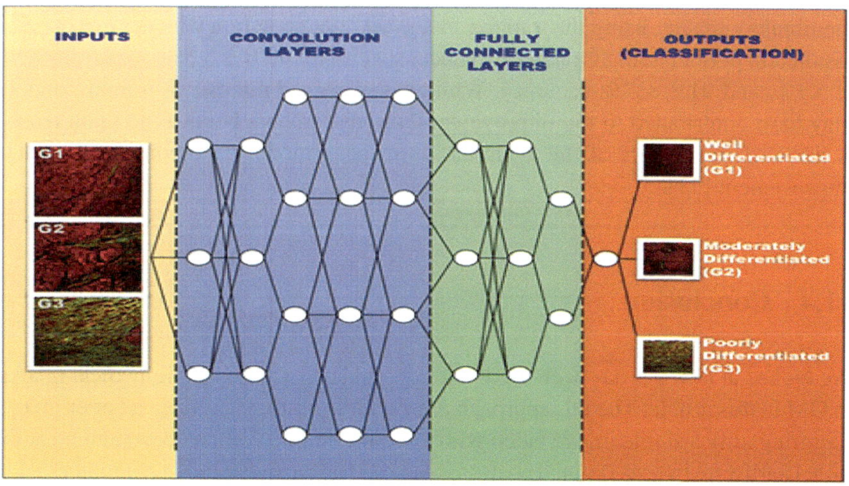

Fig. 11.7 A demonstration of the combination of multiphoton microscopy and a deep-learning algorithm for classifying HCC differentiation to create a unique computer-aided diagnostic tool. The figures are reproduced with the kind permission from [1]

out [20]. CNN is used to segment in vivo multiphoton images of vasculature in mouse models [10]. Variations in signals to noise as a function of imaging depth in MPM cause changes in image quality between image slices. These findings show that our model can be applied to new MPM images captured from various mouse models with varying image quality. Segmenting 3D vasculature images takes time, which stymies biomedical research and limits the use of imaging in clinical practice. The images of these examples were segmented using CNN from in vivo multiphoton images of vasculature in mouse models. Images from deeper within the brain tissue often have lower image contrast and higher noise levels due to the nature of MPM, resulting in more segmentation errors.

Similar to many ML solutions, DeepVess' performance is dependent on specific image features, and it degrades when tissues are labeled differently or image intensities are concentrated in a small portion of the intensity range [10]. To create a composite RGB image, the processed SHG and TPF images were inserted into the blue and red channels, respectively. The regions imaged with MPM were also imaged with BM (Basement Membrane) at high magnification for ground-truth availability [21]. The 358 MPM images from healthy and dysplastic unstained tissue sections that were collected to contain the DEJ were used to train the DL image classification method, which was inspired by a recent study. At random, images were divided into validation and training sets, with the latter enhanced using two strategies: horizontal and vertical reflections of the original images, and repeatable blurring of these horizontal and vertical reflections using a five-layer Gaussian image pyramid. The first augmentation strategy resulted in 1000 training images while combining the two yielded a total of 5000 images [21]. Combining TPF and SHG signals in MPM images of tissues improves classification accuracy when compared to addressing only TPF or SHG images, according to previous research. In the BM (basement membrane) images, using the training image set in conjunction with the first scheme resulted in poor classification performance: specificity $90.6 \pm 5.5\%$, sensitivity $93.9 \pm 5.1\%$, and $91.8 \pm 4\%$ accuracy. While the increased number of training images may have contributed to the improvement, the researchers believe the main reason is that different layers of the Gaussian impedance model have contributed to the improvement [21].

11.4 Conclusion

We looked at how DL has been used to transform and reconstruct biomedical images in OM in this article. The DL approach is extremely quick to produce a better image, requiring only a single image taken with a standard optical microscope and no additional hardware or user-defined post-processing. The current trend in computer vision and image analysis indicates that hand-crafted features and pipelines are giving way to data-driven DL approaches. Non-traditional data augmentation strategies, such as creating scale-space representations of images and then training on these representations, may also prove useful in assisting the deep method in performing recognition at multiple scales. This framework and its derivatives may apply to other forms of OM and imaging techniques after appropriate training and can be used to convert images acquired with low-resolution systems into HR and wide-field images, significantly increasing the output images' space bandwidth product. This framework can be applied to other computational imaging approaches, spanning different parts of the electromagnetic spectrum, and can be used to design computational imagers with improved resolution to OM. These methods can be applied to a variety of systems and optical modalities that have images improved, reconstructed, or transformed images. Image super-resolution systems can lower the cost of imaging systems while also allowing new samples to be characterized. Virtual staining, for example, is another technique for histopathology that can be used to improve the speed and accuracy of diagnoses. Microscopy is an ideal field to apply deep learning, as experimental data sets are typically obtained under highly controlled conditions. Especially now that powerful unsupervised learning approaches have emerged, providing the deep learning field with unique capabilities and enabling applications that are simply not possible with current OM technologies.

Acknowledgements We thank the Department of Science and Technology (DST), Government of India, India (Project Number- GITA/DST/TWN/P-95/2021) and Indian Council of Medical Research, Government of India, India (Project Number-ITR/Ad-hoc/43/2020-21) for the financial support. We thank Manipal School of Life Sciences, Manipal Academy of Higher Education, Manipal, Karnataka, India for providing the infrastructure and facilities.

References

1. J.R. Lakowicz, (Ed.) *Topics in Fluorescence Spectroscopy: Nonlinear and two-photon-induced fluorescence,* vol. 22, issue no. 3 (2019), pp. 132–219
2. U. Tauer, Advantages and risks of multiphoton microscopy in physiology. Exp. Physiol. **87**(6), 709–714 (2002)
3. R.M. Martínez-Ojeda, M.D. Pérez-Cárceles, L.C. Ardelean, S.G. Stanciu, J.M. Bueno, Multiphoton microscopy in imaging. Front. Phys. **1**(8), 128 (2020)
4. B. Mahesh, Machine learning algorithms-a review. Int. J. Sci. Res. (IJSR) **12**(9), 381–392 (2020)
5. S. Uddin, A. Khan, M.E. Hossain, M.A. Moni, Comparing different supervised machine learning algorithms for disease prediction. BMC Med. Inform. Decis. Mak. **19**(1), 1–16 (2019)
6. C. Stosiek, O. Garaschuk, K. Holthoff, A. Konnerth, In vivo two-photon calcium imaging of neuronal networks. Proc. Natl. Acad. Sci. **100**(12), 7319–7324 (2003)
7. S. Ray, A quick review of machine learning algorithms. in *2019 International conference on machine learning, big data, cloud and parallel computing,* vol. 22, issue no. 2 (2019, February), pp. 35–39
8. A. Dertat, Applied deep learning-part 1: Artificial neural networks, 2017. Towardsdatascience. d7834f67a4f6 (2018)
9. S.K. Melanthota, D. Gopal, S. Chakrabarti, A.A. Kashyap, R. Radhakrishnan, N. Mazumder, Deep learning-based image processing in optical microscopy. Biophys. Rev. **14**(2), 463–481 (2022)
10. M. Haft-Javaherian, L. Fang, V. Muse, C.B. Schaffer, N. Nishimura, M.R. Sabuncu, Deep convolutional neural networks for segmenting 3D in vivo multiphoton images of vasculature in Alzheimer disease mouse models. PLoS ONE **14**(3), e0213539 (2019)

11. D.X. Zhou, Universality of deep convolutional neural networks. Appl. Comput. Harmon. Anal. **48**(2), 787–794 (2020)
12. Z. Wang, J. Chen, S.C. Hoi, Deep learning for image super-resolution: A survey. IEEE Trans. Pattern Anal. Mach. Intell. **22**(2), 124–132 (2020)
13. V. Mannam, Y. Zhang, X. Yuan, C. Ravasio, S.S. Howard, Machine learning for faster and smarter fluorescence lifetime imaging microscopy. J. Phys.: Photonics **2**(4), 042005 (2020)
14. Z. Hang, L. Shiwei, H. Qing, L. Shijie, Q. Tingwei, R. Cai, Z. Shaoqun, A 3D high resolution generative deep-learning network for fluorescence microscopy image. BioRxiv **10**(2), 743179 (2019)
15. S.E. Crowe, G.C. Ellis-Davies, Longitudinal in vivo two-photon fluorescence imaging. J. Comp. Neurol. **522**(8), 1708–1727 (2014)
16. S. Soltanian-Zadeh, K. Sahingur, S. Blau, Y. Gong, S. Farsiu, Fast and robust active neuron segmentation in two-photon calcium imaging using spatiotemporal deep learning. Proc. Natl. Acad. Sci. **116**(17), 8554–8563 (2019)
17. L. Xiao, C. Fang, L. Zhu, Y. Wang, T. Yu, Y. Zhao, P. Fei, Deep learning-enabled efficient image restoration for 3D microscopy of turbid biological specimens. Opt. Express **28**(20), 30234–30247 (2020)
18. Y.V. Kistenev, V.V. Nikolaev, O.S. Kurochkina, A.V. Borisov, D.A. Vrazhnov, E.A. Sandykova, Application of multiphoton imaging and machine learning to lymphedema tissue analysis. Biomed. Opt. Express **10**(7), 3353–3368 (2019)
19. H. Lin, C. Wei, G. Wang, H. Chen, L. Lin, M. Ni, J. Chen, S. Zhuo, Automated classification of hepatocellular carcinoma differentiation using multiphoton microscopy and deep learning. J. Biophotonics **12**(7), e201800435 (2019)
20. M.J. Huttunen, A. Hassan, C.W. McCloskey, S. Fasih, J. Upham, B.C. Vanderhyden, S. Murugkar, Automated classification of multiphoton microscopy images of ovarian tissue using deep learning. J. Biomed. Opt. **23**(6), 066002 (2018)
21. M.J. Huttunen, R. Hristu, A. Dumitru, I. Floroiu, M. Costache, S.G. Stanciu, Multiphoton microscopy of the dermoepidermal junction and automated identification of dysplastic tissues with deep learning. Biomed. Opt. Express **11**(1), 186–199 (2020)

Chapter 12
U-Net: A Versatile Deep Learning Architecture for Multi-Disease Detection

Snehal Rajput, Rupal Kapdi, Mohendra Roy, and Mehul S. Raval

Abstract Imaging has revolutionized the medical domain due to its non-invasiveness and ability to capture internal organs accurately. The digitization of images led to the penetration of Artificial Intelligence, which improved the objectivity of analysis. The emergence of Convolution Neural Networks (CNN) allowed semantic segmentation of medical images, allowing pixel-level understanding. U-Net is a CNN, which is among the most robust architectures researchers use in state-of-the-art semantic segmentation methods. The chapter aims to discuss U-Net and its ability to work effectively across different imaging modalities. The chapter highlights variants of U-Nets and presents various data augmentation techniques to improve the performance. The chapter covers the segmentation of brain tumors, lung nodules, and liver lesions.

12.1 Introduction

Medical imaging allows visualization of the internal structure of biological organs. Clinicians mainly use this for prognostic, diagnostic, monitoring progress or recovery, and therapeutic purposes. It is a non-invasive technique without causing pain or discomfort to the patient. The essential medical imaging methods include Computed Tomography (CT), X-rays, Magnetic Resonance Imaging (MRI), and Ultrasound. Accurately detecting and diagnosing the disease is crucial for effective treatment planning based on the patient's condition. It is also vital to ensure patient

S. Rajput
School of Computer Science Engineering & Technology, Bennett University, Greater Noida, India

R. Kapdi
Institute of Technology, Nirma University, Ahmedabad, India

M. Roy
School of Technology, Pandit Deendayal Energy University, Gandhinagar, India

M. S. Raval (✉)
School of Engineering and Applied Science, Ahmedabad University, Ahmedabad, India
e-mail: mehul.raval@ahduni.edu.in

A. Gogoi and N. Mazumder (eds.), *Biomedical Imaging*, Biological and Medical Physics, Biomedical Engineering, https://doi.org/10.1007/978-981-97-5345-1_12

safety; effective diagnostics using medical imaging ensures improved chances of recovery. Each imaging modality detects different diseases; e.g., X-ray can detect bone fractures, dental problems, and lung disease; CT detects cancer in organs, brain, and cardiac issues; MRI usually detects soft tissue diseases like brain tumors and kidney disease; ultrasound detects kidney stones and heart conditions.

Each modality captures different aspects of the internal body structure; therefore, images generated by them have different characteristics. The images have different resolutions, contrast, and sensitivity. Medical imaging allows the development of autonomous or semi-autonomous diagnostic systems. It is necessary as traditional diagnostic through manual processing is slow, results may not be reproducible, and there is an inter-rater variability. Also, there can be bias in the decision-making, and experts may need more subtle diagnostic features. Automation can lead to diagnostics at a scale, faster processing and turnaround time, reproducible and standardized results, objectivity in the analysis, and easily integrated technological advances.

The disease manifests differently for each individual, with person-centric changes in symptoms and severity. Therefore, the patient responds to treatment very differently. The disease is a temporal phenomenon that evolves, which is a challenge during diagnostics. Apart from the inter-disease variability, there is also an intra-disease variability, and subtypes of the same disease may exhibit different symptoms. Specific diseases in the early stages may exhibit the same symptoms, e.g., fever in the body could be due to several different diseases. Genetic factors, environmental factors, age, and comorbid conditions are crucial during diagnostics. It indicates that patients need personalized treatment depending on their physical conditions and other factors.

Considering the variability in the imaging modalities and disease manifestation, developing robust machine learning (ML) and deep learning (DL) based algorithms is vital. They must learn well from diverse datasets, generalize across the disease and stages, be scalable, and be adaptive as new challenges emerge. Timely and accurate diagnostics are critical to effective treatment delivery. Early interventions allow better planning, lower complications, increase the survival rate in diseases like cancer, improve the patient's life as psychologically they are better off, reduce cost, and, most importantly, reduce the intensity of the treatment.

The emergence of DL has a significant impact on imaging and disease diagnostics. The effectiveness of DL is due to the following factors: automated hierarchical feature extraction from the raw data establishing relationships between different levels of abstraction; end-to-end learning, which takes images as inputs and produces predictions; it is scalable, does not suffer from feature selection bias; transfer learning allows faster convergence and also cross-domain adaptations, e.g., between X-rays and CT is possible; robust to variations and facilitates the real-time implementation. Therefore, DL's ability to select features and autonomously learn patterns has provided healthcare systems accuracy, efficiency, and generalizability.

12.2 Overview of U-Net and Its Applicability in Medical Imaging

Convolutional Neural Network (CNN) processes the spatial/spatiotemporal data like images or videos using a convolutional layer to pick out hierarchical features using filters. The features are concatenated and fed to the next layer to complete tasks like classification or segmentation—breaking the image into its subparts and labeling them. In medical imaging, it is essential to conduct semantic segmentation [17], providing a label for each pixel. It allows pixel-level understanding of the image, which is the key in medical imaging as it allows easy differentiation between different objects in the image.

U-Net is a CNN to perform effective semantic segmentation of the image or video frame [24]. The CNN shown in Fig. 12.1 has a "U" shape from which it derives its name—U-Net. It is a fully convolutional neural network [17]. The CNN with contracting structure is complemented with an upsampling layer, increasing the resolution; the contracting and expansive path results in a U shape. As we observe in Fig. 12.1, the contracting structure known as the encoder has a series of convolution—ReLU—max pooling operations, reducing the spatial information. It learns the context, and simultaneously, the number of channels in the encoder increases, causing richness in features. The expansive path, known as the decoder, upsamples the spatial information and concatenates it with higher resolution features from the encoder via skip connection. It connects corresponding contractive and expansive path layers and preserves spatial information for generating fine details. The network propagates the contextual information by having many feature channels at the decoder, generating a high-resolution segmentation map. The number of channels decreases during each upsampling stage within the decoder.

The middle layer of the U-Net (Down conv4 in Fig. 12.1) is known as the bottleneck layer. It has the highest channel dimension, showing maximal richness for the feature and capturing very high-level information. U-Net architecture has become a workhorse for medical imaging as it is well suited for segmenting the anatomical structure of the organs. It has been successfully applied across various imaging modalities—X-rays, CT, MRI, and ultrasound showcasing its robustness and flexibility. The skip connection allows the reconstruction of high-frequency content in the images, adding richness and ease of image interpretation for diagnostics. Many imaging modalities like MRI generate a 3D volume. U-Nets can quickly process slice by slice or as a 3D volume. Semantic segmentation allows it to segment and classify multiple regions simultaneously. As a CNN, it can perform transfer learning, speeding up the convergence. Therefore, with its unique architecture, U-Net has become a defacto DL tool for processing medical images, providing a wide range of diagnostics.

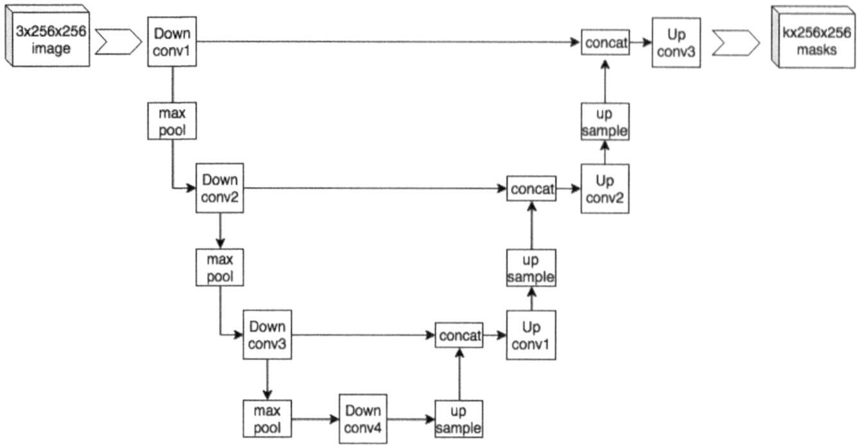

Fig. 12.1 A U-Net architecture designed to generate *k* 256-by-256 image masks for a 256-by-256 RGB image. Image Credit: By Mehrdad Yazdani—Own work, CC BY-SA 4.0, (Wikimedia Commons. 2019b, August 5)

12.2.1 Loss Functions and Optimization

The loss function assesses the performance of a machine learning algorithm by quantifying the disparity between the observed (actual) outcomes and predicted outcomes. It keeps track of the optimization process and guides it toward the best values of the parameters. Essential characteristics of loss function for medical imaging are as follows:

- They must adapt to the anatomical variations of the body organ.
- They must maintain performance for different shapes, sizes, and views of the structure, e.g., brain tumors.
- It must handle class imbalance. Most datasets will have unequal distribution for different classes, leading to a bias toward classes with fewer samples and poor learning.
- It should ensure that the segmentation results match the true anatomical structure's shape, size, and location.
- It must generate precise contours and boundaries for the region of interest (RoI).
- It guides the optimization process smoothly toward the extremum, preventing abrupt changes.
- It should allow the clinician to comprehend the ML process by emphasizing specific patterns in the decision.

Table 12.1 summarizes the loss functions popular in medical image processing and how they handle the class imbalance problem.

Table 12.1 Important loss function for addressing the issue of class imbalance

Loss	Expressions	Explanation
Dice	Dice Loss = 1–(2 * Intersection)/(Union + Intersection) Intersection refers to the pixel-wise multiplication of ground truth and predicted masks. At the same time, the union represents the summation of pixels within both the ground truth and predicted masks	The Dice Loss quantifies the dissimilarity between the ground truth and predicted segmentation masks. It spans a scale from 0 to 1, with 1 indicating perfect similarity and 0 indicating no similarity between the two masks. The aim is to minimize the dice loss or maximize the overlap between the ground truth and the predicted segmentation mask. As it balances precision and recall, it is well suited for handling data imbalance. Also, it has a high sensitivity to small regions, making it a good choice for medical imaging
Jaccard	JaccardLoss [prediction, target]: = 1–(Total [prediction * target]/ Total [(prediction + target)–(prediction * target)]) The prediction is the predicted binary, the target is the ground truth binary mask, and the total is the summation operation	It is an intersection over union (IoU) loss. It measures the similarity in the predicted and ground truth binary masks with either 0 or 1 pixels. The loss function is then 1—intersection/Union, pushing the model to follow the ground truth mask (maximize IoU). It handles class imbalance by focusing on overlapping areas and measures each pixel's contributions to the prediction. It also maintains the localized boundary or contour and the segmentation's integrity
Focal	FocalLoss [prediction, target, alpha, gamma]: = –alpha * (1–prediction)^gamma * Log[prediction] Where prediction is the probability of the correct class, alpha is a class-specific parameter or weight to handle class imbalance, and gamma is a focusing parameter to weigh complex examples	The focal loss seeks to reduce the influence of accurately classified examples on predictions while giving greater importance to challenging instances. The parameters alpha and gamma allow it to weigh and adjust loss for each class and provide control over accuracy. It also minimizes the chances of gradient saturation and effectively trains the ML algorithm

12.2.2 Evaluation Metrics

There are multiple metrics [21] for evaluating U-Net performance in disease detection tasks. Some of the key metrics include

$$Sensitivity\ (\text{True positive rate}) = \frac{TP}{(TP + FN)}$$

$$Specificity\ \text{(True negative rate)} = \frac{TN}{(TN + FP)}$$

Sensitivity is the proportion of individuals who test positive (TP) to the total number of truly positive individuals ($TP + FN$). In contrast, *Specificity* (True Negative Rate) is the fraction of individuals who test negative (TN) out of the total number of truly negative individuals ($TN + FP$). Finding the proper equilibrium between sensitivity and specificity poses a challenge in medical segmentation. Enhancing sensitivity can increase the identification of true positives but may also raise the risk of false positives. Conversely, improving specificity might reduce false positives but carries the potential of missing specific structures. Striking this balance relies on the clinical context and the consequences of false positives and negatives.

Similarly, *precision quantifies the relationship between positive predictions and all* positive predictions ($TP + FP$) set. It reflects the accuracy of the model's positive predictions. On the other hand, *recall* calculates the proportion of true positive predictions to the total count of actual positive instances (true positives + false negatives). It indicates the model's ability to identify all actual positive instances. The *F1 score* integrates precision and recall into a single metric, offering a balanced assessment of the model's effectiveness in positive and negative predictions.

Depending on the specific disease, other metrics can be the Area under the Curve for the Receiver Operating Characteristic Curve (AUC-ROC), Haudorff distance (HD), and Area Under the curve for the Precision-Recall Curve (AUC-PR). When evaluating the performance of U-Net in disease detection, it's often a good practice to consider multiple metrics to gain a detailed understanding of how well the model is performing, given the task. The selection of metrics relies on the particular disease, the clinical circumstances, and the desired tradeoffs between sensitivity, specificity, and overall segmentation accuracy.

12.2.3 U-Net Variants and Modifications

The U-Net architecture finds extensive application in the semantic segmentation of medical images, attributing each pixel to its corresponding class. As mentioned in Sect. 12.2, this architecture works as a primary deep learning network for segmentation in the medical domain. After the architecture was initially proposed by [24], many variations and improvements have been proposed for this architecture over time. Architecture is widely used for both 2D and 3D inputs. The basic architecture remains the same in 3D U-Net [6]. However, 3D operations are performed instead of 2D, namely 3D convolution and 3D max pooling. 3D U-Net [6] enables volumetric segmentation. All the 2D/3D U-Net architectures differ in the number of layers in the U-Net architecture. Figure 12.1 shows a four-layer architecture. Similarly, there may be architectures that vary in depth. The architectures can also differ in the count of convolution layers in each block, and this configuration of convolution layers enables the block to encompass the scope needed to capture features. Besides these changes, the network has the following variants:

a. **U-Net++** : An encoder-decoder network with nested skip connections and deep supervision [32]. In U-Net++ , skip connections form a hierarchical structure, unlike the original U-Net, which directly concatenates encoder and decoder feature maps. These nested skip connections link feature maps at various depths in the encoder to those in the decoder, enabling efficient multi-scale information capture. Within nested skip connections, U-Net ++ uses varying convolution layers based on pyramid levels, reducing the semantic gap between encoder and decoder feature maps for improved multi-scale information transfer. It employs deep supervision through auxiliary heads in the decoder, producing intermediate segmentation predictions. These intermediates, used in training alongside the final prediction, enhance convergence and segmentation accuracy by providing gradients at multiple stages.
b. **Attention U-Net**: With an attention mechanism, U-Net allows the network to focus only on the crucial features by removing irrelevant features from the ongoing training task. Each layer of the decoder path typically has an attention gate through which the features from the corresponding layer of the encoder path pass before concatenation. This mechanism improves the segmentation performance while maintaining the computational efficiency of the model [19].
c. **Residual U-Net**: Diakogiannis et al. [7] proposed residual U-Net, which uses the U-Net as a backbone and residual connections. Further, each residual building block incorporates multiple parallel atrous convolutions with varying dilation rates to enhance comprehension across different scales. The pyramid scene parsing pooling layer was also included to extract background information. This configuration effectively addresses the challenge of handling significant within-class variance and minimal between-class variance in pixel values for the objects of interest. Such networks find the boundary of the objects. Each task is conditioned on the inference of the previous ones, thus establishing a conditioned relationship between the various tasks, as described through the architecture's computation graph.
d. **Inception U-Net**: As per the study by Punn et al. in 2020, modifications have been applied to the Inception U-Net architecture, including incorporating switch normalization layers, convolutional layers, and inception layers. These inception layers involve the fusion of $1 \times 1, 3 \times 3,$ and 5×5 convolutions, along with hybrid pooling layers that combine elements of both max pooling and Hartley spectral pooling techniques. The inception modules enable the network to acquire deeper features from input. The network does not face a vanishing gradient problem due to the unique arrangement of the convolution layers of the inception layer.
e. **Dense U-Net**: The Vanilla U-Net has millions of parameters, increasing the training time. In addition, it suffers from vanishing gradient problems. To overcome these problems, the convolution modules at both the encoding and decoding sides are replaced with dense modules [5]. This particular variant of the U-Net architecture accomplishes a reduction in the parameter count and effectively

addresses the vanishing gradient issue. It promotes the reuse and propagation of features by amalgamating feature maps from previous layers with those of subsequent layers. The dense connections have a regularizing effect, which overcomes overfitting issues during training. The number of parameters in a dense connection is also reduced without decreasing the depth of the network.

12.2.4 Data Augmentation Strategies to Improve U-Net Performance

The term “data augmentation” (DA) refers to techniques for increasing the amount of training data by performing various transformations on the original dataset [27]. The goal of DA is to broaden the diversity of the training dataset, which can help improve the model’s generalization ability and reduce overfitting. These techniques are beneficial when available training data is limited. By introducing these variations, the model becomes more robust and capable of handling different variations of the input data that it might encounter during testing or deployment. Some of the most widely used transformations are as follows:

a. **Rotation and flipping**: In this technique, rotation involves rotating an image either clockwise (positive/right) or anticlockwise (negative/left) around an axis within a range of 0° to 360°. The rotation can be random rotation within the range or as specified by the user. Similarly, flipping can be random or user-defined, which involves horizontal (across the Y-axis) or vertical flipping (across the X-axis) of an image. Flipping can be particularly useful when objects have symmetry, which can help the model to detect objects irrespective of the orientation, thereby improving its ability to detect symmetrical patterns.
b. **Scaling and cropping**: The scaling technique resizes the image by rescaling each pixel value of an input image by multiplying it by the scale factor and adding to the offset value. Scaling helps the model locate objects precisely at different scales for object detection and segmentation. In contrast, cropping proves advantageous when dealing with sufficiently large image sizes, and it facilitates the selection of a specific height, width, and position for the cropped region.
c. **Elastic deformations and affine transformations**: Object deformations are typical in image acquisition. Elastic deformation involves applying small displacements to pixels within an image. These shifts are drawn from a Gaussian distribution with a specific pixel standard deviation. In contrast, an affine transformation refers to a linear operation that maintains the positions of points, the straightness of lines, and the relative distances between points. It includes rotation, translation, scaling, and shearing operations, which can help identify objects in diverse orientations.
d. **Color augmentations and noise injection**: Color augmentation is a set of techniques that alter the color distribution of images. Some standard color augmentation techniques include Brightness, contrast, gamma, hue, and saturation. Color augmentations can aid in alleviating biases caused by disparities in lighting,

camera configurations, and environmental circumstances. In contrast, noise injection involves adding some random variation in the pixel values of an image, which can occur during the image acquisition process. Introducing noise enhances the model's ability to withstand the presence of that specific noise in the input data. Furthermore, it encourages the model to comprehend various aspects of each class by including random features.

12.3 Summary of Important Datasets

Table 12.2 summarizes essential challenges, tasks, and available annotated datasets.

Table 12.2 Various publicly available biomedical datasets

Challenge	Task	Dataset description
Brain tumor Segmentation (BraTS) RSNA-ASNR-MICCAI challenge [1, 25]	Brain tumor segmentation into multiple lesions	BraTS2021 consists of 3D multiple modal MRI images. The training set includes images from 1251 patients, while the validation and test datasets contain images from 219 and 530 patients, respectively. However, truth labels from validation and test sets are not available to the user
Lung Nodule Analysis (LUNA) [26]	Challenge tracks: 1. Nodule detection (NDET) 2. False positive reduction (FPRED)	2D images-888 CT scans for training and test sets total
Liver Tumor Segmentation (LiTS) [2]	Segment liver lesions	The training dataset consists of 130 3D CT scans, while the test dataset comprises 70 CT scans
Kidney and Kidney-Tumor Segmentation challenge (KiTS23) MICCAI challenge [10]	Semantic segmentation of kidneys, renal tumors, and renal cysts	The training dataset includes 489 cases of 3D CT scans, while the test dataset contains 110 cases of 3D CT scans
Fast, low resource and accurate organ and Pan-cancer segmentation (FLARE)–in the abdomen CT-MICCAI challenge [18]	It focuses on pan-cancer segmentation, which covers various abdominal (13) cancer types	The training set includes 4,000 3D CT scans from 30 + medical centers. 2200 cases have partial labels, and 1800 cases are unlabeled
Diabetic Retinopathy Segmentation (DRIVE) [23]	Retinal vessel segmentation from the images	40 2D images have been split into two sets: a training set comprising 20 images and a test set containing 20 images

12.4 U-Net Applications in Multi-Disease Detection

In addition to the abovementioned variation, U-Net has numerous other variations, including cascaded U-Net arrangements, parallel U-Net arrangements, and a combination of two or more modules, as discussed above. The success of U-Net for biomedical image segmentation is apparent from its numerous biomedical applications, as discussed in this section.

12.4.1 U-Net for Tumor Segmentation in Brain MRI Scans

Glioblastoma is the most common and highly aggressive malignant primary tumor in the adult central nervous system. This tumor exhibits intrinsic variation in intensities, appearance, regions, and structure; hence, patients have poor survival chances. The brain tumor segmentation (BTS) task comprises delineating tumor tissues from healthy brain tissues. The goal is to develop machine learning algorithms and models that can accurately segment and classify different tumor regions within the brain scans. This can aid medical experts in diagnosing and treating brain tumors more effectively. Over the last 11 years, the Brain Tumor Segmentation (BraTS) challenge [1, 25] has been conducting a competition that provides the largest dataset of 3D MRI scans for participants to work with. In this challenge, participants are tasked with accurately delineating tumor areas into three distinct regions: the whole tumor (WT), the tumor core (TC), and the enhanced or active tumor (ET) (Fig. 12.2).

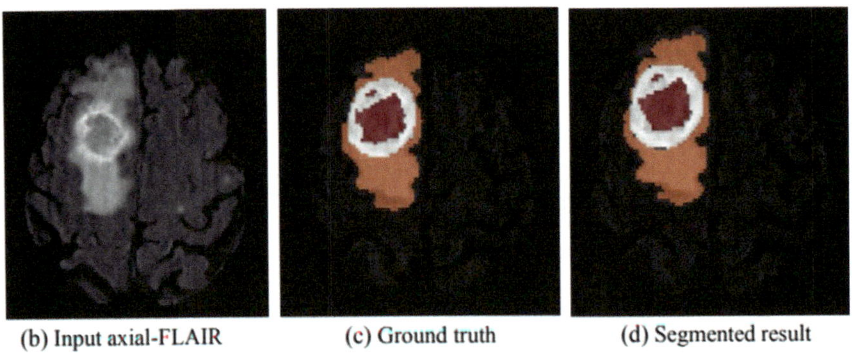

Fig. 12.2 Images from the BraTS2019 dataset, Input FLAIR image, ground truth, and segmented results derived by Snehal Rajput et al.

12.4.1.1 Data Preparation and Pre-Processing

Radiological images acquired from different institutions have different protocols and scanning machines. Integration of all the images requires some preliminary adjustments to align them with a uniform standard. Some widely used techniques include mapping to the same anatomical structure, stripping the skull and other non-brain structures, interpolating the voxel or pixel resolution, normalizing pixels' intensity values, and histogram matching.

12.4.1.2 Customization and Training of U-Net for Brain Tumor Segmentation

Due to advancements in computational capabilities, there has been a surge in the development of more complex U-Net models for segmenting medical images. Some of the best-performing architectures are: Isensee et al. [11] proposed a 5-stage encoder-decoder architecture with the skip connection. Their approach emphasizes training based on region-based training, extensive data augmentation post-processing, and ensemble predictions from multiple 3D U-Net models. Jia et al. [13] proposed a two-stage cascaded approach to segment brain regions from coarser to finer. They employ a single model to segment the more significant brain regions and then cascade additional models to segment brain tumors into more detailed sub-regions. Yuan et al., [31] proposed a full-scale attention network where the skip connection at each layer is replaced with a skip connection from all the layers. In this arrangement, each decoding layer considers the output feature maps from all the encoding layers, which allows the capture of detailed and broader contexts at different scales. Jia et al. [13] incorporated a transformer and self-attention mechanism into the 3D U-Net to capture the long-range dependencies.

12.4.1.3 Results, Performance, and Validation

The standard metric used for assessment is the Dice similarity coefficient (DSC), which measures the overlapping pixel areas between predicted and ground truth areas. The DSC ranges from 0 to 100, where 0 indicates no overlap, and 100 represents perfect agreement between the two sets. The results of the top methods on BraTS 2020 and 2021 datasets are shown in Table 12.3.

12.4.2 U-Net for Lung Nodules Detection Using CT Scans

Lung cancer ranks among the most severe respiratory disorders. It continues to become the most commonly diagnosed cancer. As per Sung et al. [28], there were approximately 17 lacs mortalities due to lung cancer, 18% out of all the cancel

Table 12.3 BTS results of top BraTS challenge (2020 and 2021) methods. Results are derived from the test set

Method	Dataset	DSC
Isensee et al. [11]	Training set—369 Test set—166	88.95(WT), 85.06(TC) and 82.03(ET)
Jia and Shu [14]	Training set—369 Test set—166	91.29(WT), 85.46(TC) and 0.787(ET)
Yuan et al. [31]	Training set—369 Test set—166	88.28(WT), 84.33(TC) and 81.77(ET)
Jia et al. [13]	Training set—1, 251 Test set—530	92.57(WT), 93.50(TC) and 88.74(ET)

mortalities, the world's leading cause of death of cancer-associated mortality. When lung cancer is detected in its early stages, the likelihood of a cure dramatically rises, increasing from 14 to 49% within five years. Non-calcified lung nodules serve as an early indicator of lung cancer. Hence, accurate detection of lung nodules is highly recommended in lung cancer treatment. Computed Tomography (CT) is a commonly employed method for lung nodule detection due to its high sensitivity, rapid acquisition, and low cost. Identifying nodules in CT images can be time- and labor-intensive, primarily because of their small, rounded, or irregular shapes.

Numerous Computer-Aided Detection (CAD) systems have been created to aid radiologists in the early identification and delineation of lung nodules.

12.4.2.1 Data Preparation and Pre-Processing

Vendt [3] has proved that the LUNA 16 dataset[1] is widely used to address lung nodule detection and segmentation tasks. Around 888 CT images are included in the LIDC-IDRI dataset. The thickness of each slice is less than 2.5 mm. Four experienced radiologists have annotated the dataset where identified lung nodules have size > 3 mm.

12.4.2.2 Customization and Training of U-Net for Lung Nodule Detection

In [15], authors have proposed the modified 3D U-Net ++ architecture, where each region map of the first layer encoder blocks and last decoder block are given to the loss function. A customized dataset is used to generate the segmentation and compared using the Dice similarity coefficient and intersection over union. Authors [4] have proposed the five-layer deep 2D Residual U-Net for lung nodule segmentation. Also, two U-Net ++ networks are named shallow U-DL network and deep

[1] https://www.kaggle.com/datasets/fanbyprinciple/luna-lung-cancer-dataset/code.

Table 12.4 Results of lung nodule detection by U-Net

Method	Dataset	DSC	IoU
Kido et al. [15]	Custom	0.845	0.738
Bruntha et al. [4]	LIDC-IDRI [3]	0.870	0.770
Wang et al. [30]—Shallow U-DL	LIDC-IDRI	0.745	–
Wang et al. [30]—Deep U-DL	LIDC-IDRI	0.739	–

U-DL network [30]. The shallow U-DL network contained 16 layers, whereas the deep U-DL network had 201 layers. The comparison for lung nodule detection is shown in Table 12.4. The results on LIDC-IDRI are encouraging, but they need significant improvement before algorithms are deployed in the field.

12.4.3 U-Net for Liver Lesion Segmentation in Abdominal MRI Scans

The liver, the largest solid organ in the human body, plays a vital role in digestion and metabolism. Liver anomalies and essential biomarkers are crucial for diagnosis and hepatic tumor disease. Such biomarkers can be visualized using MRI or CT images. Again, manual segmentation of the lesion in 3D CT images is a time-intensive process, lacks reproducibility, and heavily relies on the expertise of individuals [2]. Many datasets are publicly available for liver tumor segmentation. One such dataset is the Liver Tumor Segmentation (LiTS) challenge dataset [2]. This dataset provides 201 CT volumes of liver tumor patients. The tumors exhibited diverse lesion-to-background ratios, encompassing both hyper-dense and hypo-dense characteristics. These images comprised a combination of abdominal CT scans taken before and after therapy. The imaging process involved various CT scanners and acquisition protocols, encompassing common real-world clinical data artifacts such as metal artifacts. Therefore, the dataset was characterized by significant resolution and image quality diversity. The in-plane (x–y plane) image resolution varies, spanning from 0.56 mm to 1.0 mm and 0.45 mm to 6 mm.

12.4.3.1 Data Preparation and Pre-Processing

CT images are 3D images where pre-processing can be done either slice-wise or image-wise. Generally, the slices which do not include the organ part are chopped off from the images. The intensity normalization follows in the next step. Many approaches apply histogram equalization. Image augmentation is necessary to increase the dataset size for training deep learning models.

Table 12.5 Result of liver lesion segmentation by U-Net

Method	Dataset	Accuracy	Recall	Precision	Mean IoU
Özcan et al. [20]	LiST	99.48	95.12	96.78	95.82
Jeong et al. [12]	Custom	–	0.923	0.923	0.855

12.4.3.2 Customization and Training of U-Net for Liver Lesion Segmentation

Özcan et al. [20] introduces an inception module-based U-Net model. This module has been added to the encoder-decoder concatenation skip links. The module contains two sequential inception blocks. Features from the encoder path are fed into four parallel paths of one inception module, which are then concatenated and given to the similar subsequent inception module of the same configuration. These features were then combined with the features of the decoder path for further process. Jeong et al. [12] applied U-Net with an attention mechanism where more complex hierarchical features are extracted.

Additionally, their architecture incorporated the concatenation of hierarchical features through skip connections at each encoder and decoder level. The loss is calculated by combining the decoder blocks' segmentation map. They used a custom dataset for the designed architecture [8]. Table 12.5 shows essential results.

12.4.4 Transfer Learning in U-Net

We can readily utilize the acquired weights and architecture to address our specific problem by employing pre-trained models that have previously undergone training on extensive datasets. This approach is referred to as transfer learning (TL) [29], adapting the knowledge gained from the pre-trained model to our unique problem statement. As per the requirement of our unique problem, we introduce alterations to the existing model through a process known as fine-tuning.

Given our assumption that the pre-trained network has undergone substantial training, we exercise caution in altering the weights excessively and prematurely. Given that the pre-trained model has been trained on a distinct dataset, it becomes necessary to undertake a fine-tuning process when applying it to a new dataset. This new dataset might exhibit similarities to the original dataset used for pre-training, or it could present an entirely different set of data. Various methods for fine-tuning are as follows:

1. Feature extraction by removing the output layer and using it as feature extraction for the new dataset.
2. Selective layer training while keeping others frozen: In this strategy, the initial layers' weights remain frozen while only the upper layers are retrained. The tradeoff between frozen and trainable layers must be tested and validated.

In their study of brain lesion segmentation, Litjens et al. [16] highlighted that a transfer learning model trained on MRI brain data faced difficulties when applied to a comparable dataset but with a distinct acquisition protocol. The underlying explanation for this challenge was attributed to variations in the dataset distributions. Additionally, after fine-tuning the transfer learning model on the target dataset, it showcases the capacity to achieve strong generalization with fewer training epochs as opposed to training it from scratch. The TL-based approach can also be used for cross-modality domains, e.g., cardiac segmentation, where the TL model was trained on MRI images [9]. In contrast, the target dataset was CT images using adversarial learning to learn the feature mapping between cross-domains. The study has reported effective segmentation performance.

Utilizing TL for medical image segmentation proves advantageous in enhancing the precision of deep learning models, particularly in scenarios where available medical images are scarce. It reduces training time for the specific problem. However, the effectiveness is greatly influenced by the target problem, size of the pre-trained dataset, image modality, acquisition protocol, image resolution, and fine-tuning [16]. Wang et al. [30] conducted a study focused on segmenting ribs from lungs using ultrasound images. Their approach involved employing two pre-trained U-Net models: V-U-Net, which uses a VGG architecture pre-trained on the ImageNet dataset, and X-UNet, which employs a U-Net architecture pre-trained on a grayscale natural salient object dataset. The authors adopted different strategies for these two models regarding the fine-tuning process. In the case of V-U-Net, they froze the entire four-level contracting path during fine-tuning. Conversely, for X-UNet, only the bottleneck (BN) layer was kept frozen.

In contrast, the other layers were trained on the target dataset. Notably, X-UNet, fine-tuned with all layers, demonstrated superior performance compared to fine-tuning solely the bottleneck layer. Consequently, X-UNet outperformed V-U-Net in terms of segmentation accuracy and effectiveness.

Similarly, Zhou et al. [32] also highlighted in their study based on medical imaging application that utilizing a pre-trained CNN with appropriate fine-tuning yielded better or comparable performance compared to training a CNN from scratch. Fine-tuned CNNs demonstrated increased robustness to training set sizes compared to scratch-trained CNNs. The optimal choice between shallow and deep fine-tuning varied based on the specific application. The proposed layer-wise fine-tuning approach offers a practical means to achieve optimal performance for the given application, contingent on the available data volume.

12.5 Conclusion and Future Directions

The U-Net provides a pixel-level classification, which is extremely useful in identifying anatomical structures in medical images. This allows predicted contours to follow the organ's shape, size, and location, resulting in excellent accuracy. Skip connections allow it to learn at different scales and preserve the fine details. It provides a mechanism to fuse information from different MRI modalities and allows 2D and 3D processing of the images. Being a CNN enables transfer learning, leading to faster network convergence.

However, there are several implementation challenges to work through. Usually, medical datasets involve several complexities, such as patient privacy, variation in the acquisition protocols, and, most importantly, a lack of reliable annotations. This results in small and unreliable datasets. Collecting a massive amount of data, getting it annotated by experts, and opening it up to the public domain requires considerable human resources and funding. Moreover, having robust prediction annotations by only one expert is insufficient, and multiple experts would need to annotate the data with consensus. This shortage of datasets reflects poor generalization by the ML model. Architecture like U-Net has millions of parameters, so they can easily overfit onto a small dataset. So, overfitting and lack of quality datasets go hand-in-hand. There are solutions to these problems, and solutions like data augmentation, domain adaptation, and generative AI can help mitigate the challenge.

With the increasing use of AI in the medical domain, it is crucial to explain the decisions given by the algorithms to clinicians. As discussed above, U-Net excels as a "black box" in effective segmentation. However, it cannot explain "Why" and "How" it arrived at the decisions. This has not allowed U-Net to be widely accepted in clinical practice. The ML algorithm must provide insight into the functioning and decision-making process and explain it to all stakeholders in "their" language. The model's interpretability is necessary for building trust and confidence in U-Nets' practical use. The outcomes of the U-Net model must be validated and vetted by medical practitioners. The interpretability also fuses trust in the patients as medical practitioners can confidently explain the treatment plan and diagnostics. The interpretability also allows the model to be debugged and provides checkpoints to analyze its failure. The limitations of the U-Net can now be picked up, and the model can be refined. The decisions by medical practitioners can be full of bias, and they need to be mitigated. Interpretability uncovers the source of data or algorithmic bias and allows better training and generalization. The research directions, like attention maps, saliency maps, and relevance propagation across the layers, help locate regions and features that contribute more to the decision-making. As the applicability of U-Net cuts across other modalities, more learning toward interpretability takes place.

U-Net architecture is memory-hungry and needs heavy computational and memory resources. It will not only be a heavy model during training, but it can be heavy during deployment and testing. The training time is also longer with more contracting and expanding channels of the U-Net, which increases the latency. Moreover, training large models faster would need a "thick" GPU or their cluster to make computations faster and more efficient. This is especially necessary for medical image processing. For some researchers, this can be detrimental to their research outcome as they may lack access to state-of-the-art computations. The need for memory increases as the U-Net models deepen and the datasets get big. This would severely limit the data that can be processed at a time, limiting the minibatch size.

An accurate model is insufficient, and the algorithm must be deployed on edge devices. The large-memory models fail when deployed on memory-constrained devices like mobile phones or edge devices. The research to generate a "light-weight" U-Net without compromising segmentation results is a developing area of research. Several techniques, like self-attention mechanisms, reducing the network size, and optimizing skip connections, can balance accuracy and model size. The penetration of AI into VLSI has also accelerated the development of AI or ML accelerator boards that can support faster inference. The combination of optimization from software and hardware leads the way to deploy efficient U-Nets on the end devices for medical image-based diagnostics. The other solution is to use cloud-based resources to remove the resource constraints. Demand-based models are also easy to develop using cloud computing. Utilizing transfer learning from pre-trained models on extensive datasets can also speed up the model training process and enhance resource efficiency.

Acknowledgements M. Roy acknowledges the seed grant No.ORSP/R&D/PDPU/2019/MR/RO051$ of PDEU (for the computing facility), the core research grant No. CRG/2020/000869$ of the Science and Engineering Research Board (SERB), Govt. of India and the project grant no GUJCOST/STI/2021-22/3873$ of GUJCOST, Govt. of Gujarat, India. M. S. Raval acknowledges the grant number. GUJCOST/STI/2021-22/3858 of Gujarat Council of Science and Technology (GUJCOST), Govt. of Gujarat, India for computing facility to support the work.

References

1. B.H. Menze, A. Jakab, S. Bauer, J. Kalpathy-Cramer, K. Farahani, J. Kirby, K. Van Leemput, The multimodal brain tumor image segmentation benchmark (BRATS). IEEE Trans. Med. Imaging **34**(10), 1993–2024 (2014)
2. P. Bilic, P. Christ, H.B. Li, E. Vorontsov, A. Ben-Cohen, G. Kaissis, B. Menze, The liver tumor segmentation benchmark (lits). Med. Image Anal., **84**, 102680
3. Vendt, B, Data from The Lung Image Database Consortium (LIDC) and Image Database Resource Initiative (IDRI): A completed reference database of lung nodules on CT scans (LIDC-IDRI). Last modified 21st August, (2023)
4. P.M. Bruntha, S. Dhanasekar, L.J. Ahmed, V. Govindaraj, S.I.A. Pandian, S.S. Abraham, Lung_RUNET: A segmentation framework for lung nodules. in *2023 9th International Conference on Advanced Computing and Communication Systems (ICACCS)* ,vol. 1 (IEEE, 2023), pp. 658–661
5. Y. Cao, S. Liu, Y. Peng, J. Li, DenseUNet: densely connected UNet for electron microscopy image segmentation. IET Image Proc. **14**(12), 2682–2689 (2020)
6. O. Çiçek, A. Abdulkadir, S.S. Lienkamp, T. Brox, O. Ronneberger, 3D U-Net: learning dense volumetric segmentation from sparse annotation. in *Medical image computing and computer-assisted intervention–MICCAI 2016: 19th international conference, Athens, Greece, October 17–21, 2016, Proceedings, Part II 19* (Springer International Publishing, 2016), pp. 424–432
7. F.I. Diakogiannis, F. Waldner, P. Caccetta, C. Wu, ResUNet-a: A deep learning framework for semantic segmentation of remotely sensed data. ISPRS J. Photogramm. Remote. Sens. **162**, 94–114 (2020)
8. T. Fan, G. Wang, X. Wang, Y. Li, H. Wang, MSN-Net: a multi-scale context nested U-Net for liver segmentation. Signal, Image Video Process., **15**, 1089–1097 (2021)
9. M. Ghafoorian, A. Mehrtash, T. Kapur, N. Karssemeijer, E. Marchiori, M. Pesteie, C.R. Guttmann, F.E. de Leeuw, C.M. Tempany, B. Van Ginneken, A. Fedorov, Transfer learning for domain adaptation in MRI: Application in brain lesion segmentation. in *Medical image*

computing and computer assisted intervention—MICCAI 2017: 20th International conference, Quebec City, QC, Canada, September 11–13, 2017, Proceedings, Part III 20 (Springer International Publishing, 2017), pp. 516–524
10. N. Heller, F. Isensee, K.H. Maier-Hein, X. Hou, C. Xie, F. Li, C. Weight, The state of the art in kidney and kidney tumor segmentation in contrast-enhanced CT imaging: Results of the KiTS19 challenge. Med. Image Anal. **67**, 101821 (2021)
11. F. Isensee, P. Kickingereder, W. Wick, M. Bendszus, K.H. Maier-Hein, No new-net. in *Brainlesion: Glioma, Multiple sclerosis, stroke and traumatic brain injuries: 4th International workshop, BrainLes 2018, Held in conjunction with MICCAI 2018, Granada, Spain, September 16, 2018, Revised selected papers, Part II 4* (Springer International Publishing, 2019), pp. 234–244
12. J.G. Jeong, S. Choi, Y.J. Kim, W.S. Lee, K.G. Kim, Deep 3D attention CLSTM U-Net based automated liver segmentation and volumetry for the liver transplantation in abdominal CT volumes. Sci. Rep. **12**(1), 6370 (2022)
13. H. Jia, W. Cai, H. Huang, Y. Xia, H^ 2 2 NF-net for brain tumor segmentation using multimodal MR imaging: 2nd place solution to BraTS challenge 2020 segmentation task. in *Brainlesion: Glioma, multiple sclerosis, stroke and traumatic brain injuries: 6th International workshop, BrainLes 2020, Held in conjunction with MICCAI 2020, Lima, Peru, October 4, 2020, Revised Selected Papers, Part II 6* (Springer International Publishing, 2020), pp. 58–68
14. Q. Jia, H. Shu, Bitr-unet: a cnn-transformer combined network for mri brain tumor segmentation. In *International MICCAI brainlesion workshop* (Springer International Publishing, Cham, 2021), pp. 3–14
15. S. Kido, S. Kidera, Y. Hirano, S. Mabu, T. Kamiya, N. Tanaka, N. Tomiyama, Segmentation of lung nodules on ct images using a nested three-dimensional fully connected convolutional network. Front. Artif. Intell. **5**, 782225 (2022)
16. G. Litjens, T. Kooi, B.E. Bejnordi, A.A.A. Setio, F. Ciompi, M. Ghafoorian, C.I. Sánchez, A survey on deep learning in medical image analysis. Med. Image Anal. **42**, 60–88 (2017)
17. J. Long, E. Shelhamer, T. Darrell, Fully convolutional networks for semantic segmentation. in *Proceedings of the IEEE conference on computer vision and pattern recognition* (2015), pp. 3431–3440
18. J. Ma, Y. Zhang, S. Gu, C. Ge, S. Ma, A. Young, ... B. Wang, *Unleashing the strengths of unlabeled data in pan-cancer abdominal organ quantification: The FLARE22 challenge.* (2023). arXiv preprint arXiv:2308.05862
19. O. Oktay, J. Schlemper, L.L. Folgoc, M. Lee, M. Heinrich, K. Misawa, D. Rueckert, *Attention u-net: Learning where to look for the pancreas.* (2018). arXiv preprint arXiv:1804.03999
20. F. Özcan, O.N. Uçan, S. Karaçam, D. Tunçman, Fully automatic liver and tumor segmentation from CT image using an AIM-Unet. Bioengineering **10**(2), 215 (2023)
21. D.M. Powers, Evaluation: from precision, recall and F-measure to ROC, informedness, markedness & correlation. J. Mach. Learn. Technol. **2**(1), 37–63 (2011)
22. N.S. Punn, S. Agarwal, Inception u-net architecture for semantic segmentation to identify nuclei in microscopy cell images. ACM Trans. Multimed. Comput., Commun., Appl. (TOMM), ACM Trans. Multimed. Comput., Commun., Appl. (TOMM), **16** (1), 1–15 (2020)
23. T.A. Qureshi, M. Habib, A. Hunter, B. Al-Diri, A manually-labeled, artery/vein classified benchmark for the DRIVE dataset. in *Proceedings of the 26th IEEE international symposium on computer-based medical systems* (IEEE, 2013), pp. 485–488
24. O. Ronneberger, P. Fischer, T. Brox, U-net: Convolutional networks for biomedical image segmentation. in *Medical image computing and computer-assisted intervention–MICCAI 2015: 18th International conference, Munich, Germany, October 5–9, 2015, Proceedings, Part III 18* (Springer International Publishing, 2015), pp. 234–241
25. S. Bakas, H. Akbari, A. Sotiras, M. Bilello, M. Rozycki, J.S. Kirby, C. Davatzikos, Advancing the cancer genome atlas glioma MRI collections with expert segmentation labels and radiomic features. Sci. Data **4**(1), 1–13 (2017)
26. A.A.A. Setio, A. Traverso, T. De Bel, M.S. Berens, C. Van Den Bogaard, P. Cerello, C. Jacobs, Validation, comparison, and combination of algorithms for automatic detection of pulmonary nodules in computed tomography images: the LUNA16 challenge. Med. Image Anal. **42**, 1–13 (2017)

27. C. Shorten, T.M. Khoshgoftaar, A survey on image data augmentation for deep learning. J. Big Data **6**(1), 1–48 (2019)
28. H. Sung, J. Ferlay, R.L. Siegel, M. Laversanne, I. Soerjomataram, A. Jemal, F. Bray, Global cancer statistics 2020: GLOBOCAN estimates of incidence and mortality worldwide for 36 cancers in 185 countries. CA: Cancer J. Clin., **71**(3), 209–249 (2021)
29. N. Tajbakhsh, J.Y. Shin, S.R. Gurudu, R.T. Hurst, C.B. Kendall, M.B. Gotway, J. Liang, Convolutional neural networks for medical image analysis: Full training or fine tuning? IEEE Trans. Med. Imaging **35**(5), 1299–1312 (2016)
30. Y. Wang, C. Zhou, H.P. Chan, L.M. Hadjiiski, A. Chughtai, E.A. Kazerooni, Hybrid U-Net-based deep learning model for volume segmentation of lung nodules in CT images. Med. Phys. **49**(11), 7287–7302 (2022)
31. Y. Yuan, Automatic brain tumor segmentation with scale attention network. in *Brainlesion: Glioma, multiple sclerosis, stroke and traumatic brain injuries: 6th international workshop, BrainLes 2020, Held in conjunction with MICCAI 2020, Lima, Peru, October 4, 2020, Revised Selected Papers, Part I 6* (Springer International Publishing, 2021), pp. 285–294
32. Z. Zhou, M.M. Rahman Siddiquee, N. Tajbakhsh, J. Liang, Unet++: A nested u-net architecture for medical image segmentation. in *Deep learning in medical image analysis and multimodal learning for clinical decision support: 4th International workshop, DLMIA 2018, and 8th International workshop, ML-CDS 2018, Held in conjunction with MICCAI 2018, Granada, Spain, September 20, 2018, Proceedings 4* (Springer International Publishing, 2018), pp. 3–11
33. File: Example architecture of U-Net for producing k 256-by-256 image masks for a 256-by-256 RGB image.png—Wikimedia Commons. (2019b). https://commons.wikimedia.org/w/index.php?curid=81055729

Chapter 13
Memristor-Based Neuromorphic Computing and Artificial Neural Networks for Computer Vison and AI—Applications

Prince Patel, Mansi Patel, Ankur Solanki, and Mohendra Roy

Abstract The traditional von Neumann architecture seen in digital processors generally suffers from the data transfer rate bottleneck and inefficiency in terms of energy consumption. The objective is to emulate the computational functions of the human brain, with special attention to attaining high compactness and energy efficiency. On the way to achieving this goal, there are a number of formidable obstacles. Because the biological brain uses extremely low power (approx. 40–60 W), high-density neural networks. Therefore new device concepts that combine low power consumption, scalability, and advanced computing functionality are needed. In this article, we will try to summarize the memristive-based neuromorphic computing. Here, the physics and functioning of CMOS-based floating-gate memory devices in artificial neural networks are explored, and then different memristive concepts relevant to deep neural network and spiking neural network architectures are reviewed and discussed. The final section discusses the main technological obstacles and potential directions for neuromorphic computing.

13.1 Introduction

The remarkable advancements in computer technology over the past half-century have been founded in Von Neumann's 1940s architecture [1]. Higher levels of data communication are required between data processing and storage because this stored-programme computer model divides them into separate domains. Gordon Moore predicted in the 1960s that the introduction of integrated circuits would lead to a

P. Patel · M. Roy (✉)
Department of Information and Communication Technology, School of Technology, Pandit Deendayal Energy University, Gandhinagar, India
e-mail: mohendra.roy@ieee.org

M. Patel · A. Solanki
Flextronics Lab, Department of Physics, School of Energy Technology, Pandit Deendayal Energy University, Gandhinagar, India

A. Gogoi and N. Mazumder (eds.), *Biomedical Imaging*, Biological and Medical Physics, Biomedical Engineering, https://doi.org/10.1007/978-981-97-5345-1_13

doubling of transistors in integrated circuits every 18–24 months. As a result of this exponential growth, computing systems became more efficient, as evidenced by faster processing and lower power consumption. But current semiconductor manufacturing technologies are getting close to the boundaries set by Moore's law. As a result, different approaches are being investigated to propel the advancement of processing systems in the future. Two different approaches suggest creating novel devices and new processing paradigms that go beyond the limitations of traditional complementary metal–oxide–semiconductor (CMOS) technologies [2, 3].

Concurrent with the advancement of computing platforms, some researchers began modelling neural systems in the 1960s, from individual neurons to complex networks, by utilizing newly developed electronic technologies. A processing paradigm that differs greatly from traditional computer architecture and offers superior performance was made evident by a deeper understanding of the structure and basic principles of the human brain. The human brain, with its 20 W of power consumption, is capable of handling novelty, complexity, and ambiguity better than even the most advanced supercomputers, which are known for their speed and accuracy. This is especially true for practical tasks like motion control and visual recognition. The field of neuromorphic computing emerged as a result of this realization, and it was in this field that Carver Mead first used the term "neuromorphic engineering" to refer to the creation of microelectronic information processing systems that function similarly to their biological counterparts [4–6]. Mead drew attention to the similarities between the physics of biological neurons and the behaviour of transistors in the sub-threshold regime in the 1980s, which sparked the development of silicon retinas and a new paradigm for computing that drew inspiration from biological neural systems.

In order to improve speed and energy efficiency, modern engineers are investigating in-memory computing architectures, in addition to trying to mimic the highly parallel architecture of biological brains. This has led to an expansion of the inspiration for neuromorphic engineering over time. In spite of depending on basic and erratic computation units, researchers achieve impressive performance in adapting to complex and unexpected environments with quick reaction times and low power consumption by delving into signal information encoding, computational principles, and learning paradigms seen in simple biological brains. In contrast, a number of beyond CMOS technologies have been put forth in response to Moore's law's limitations. Among these is the nanoscale memristor, which is one of the most promising instruments. Chua proposed the memristor theory in the 1970s, and in 2008, a group at HP Labs conducted an experimental demonstration. The memristor functions as a variable resistor, allowing its value to be changed by applying particular voltages or currents. Memristors are primarily binary devices that can switch between two resistance values: High-Resistance State (HRS) and Low-Resistance State (LRS). They are typically constructed as a union of metal-dielectric(s)-metal, with the dielectric layer being only a few nanometers thick [7–9].

In the field of neuromorphic engineering, memristors have attracted attention because of their peculiar plasticity behaviour, which is similar to the adaptation rules found in biological synapses. According to different patterns of stimulation, these devices can adapt and alter their behaviour over time; this behaviour is similar to the brain's plasticity. Potentiation and depression (P&D), which is seen in biological neurons, is a learning rule that memristors exhibit when stimulated with pulse-trains that mimic input from spiking neurons. This makes memristors regarded as synthetic inorganic synapses, demonstrating their potential applications in neuromorphic computing.

All things considered, the interwoven tales of the development of conventional computing, neuromorphic engineering, and beyond CMOS technologies most notably memristors vividly depict the dynamic nature of technological progress. Moore's law faces obstacles, but the search for novel processing paradigms and creative devices keeps pushing computing systems forward into uncharted territory. These developments are motivated by the complex and well-oiled mechanisms found in the human brain.

Memristors are replacing CMOS platforms in the implementation of neuromorphic systems. This article examines this trend and presents a new paradigm that gets around the current drawbacks of conventional processing systems.

13.1.1 Memristors

Memristors, a contraction of "memory + resistors," represent a distinct class of two-terminal non-volatile memory devices where resistance is intricately linked to the historical application of voltage and current. First conceptualized by Leon Chua in 1971 as the fourth fundamental circuit element alongside resistors, capacitors, and inductors, memristors are passive two-terminal electronic components that sustain a relationship between the time integrals of current and voltage across them [10, 11]. Their unique characteristic lies in the ability to alter and retain resistance, differentiating them from traditional resistors and establishing them as a form of non-volatile memory. Typically composed of a thin metal oxide layer sandwiched between two metal electrodes, memristors determine resistance based on the concentration of oxygen vacancies in the metal oxide layer. This concentration can be manipulated by applying voltage, leading to a change in resistance and, consequently, a shift in the device's state, facilitating information storage and processing. A crucial aspect of memristors is their pivotal role in neuromorphic computing, mimicking synapses found in biological systems [12–14] (Fig. 13.1).

In biological neural networks, synapses transmit signals between neurons, a function emulated by memristors. Their ability to modulate resistance enables the simulation of synaptic connections' strengthening or weakening, making them integral to brain-inspired computing systems. Memristors offer advantages such as low power consumption, high density, and the capability to perform memory storage and processing tasks within the same device. The I–V (current–voltage) characteristics of memristors define the relationship between current flow and applied voltage. Initially displaying non-linear behaviour, the resistance changes as voltage is applied,

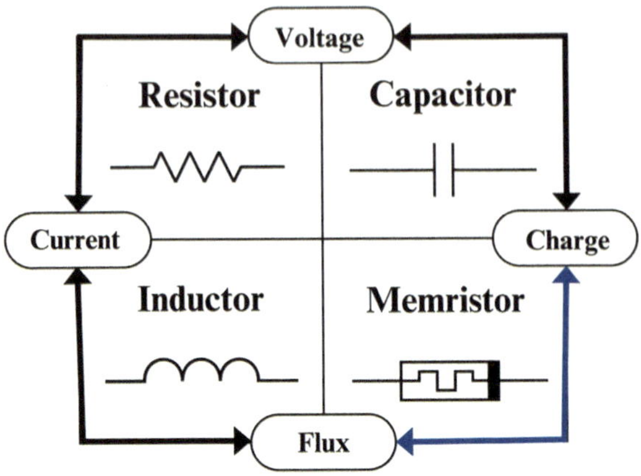

Fig. 13.1 Current I, charge Q, voltage V, and magnetic flux Φ are the four fundamental circuit quantities that are explained. The missing memristor in the circuit [15]

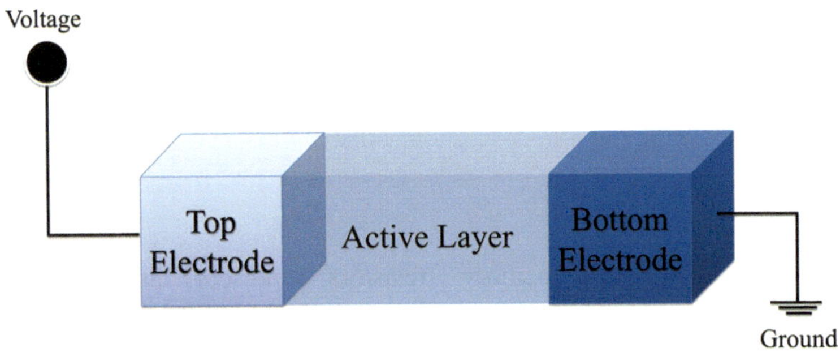

Fig. 13.2 Memristor structure [15]

and the resulting I-V curve provides insights into the memristor's different states and switching behaviour. The hysteresis loop in the I-V curve signifies the device's memory aspect [16, 17] (Fig. 13.2).

The unique combination of features positions memristors as standout components in neuromorphic computing. Their seamless integration of processing and memory functions at a single site mirrors the behaviour of synapses in biological neural networks, enhancing efficiency by reducing energy consumption and data movement. The non-volatile nature of memristors contributes to improved energy efficiency and faster boot-up times. Their analogue behaviour, allowing a continuous range of resistance values, closely resembles the complexity of biological synapses, making them well-suited for accurately simulating neural processes. In applications demanding small size and energy efficiency, such as Internet of Things (IoT) devices and portable gadgets, memristors excel due to their high density, scalability, and low power consumption. The parallel processing capability aligns with the inherent parallelism of the brain, and their flexibility facilitates the simulation of synaptic plasticity, a vital aspect of memory and learning. Despite some drawbacks, memristors contribute significantly to the promising field of neuromorphic computing, paving the way for the development of efficient and brain-inspired computing systems [18].

Artificial neural networks (ANNs) and memristors are integrated through a methodical process that connects the computational model of the brain with physical components [19, 20]. The first step is to define the neural network's architecture, including its layers, neurons, and connectivity. Then, memristors are mapped onto neurons and synapses, where each memristor represents a synaptic connection and stores corresponding weights because of its analogue capabilities. This mapping process is followed in the fabrication and hardware platform integration of memristor devices. The resistance states of memristors are then programmed according to synaptic weights, enabling the analogue representation of neural connections. In order to take advantage of memristors' special qualities—namely, their capacity to mimic synaptic plasticity—learning algorithms have been modified to enable effective and realistic learning procedures. An analogue behaviour and the parallel processing capabilities of memristors are taken into account when configuring signal propagation within the neural network. Using MNIST datasets, the integrated system is tested and optimized, with parameters adjusted to reach target performance levels. The fields of materials science, computer science, and neuroscience must continuously collaborate and iterate in order to keep up with the rapid advancements in neural networks and memristor technologies [21–23].

13.1.2 Neuromorphic Computing

Neuromorphic computing, drawing inspiration from the architecture and computational principles of biological brains, has evolved over the years to address complex computational problems. The foundation of this field lies in Artificial Neural Networks (ANNs), first introduced in the 1940–1950s [24, 25]. ANNs comprised interconnected neurons with adjustable synaptic weights, solving intricate pattern recognition, classification, and function estimation challenges [26–28] (Fig. 13.3).

The early ANNs lacked temporal aspects, but in the 1950s–60 s, a second generation emerged with continuous analogue-valued outputs, introducing non-linear activation functions like sigmoid and ReLU. The advent of backpropagation allowed the development of Deep Neural Networks (DNNs) with multiple hidden layers, marking a significant advancement [30, 31]. Despite their success, DNNs exhibit inefficiencies in terms of speed, power consumption, and information coding when compared to the human brain (Fig. 13.4).

Unlike conventional DNNs, the human brain processes information continuously over time, utilizing spiking neurons that transmit signals through spikes. This led to the emergence of the third generation of neural networks Spiking Neural Networks (SNNs) [32–34]. SNNs aim to bridge the gap between neuroscience and machine learning by incorporating biologically-realistic models of neurons. These networks introduce the concept of time into their operations, operating asynchronously with spike trains representing information flow.

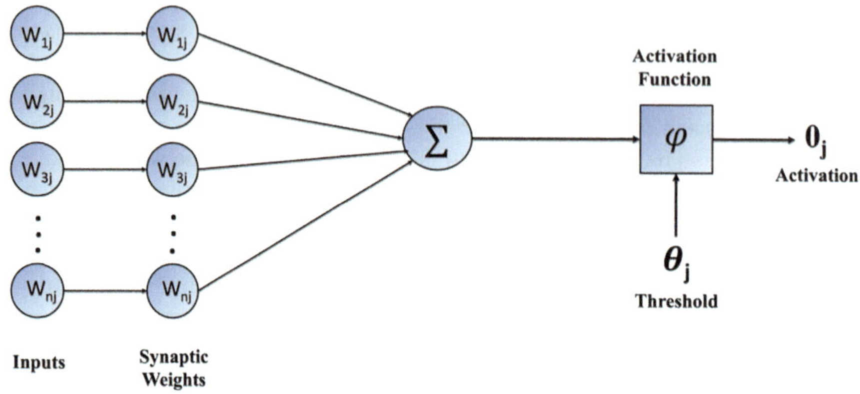

Fig. 13.3 Diagram of an artificial neuron with n inputs with their corresponding synaptic weights [29]

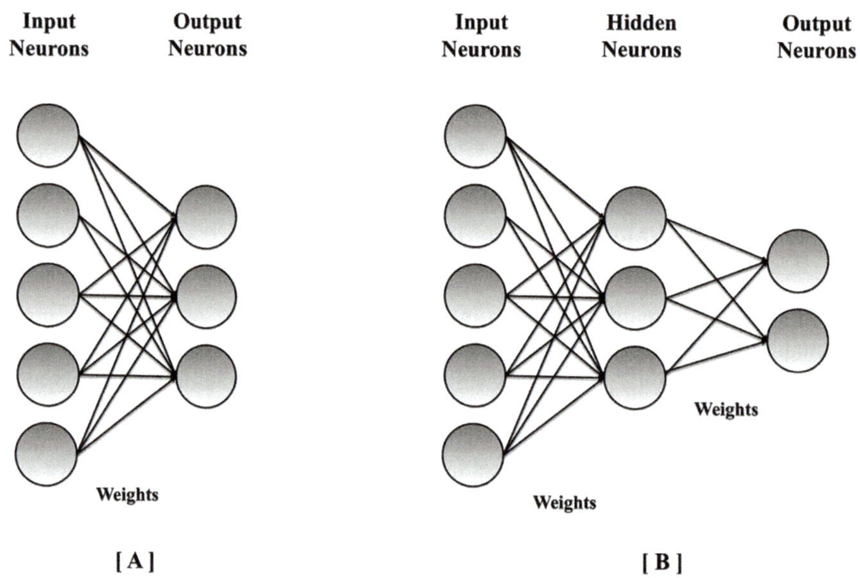

Fig. 13.4 **a** Single-layer perceptron architecture. **b** Adding more than one layer of trainable weights to a multi-layer perceptron [29]

SNNs have advantages in handling dynamic information, such as visual gesture recognition or speech recognition, more naturally than traditional artificial intelligence systems. Unlike static image-based computations, SNNs process information in a continuous time-driven manner, leveraging temporal correlations and achieving higher efficiency. Several coding methods for spike trains have been proposed, including frequency-based coding, time-to-first-spike encoding, rank order coding, and synchronous detection coding. The choice of coding method influences power efficiency and computational speed in SNNs.

Various neuron models describe the behaviour of biological neurons, ranging from complex biophysical models like Hodgkin-Huxley to simplified models like the Integrate and Fire model. While detailed biophysical models offer accuracy, simplified models are preferred for computational efficiency and hardware implementation. Connectivity in neural networks varies, with fully connected networks allowing each neuron to connect to all neurons in the next layer. Convolutional Neural Networks (ConvNets) reduce interconnections, inspired by the visual cortex's projective field. This architecture is commonly used for image processing applications.

Comparing ANNs and SNNs, SNNs exhibit "pseudo-simultaneity," achieving low latency between input and output spike flows. SNNs process spikes almost in real-time, enabling faster recognition speeds dependent on input statistics and coding strategies. Power consumption in SNNs depends on stimulus statistics and coding strategies, making efficient coding crucial for power efficiency. SNNs face challenges, including the complexity of incorporating the time variable and the need for parallel hardware resources due to real-time spike processing. While SNNs excel in applications requiring recurrent neural architectures, their learning methods differ from those of ANNs. Despite these challenges, recent work reports impressive accuracy for SNNs, with potential applications in recognizing dynamic sequences efficiently (Table 13.1).

13.2 Applications of Neuromorphic Computing

13.2.1 Medical Image Analysis and Computer Vision

The integration of memristors with neural networks heralds a revolutionary era in medical image analysis, promising unprecedented levels of precision and efficiency. In the critical realm of medical imaging, where accuracy is paramount, memristor-based neural networks bring unparalleled capabilities to the forefront, enabling more precise and efficient analysis. Medical image analysis encompasses crucial tasks for accurate diagnostics, spanning the identification of anomalies in X-rays to the scrutiny of intricate details in magnetic resonance imaging (MRI) and computed tomography (CT) scans. The continuous learning and adaptive nature of memristors play a pivotal role in addressing the inherent complexities of medical imaging [35–37] (Fig. 13.5).

A key challenge in medical image analysis is the demand for systems to adapt to varying patient conditions and imaging modalities. Memristors, emulating synaptic plasticity, empower neural networks to learn from diverse datasets and dynamically adjust their processing parameters. This adaptability ensures the system's effectiveness across different imaging scenarios, contributing to more accurate and reliable diagnostic outcomes. In tasks such as tumour detection or anomaly identification,

Table 13.1 Comparison between ANN and SNN [29]

Feature	ANN	SNN
Data processing	Frame-based	Spike-based
Latency	High	Low Pseudo-simultaneity
Time resolution	Low	High Preservation of spatio-temporal correlation
Time processing	Sampled	Continuous
Neuron model complexity	Low	High
Recognition accuracy	Higher	Lower
Hardware multiplexing	Possible	Not possible
System scale-up	Ad hoc	Adding modules
Recognition speed	Low	High
Power consumption	Independent on input stimulus Dependent on hardware resources Dependent on system complexity Determined by processor power and memory fetching Independent on input stimulus	Dependent on input statistics Not dependent on system complexity Determined by power-per-event processing in modules Dependent on stimulus statistics
Recurrent topologies	Need to iterate until converge	Instantaneous

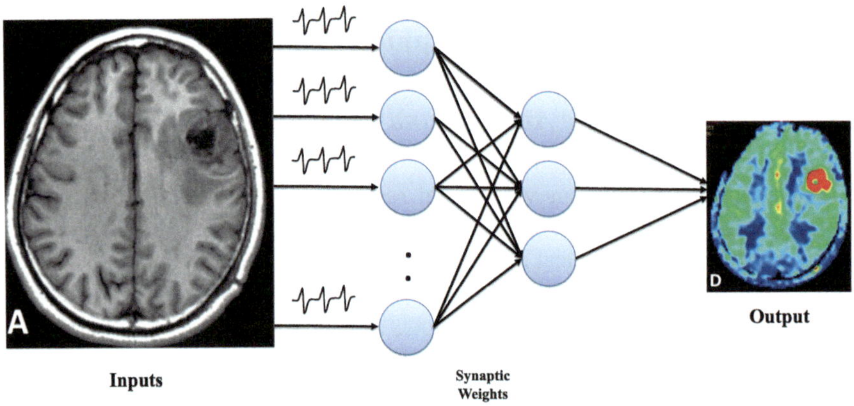

Fig. 13.5 Schematic representation of neuromorphic computing based brain tumour segmentation [38]

memristor-enhanced neural networks excel in discerning subtle patterns and abnormalities within medical images. As the system encounters new cases, it refines its understanding, continually improving diagnostic accuracy over time.

This innovation holds immense potential for expediting the diagnostic process, providing clinicians with enhanced tools to detect and analyse medical conditions more swiftly and accurately. The marriage of memristors with neural networks in medical image analysis marks a significant stride towards personalized and precise healthcare, where technology becomes an invaluable ally in the pursuit of early detection and effective treatment [39–41].

13.2.2 Neuromorphic Prosthetics and Brain-Machine Interfaces

Neuromorphic prosthetics and brain-machine interfaces (BMIs) have achieved remarkable sophistication through the incorporation of memristors into neural networks. These memristor-based networks play a pivotal role in advancing prosthetic technologies and BMIs by emulating synaptic plasticity, a key feature of dynamic learning in biological systems. This emulation translates into heightened adaptability and bidirectional communication between prosthetic devices and the human nervous system. The integration of memristors allows prosthetic limbs to intuitively respond to neural signals, granting users unprecedented control and precision in their movements. This adaptability is crucial for creating a more immersive and natural prosthetic experience, significantly enhancing the overall quality of life for users [42, 43].

In the field of BMIs, memristor-based neural networks contribute to more seamless and efficient communication between the brain and external devices. These interfaces can decode neural signals with enhanced accuracy, enabling users to control external devices, such as computers or robotic limbs, with remarkable precision. The bidirectional flow of information, enriched by memristor-enabled adaptability, fosters an intuitive and responsive interaction between the user's brain and connected technology. In summary, the ground-breaking integration of memristors with neural networks in neuromorphic prosthetics and BMIs represents a significant leap towards creating more natural and effective interfaces, ensuring devices not only respond to neural signals with precision but also continuously adapt to the user's unique needs and preferences [36, 44].

13.2.3 Disease Diagnosis and Prediction

The integration of memristors with neural networks represents a ground-breaking approach to handling complex medical data. This dynamic synergy allows for the analysis of diverse datasets, including genetic information, patient records, and intricate imaging data. Memristor-based neural networks stand out for their intrinsic ability for continuous learning and adaptability, setting them apart from traditional diagnostic tools. These networks evolve over time, refining their predictive capabilities, thanks to the memristor's unique property of mimicking synaptic plasticity, which emulates the learning process observed in biological systems [45, 46].

Harnessing this continuous learning, the system becomes increasingly adept at interpreting intricate patterns within medical data, leading to more accurate and personalized diagnoses. The memristor's capacity to retain information ensures that the neural network builds upon its knowledge base, continually improving its predictive accuracy. This innovative approach holds immense potential for revolutionizing disease diagnosis by providing real-time, data-driven insights. As the memristor-based neural network navigates through vast datasets, it identifies subtle correlations and trends that may escape conventional methods. Ultimately, this fusion of memristors and neural networks not only enhances the efficiency of diagnostic processes but also heralds a new era of personalized and adaptive healthcare solutions [47].

13.2.4 Drug Discovery and Development

The integration of memristors into neural networks marks a ground-breaking advancement, particularly in the realm of drug development. Memristor-based neural networks offer the unprecedented ability to accelerate the analysis of extensive datasets, spanning molecular structures, biological interactions, and clinical trial results. At the core of this innovation lies the adaptive learning capabilities of memristors, addressing a significant limitation in traditional drug development methodologies characterized by static models and algorithms. Memristors introduce a dynamic dimension to the process, enabling real-time adjustments and continuous learning from new data, thereby adapting to evolving insights. This real-time adaptability proves to be instrumental in optimizing the identification of potential drug candidates for various diseases [48, 49].

Memristor-enhanced neural networks exhibit the capacity to swiftly and accurately analyse large datasets, discerning subtle patterns and relationships that may elude conventional methods. The system's ability to evolve over time ensures it remains at the forefront of scientific advancements, incorporating the latest findings into its decision-making processes. The integration of memristors into drug discovery processes not only expedites the identification of promising compounds but also enhances the efficiency of clinical trial design. The adaptive nature of memristor-based neural networks holds the promise of a more agile and responsive drug development pipeline, ultimately leading to faster and more effective solutions for a myriad of diseases [50, 51].

13.2.5 Brain Disorders and Cognitive Health

The collaboration between memristors and neural networks emerges as a potent tool for the in-depth study and simulation of complex neural circuits, with a particular focus on disorders such as Alzheimer's, Parkinson's, and epilepsy. Memristor-based neural networks offer a dynamic platform that significantly advances our comprehension of these conditions. Central to this advancement is the adaptive learning capabilities of memristors, which play a pivotal role in unravelling the intricate nature of neural circuits associated with brain disorders. These networks can simulate dynamic changes observed in synaptic connections, mirroring the synaptic plasticity inherent in biological systems. This unique capability allows researchers and clinicians to study the evolving nature of neural networks within the context of different disorders. By delving into the underlying mechanisms of brain disorders, memristor-based neural networks contribute to the development of innovative therapies. The ability to replicate and manipulate neural circuitry provides a virtual laboratory for testing potential interventions and treatment strategies. This not only accelerates the pace of research but also opens new avenues for precision medicine in the realm of neurological and cognitive health [52, 53].

Furthermore, these simulated neural networks serve as valuable models for testing the efficacy of therapeutic interventions. Researchers can utilize memristor-enhanced systems to explore the impact of various drugs, electrical stimulations, or other treatments on simulated neural circuits, providing insights that can inform real-world clinical practices. In essence, the union of memristors and neural networks holds the promise of transformative breakthroughs in understanding, treating, and eventually preventing brain disorders, ushering in a new era of precision medicine for cognitive health [54, 55].

13.2.6 Healthcare Robotics and Assistance

The integration of memristors into neural networks marks a transformative shift in healthcare robotics and assistance, enhancing the adaptability and learning capabilities of robotic systems. Memristor-enhanced neural networks serve as the cognitive backbone for healthcare robots and assistive devices, propelling them into a realm of personalized and responsive care. The dynamic nature of memristors, mirroring synaptic plasticity, introduces a game-changing quality to neural networks, particularly for healthcare robots. This adaptability enables these robots to learn from patient interactions and adjust their behaviour accordingly, tailoring responses to individual needs and fostering a more personalized and empathetic interaction with patients [56, 57].

Equipped with memristor-based neural networks, healthcare robots overcome the limitations of static programming. They can adapt to changing environments, evolving alongside the dynamic healthcare landscape. Whether in a hospital setting or a home environment, these robots exhibit agility, enhancing their utility across various healthcare applications. The continuous learning facilitated by memristors empowers healthcare robots to provide not just routine assistance but also to anticipate and respond to evolving patient needs. From reminding patients to take medication to assisting with mobility challenges, these robots become proactive partners in healthcare, contributing to improved patient outcomes and overall well-being. In essence, the marriage of memristors with neural networks propels healthcare robotics into a new era of adaptability and personalized care, where robots serve as intelligent companions in the journey towards better health and an enhanced quality of life [58, 59].

13.2.7 Object Recognition

In the realm of computer vision, memristor-based neural networks are reshaping the landscape of object recognition and classification. These networks boast unmatched accuracy and adaptability, thanks to the continuous learning capabilities of memristors. This adaptability proves crucial in dynamic or evolving environments, enabling the system to evolve and refine its recognition accuracy over time. Whether tasked with identifying objects in changing scenes or adapting to new visual contexts, memristors ensure a more efficient and precise object recognition process. Facial recognition undergoes a revolutionary transformation with the integration of memristors into neural networks. Memristors elevate the performance of facial recognition systems by enhancing their adaptive learning capabilities. This enhancement enables the system to excel in recognizing and differentiating faces even under challenging conditions such as varying lighting or changes in facial expressions [60, 61] (Fig. 13.6).

The result is improved accuracy and reliability in facial recognition applications, contributing to advancements in security, user authentication, and personalized interactions. In the domain of pattern recognition, memristor-enhanced neural networks emerge as pioneers in deciphering complex visual patterns within images. Whether

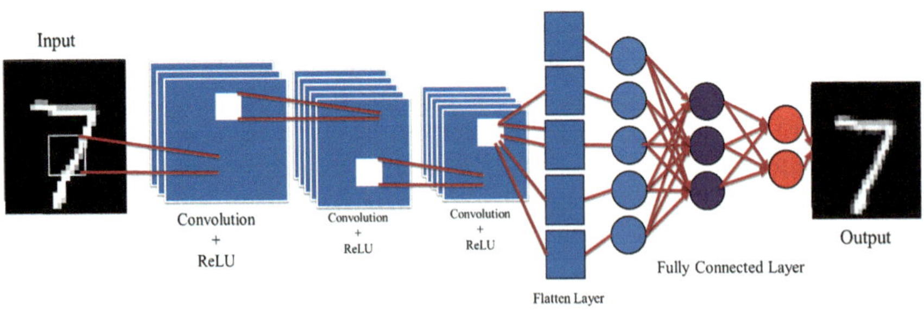

Fig. 13.6 The developed architecture for character recognition by simulating the memristor property for each neuron [62]

tasked with identifying textures, shapes, or intricate visual features, the adaptive nature of memristors proves instrumental in refining pattern recognition algorithms. This capability extends across diverse applications, from medical imaging to industrial quality control, where the system continuously learns and adapts to discern subtle patterns. The result is precision and efficiency in pattern recognition tasks, revolutionizing industries reliant on accurate visual analysis [63].

13.2.8 Augmented Reality (AR) and Virtual Reality (VR)

The convergence of memristors with neural networks represents a pivotal advancement in the realms of Augmented Reality (AR) and Virtual Reality (VR), where image recognition serves as a foundational element for immersive experiences. Memristor-enhanced image recognition technology provides a transformative boost to AR and VR applications by empowering neural networks to identify and track objects in real-time. In the context of AR, where digital elements are overlaid onto the physical world, memristor-based neural networks excel in recognizing and augmenting real-world objects. The adaptive learning capabilities of memristors enable AR systems to continually enhance their ability to seamlessly integrate virtual elements into the user's environment. Whether placing virtual furniture in a room or overlaying information onto real-world objects, the accuracy and responsiveness of memristor-enhanced image recognition significantly elevate the overall AR experience [64, 65].

Similarly, in the immersive realm of VR, where users are surrounded by entirely virtual environments, memristor-based neural networks contribute to a heightened sense of realism. The capability to identify and track objects in real-time enhances interactive possibilities within the virtual space. Users can engage in more dynamic and responsive interactions, such as manipulating virtual objects or navigating virtual environments with a higher degree of precision. This integration of memristors with image recognition not only elevates the visual fidelity of AR and VR but also opens the door to more sophisticated and interactive applications. Whether applied in gaming, education, or professional simulations, the marriage of memristors with neural networks in AR and VR lays the foundation for a more immersive and adaptive digital experience [66].

13.3 Conclusion

The journey from traditional computing to neuromorphic computing and beyond CMOS technologies, particularly memristors, reveals a dynamic quest for innovation. As Moore's law faces limitations, the fusion of artificial neural networks and memristors in neuromorphic computing emerges as a promising paradigm. Memristors, with their unique ability to seamlessly integrate processing and memory, mimic synaptic plasticity, offering energy-efficient and parallel processing. Despite challenges, the evolution towards Spiking Neural Networks signifies a stride towards biologically-inspired efficiency. The collaborative efforts of materials science, computer science, and neuroscience underscore the continuous pursuit of advancing neural networks and memristor technologies.

References

1. D. Ielmini, H-S.P.J.N.E. Wong, In-memory computing with resistive switching devices. Nat. Electron. **1**, 333–343 (2018)
2. A.J.J.J.O.A.P. Chen, Beyond-CMOS roadmap—from Boolean logic to neuro-inspired computing. Jpn. J. Appl. Phys. **61**, SM1003 (2022)
3. M.R. Azghadi, B. Linares-Barranco, D. Abbott, P.H.J.I.T.O.B.C. Leong, and systems: A hybrid CMOS-memristor neuromorphic synapse. IEEE Trans. Biomed. Circuits Syst. **11**, 434–445 (2016)
4. W. Olin-Ammentorp, N.J.S.P. Cady, Biologically-inspired neuromorphic computing. **102**, 261–276 (2019)
5. M.P. Fok, D. Rosenbluth, K. Kravtsov, P.R.J.I.S.P.M. Prucnal, Lightwave neuromorphic signal processing [in the spotlight]. **27**, 160–158 (2010)
6. F. Zenke et al., Visualizing a joint future of neuroscience and neuromorphic engineering. **109**, 571–575 (2021)
7. S. Kim, S. Choi, W.J.A.N. Lu, Comprehensive physical model of dynamic resistive switching in an oxide memristor. ACS Nano **8**, 2369–2376 (2014)
8. S.M. Hus et al., Observation of single-defect memristor in an MoS2 atomic sheet. Nat. Nanotechnol., **16**, 58–62 (2021)
9. Y. Yang, W.J.N. Lu, Nanoscale resistive switching devices: mechanisms and modeling. **5**, 10076–10092 (2013)
10. D.B. Strukov, G.S. Snider, D.R. Stewart, R.S.J.N. Williams, The missing memristor found. **453**, 80–83 (2008)
11. X. Xiao et al., Recent advances in halide perovskite memristors: materials, structures, mechanisms, and applications. Adv. Mater. Technologies. **5**, 1900914 (2020)
12. P. Krzysteczko, J. Münchenberger, M. Schäfers, G. Reiss, A.J.A.M. Thomas, The memristive magnetic tunnel junction as a nanoscopic synapse-neuron system. **24**, 762–766 (2012)
13. S. Hu et al., SPICE study of STDP characteristics in a drift and diffusive memristor-based synapse for neuromorphic computing. **10**, 6381–6392 (2022)
14. H. Bian, Y.Y. Goh, Y. Liu, H. Ling, L. Xie, X.J.A.M. Liu, Stimuli-responsive memristive materials for artificial synapses and neuromorphic computing. **33**, 2006469 (2021)
15. M. Patel, J. Gosai, N. Chaudhari, A. Solanki, Hybrid perovskite-based memristor devices. in *Information systems management* (IntechOpen, 2023)
16. M. Patel, J. Gosai, A. Lokhandwala, A.J.A.A.E.M. Solanki, Slow migration-controlled resistive switching in stable dion–jacobson hybrid perovskites for flexible memristive applications. (2024)
17. M. Patel, D.D. Kumbhar, J. Gosai, M.R. Sekhar, A.T. Mallajosyula, A.J.A.E.M. Solanki, Hybrid perovskite-based flexible and stable memristor by complete solution process for neuromorphic computing. **9**, 2200908 (2023)
18. M. Patel et al., MXenes: promising 2D memristor materials for neuromorphic computing components. (2022)
19. J. Wang et al., Handwritten-digit recognition by hybrid convolutional neural network based on HfO2 memristive spiking-neuron. Scientific Reports **8**, 12546 (2018)

20. Q. Duan et al., Spiking neurons with spatiotemporal dynamics and gain modulation for monolithically integrated memristive neural networks. **11**, 3399 (2020)
21. X. Ji, Z. Dong, C.S. Lai, D.J.I.C.M. Qi, A brain-inspired in-memory computing system for neuronal communication via memristive circuits. **60**, 100–106 (2022)
22. K. Liao et al., Memristor based on inorganic and organic two-dimensional materials: mechanisms, performance, and synaptic applications. **13**, 32606–32623 (2021)
23. V. Vakharia, M. Shah, V. Suthar, V.K. Patel, A.J.P.S. Solanki, Hybrid perovskites thin films morphology identification by adapting multiscale-SinGAN architecture, heat transfer search optimized feature selection and machine learning algorithms. **98**, 025203 (2023)
24. S. Agatonovic-Kustrin, R.J.J.O.P. Beresford, Basic concepts of artificial neural network (ANN) modeling and its application in pharmaceutical research. J. Pharm. Biomed. Anal. **22**, 717–727 (2000)
25. J.V.J.J.O.C.E. Tu, Advantages and disadvantages of using artificial neural networks versus logistic regression for predicting medical outcomes. **49**, 1225–1231 (1996)
26. S.M. Siniscalchi, T. Svendsen, C.-H.J.N. Lee, An artificial neural network approach to automatic speech processing. **140**, 326–338 (2014)
27. I. Litvan et al., What can artificial neural networks teach us about neurodegenerative disorders with extrapyramidal features? **119**, 831–839 (1996)
28. S.H. Reshak, Drugs designing using artificial intelligence based pharmaceutical systems, in *2022 International Conference for Advancement in Technology (ICONAT)* (IEEE, 2022), pp. 1–4
29. L.A. Camuñas-Mesa, B. Linares-Barranco, T.J.M. Serrano-Gotarredona, Neuromorphic spiking neural networks and their memristor-CMOS hardware implementations. **12**, 2745 (2019)
30. L. Shen, Z. Lin, Q. Huang, Relay backpropagation for effective learning of deep convolutional neural networks, in *Computer Vision–ECCV 2016: 14th European Conference, Amsterdam, The Netherlands, October 11–14, 2016, Proceedings, Part VII 14* (Springer, 2016), pp.467–482
31. T.P. Lillicrap, A. Santoro, L. Marris, C.J. Akerman, G.J.N.R.N. Hinton, Backpropagation and the brain. **21**, 335–346 (2020)
32. S.R. Kulkarni, A.V. Babu, B. Rajendran, Spiking neural networks—Algorithms, hardware implementations and applications. in *2017 IEEE 60th International Midwest Symposium on Circuits and Systems (MWSCAS)* (IEEE, 2017), pp. 426–431
33. B. Cramer, Y. Stradmann, J. Schemmel, F.J.I.T.O.N.N. Zenke, The heidelberg spiking data sets for the systematic evaluation of spiking neural networks. IEEE Trans. Neural Netw. Learn. Syst. **33**, 2744–2757 (2020)
34. S.B. Furber, F. Galluppi, S. Temple, L.A.J.P.O.T.I. Plana, The spinnaker project. **102**, 652–665 (2014)
35. M. Prezioso, F. Merrikh-Bayat, B.D. Hoskins, G.C. Adam, K.K. Likharev, D.B.J.N. Strukov, Training and operation of an integrated neuromorphic network based on metal-oxide memristors. **521**, 61–64 (2015)
36. P. Yao et al., Fully hardware-implemented memristor convolutional neural network. **577**, 641–646 (2020)
37. C. Zhang et al., Bioinspired artificial sensory nerve based on nafion memristor. **29**, 1808783 (2019)
38. W-Z. Hu et al., Differentiation of neoplastic and non-neoplastic intracranial enhancement lesions using three-dimensional pseudo-continuous arterial spin labeling. Front. Neurosci. **16**, 812997. https://doi.org/10.3389/fnins.2022.812997
39. J. Chen et al., An efficient memristor-based circuit implementation of squeeze-and-excitation fully convolutional neural networks. **33**, 1779–1790 (2021)
40. F. Yu, X. Kong, A.A.M. Mokbel, W. Yao, S.J.I.T.O.C. Cai, S.I.E, Briefs: Complex dynamics, hardware implementation and image encryption application of multiscroll memeristive Hopfield neural network with a novel local active memeristor. **70**, 326–330 (2022)
41. Q. Lai, Z. Wan, H. Zhang, G.J.I.T.O.N.N. Chen, L., Systems: Design and analysis of multiscroll memristive hopfield neural network with adjustable memductance and application to image encryption. (2022)

42. A. Lebedeva et al., Development a cross-loop during adaptive stimulation of hippocampal neural networks by an artificial neural network. in *2022 Fourth International Conference Neurotechnologies and Neurointerfaces (CNN)* (IEEE, 2022), pp. 82–85
43. L. Wang et al., Artificial synapses based on multiterminal memtransistors for neuromorphic application. **29**, 1901106 (2019)
44. V.T. Pham, S. Jafari, S. Vaidyanathan, C. Volos, X.J.S.C.T.S. Wang, A novel memristive neural network with hidden attractors and its circuitry implementation. **59**, 358–363 (2016)
45. M.T. Maurano et al., Systematic localization of common disease-associated variation in regulatory DNA. **337**, 1190–1195 (2012)
46. S.A. Lambert, G. Abraham, M.J.H.M.G. Inouye, Towards clinical utility of polygenic risk scores. Hum. Mol. Genet., **28**, R133–R142 (2019)
47. R.M. Poses, R.D. Cebul, M.Collins, S.S.J.A.O.I.M. Fager, The importance of disease prevalence in transporting clinical prediction rules: the case of streptococcal pharyngitis. Ann. Intern. Med. **105**, 586–591 (1986)
48. J.J.S. Drews, Drug discovery: a historical perspective. Science **287**, 1960–1964 (2000)
49. H.S. Chan, H. Shan, T. Dahoun, H. Vogel, S.J.T.I.P.S. Yuan, Advancing drug discovery via artificial intelligence. **40**, 592–604 (2019)
50. C.S.J.D.D.T. Fishburn, Translational research: the changing landscape of drug discovery. Drug Discov. Today **18**, 487–494 (2013)
51. R.C. Mohs, N.H.J.A.S. Greig, Drug discovery and development: Role of basic biological research. Transl. Res. & Clin. Interv., **3**, 651–657 (2017)
52. A.F.J.N.N. Arnsten, Stress weakens prefrontal networks: molecular insults to higher cognition. Nat. Neurosci. **18**, 1376–1385 (2015)
53. C.-Y. Wee et al., Identification of MCI individuals using structural and functional connectivity networks. Neuroimage **59**, 2045–2056 (2012)
54. C. Iadecola et al., Vascular cognitive impairment and dementia: JACC scientific expert panel. **73**, 3326–3344 (2019)
55. P.T. Francis, N.R. Sims, A.W. Procter, D.M.J.J.O.N. Bowen, Cortical pyramidal neurone loss may cause glutamatergic hypoactivity and cognitive impairment in Alzheimer's disease: investigative and therapeutic perspectives. J. Neurochem. **60**, 1589–1604 (1993)
56. J. Broekens, M. Heerink, H.J.G. Rosendal, Assistive social robots in elderly care: a review. **8**, 94–103 (2009)
57. A. Sorrentino et al., Personalizing care through robotic assistance and clinical supervision. Front. Robot. AI. **9**, 883814 (2022)
58. M.-F.R. Lee, F.-H.S. Chiu, C.W. de Silva, C-Y.A.J.I.J.O.F.S. Shih, Intelligent navigation and micro-spectrometer content inspection system for a homecare mobile robot. **16**, (2014)
59. M.J.R. Kassler, Robotics for health care: a review of the literature. **11**, 495-516 (1993)
60. R.D. Singh, A. Mittal, R.K.J.M.T. Bhatia, Applications: 3D convolutional neural network for object recognition: a review. Multimed. Tools Appl. **78**, 15951–15995 (2019)
61. M. Jain et al., Insights into synaptic functionality and resistive switching in lead iodide flexible memristor devices. (2024)
62. A. Ali, S.Z. Kazmi, G. Shahzadi, M. Rashid, M.U. Ahsan, Demystifying CNN with mathematical insights: A prelude with application to an ai-based sustainable solution for diabetic retinopathy diagnosis. (2023)
63. Y. Gao, M.A. Maraci, J.A. Noble, Describing ultrasound video content using deep convolutional neural networks. in *2016 IEEE 13th International Symposium on Biomedical Imaging (ISBI)*, (IEEE, 2016), pp. 787–790
64. S. Moro, P. Rita, P. Ramos, J.J.J.O.H. Esmerado, Technology: Analysing recent augmented and virtual reality developments in tourism. J. Hosp. Tour. Technol. **10**, 571–586 (2019)
65. J.J. Kim, Y. Wang, H. Wang, S. Lee, T. Yokota, T.J.A.F.M. Someya, Skin electronics: next-generation device platform for virtual and augmented reality. Adv. Funct. Mater. **31**, 2009602 (2021)
66. K. Yin, Z. He, J. Xiong, J. Zou, K. Li, and S.-T.J.J.O.P.P. Wu, Virtual reality and augmented reality displays: advances and future perspectives. J. Phys.: Photonics, **3**, 022010 (2021)

Chapter 14
Generations of CT—the Evolution and Future Prospects

M. J. Akshit Aiyappa, B. Suresh Kumar Shetty, and Jini Abraham

Abstract In the year 1972, at the British Institute of Radiology Annual Congress, a distinguished research scientist named Sir Godfrey Newbold Hounsfield declared the creation of a ground-breaking imaging method, he called "computerized axial transverse scanning." This technique's core idea was very straightforward: a pencil-sized x-ray beam was used to analyze a thin cross-section of the head, known as the tomographic slice, at various angles. The scintillation detector calculated the transmitted radiation, which was then entered into a computer for mathematical algorithm analysis and reconstruction as a tomographic image. There was an extraordinary feature on this tomographic image, that had never been observed on an x-ray image previously. With the help of computer analysis, the image demonstrated radiographical differences in various soft tissues, gray and white matter. This invention came to be known as computerized axial tomography (CAT), which now is popularly called computed tomography (CT). Like many great discoveries, CT was the end product of years of work put on by various investigators. Throughout the years, CT has undergone multiple improvements. In this chapter, we will deal with the development and generations of CT, its rapid technical advances that have brought dramatic changes in the field of biomedical imaging, and how these changes are having an impact on advancement today.

M. J. Akshit Aiyappa · J. Abraham
Department of Radiodiagnosis, AJ Institute of Medical Sciences and Research Centre, Mangalore, India

B. S. K. Shetty (✉)
Department of Forensic Medicine, Kasturba Medical College, Mangalore, India
e-mail: suresh.shetty@manipal.edu

Manipal Academy of Higher Education, Manipal, Karnataka, India

A. Gogoi and N. Mazumder (eds.), *Biomedical Imaging*, Biological and Medical Physics, Biomedical Engineering, https://doi.org/10.1007/978-981-97-5345-1_14

14.1 Introduction

A novel procedure that involved imaging the brain of a live patient was carried out on October 1st, 1971. Under Sir Godfrey Newbold Hounsfield's direction, CT was carried out using a single detector. A dramatic shift in medical imaging began when an extraordinary image on a monitor screen surfaced after a long wait in computer processor reconstruction. The live visual was of a middle-aged female patient of Dr. James Ambrose, at Atkinson Morley Hospital, who had an unexplained brain lesion that was eventually removed and identified as a cystic astrocytoma (Fig. 14.1). The new image was just as good as the previous one, so even though the scanning procedure took a while, it was definitely worth the wait. When he examined the initial clinical scan, Dr. Ambrose commented that it had far more information than he had anticipated. This marked the start of a new era in medical imaging and diagnostic techniques, that would swiftly replace invasive procedures such as pneumocephalograms and ventriculograms [1, 2].

Impressive technological advances and modifications, since the inception of CT, have resulted in better image quality at a lower radiation dose. CT has completely transformed the way medicine is practiced and is still advancing our understanding of diseases and the treatment of serious health issues [3]. As a result, there is a significant rise in the indication for CT scans. There are currently about 3000 scanners in use in India, and during the next ten years, this modality could double in potential [4]. Since CT scanning is the primary source of radiation exposure, the radiology community has always been interested in technological advances for radiation dose optimization [5]. It's essential to understand how CT has developed over time and how each generation of CT has overcome obstacles. The way parts are arranged and

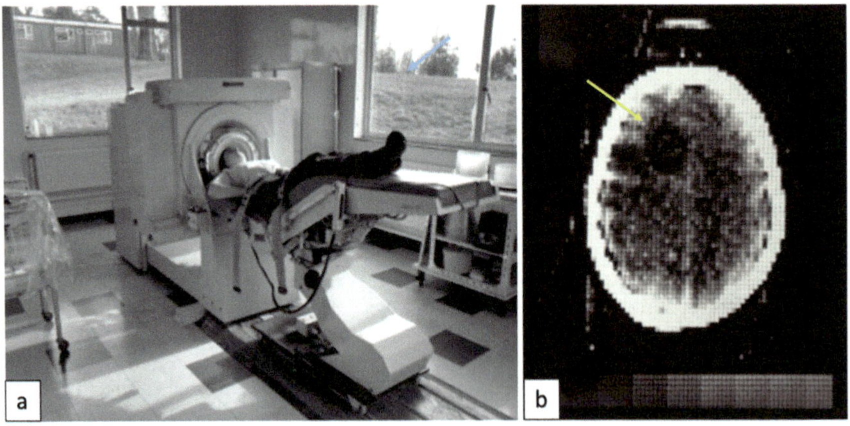

Fig. 14.1 Atkinson Morley Hospital's first CT scanner is shown in (**a**) (https://collection.sciencemuseumgroup.org.uk/objects/co134790/emi-ct-brain-scanner-ct-scanner). **b** Cystic astrocytoma is shown in the image from the initial patient scan

the amount of mechanical motion needed to gather data, determines the classification of CT. The order in which each CT scanner design has been released is what we mean when we refer to a generation of CT. Now, let's look at the generations of CT and the developments that followed during the evolution of CT.

14.1.1 First-Generation CT

Hounsfield created a pencil-beam width radiation beam by collimating the x-ray beam as he had imagined dividing the subject or the patient into axial slices. This beam width indicated the slice thickness to be scanned. The beam measured 3 mm and 13 mm in the slice plane and the subject axis respectively. On the opposite side of the patient was the x-ray tube, that was connected to a detector and tube swept a narrow beam. This sweeping motion of the tube across the patient is termed as *translation* [6]. The detector records multiple readings of x-ray transmission through the patient throughout translation (Fig. 14.2). The collection of these measurements taken during the translation is called a *view*. In the Mark I scanner created by Hounsfield, the tube and detector are rotated by 1 ° all over the subject after the translation is complete, and the process is replicated to obtain a second image and to collect 180 views over 180°. Modern machines are able to collect over 1000 views in 360°. Linear translation, in combination with increasing rotation, is termed as *translate-rotate motion*. The first-generation CT geometry required about five to six minutes to complete the scan and the results were recorded with the help of sodium iodide scintillation detector. In order to reduce the examination time, the Mark I scanner made use of two detectors, that were adjacent to each other, and a broad x-ray beam, measuring 26 mm in width, to simultaneously capture data for two slices.

In first-generation CT scanners, measurements are made at different times during the translation motion of the tube and detector. In order to collect enough information to rebuild the image, views are taken from several perspectives. There was only one

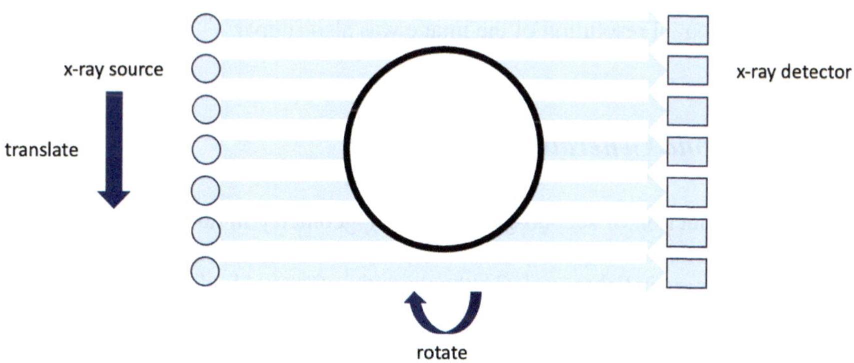

Fig. 14.2 X-ray transmission in first-generation CT scanner

detector and one x-ray source in the first-generation of CT scanners. To create an axial image, a single beam would enter the patient's body and be detected by a single detector. The detector and the x-ray source move simultaneously to collect the data. The x-ray source would need to be translated multiple times for each view to reconstruct a single-slice.

Hounsfield's idea was to divide a slice into voxels, which are three-dimensional rectangular boxes, arranged in a matrix. The X and Y directions are within the slice plane, and are determined by the slice of matrix, into which the slice is intended to be divided and the area across which the measurements are taken. The Z-direction is along the axis of the subject and corresponds to slice thickness. For example, if a scan circle, that has a diameter of 250 mm, is divided into a matrix with 250 rows by 250 columns, each voxel would measure 1×1 mm. Each voxel would measure 0.5 mm by 0.5 mm in size in a 512×512 matrix. We refer to this matrix as the reconstruction matrix. Determining each voxel's degree of narrow x-ray beam attenuation is the primary goal of CT image reconstruction. A one-dimensional grayscale image of the slice displays the calculated attenuation values. Pixels are the two voxel dimensions that are located in the plane of the slice [7].

In the initial Electric and Musical Industries (EMI) head scanner design, a rubber membrane was used to hollow out the patient's head and was placed inside a box filled with water (Fig. 14.3). In April 1972, the British Radiologic Society meeting featured a presentation of the findings from the first clinical EMI scanner [8]. A number of companies saw an explosion in research and development, as a result of the scanner's unprecedented success. A water-filled box, a single detector, and a single narrow beam were designed to spin with the device in 1° increments, so as to provide a constant path length between the patient and water for the transmission measurements. The accuracy of the attenuation coefficient measurements was maximized, thanks to the water-filled box. It was necessary to remove the water-filled box in order to increase the functionality of CT. When the beam emerges from the x-ray tube, it goes through a "bow-tie filter." Usually made of thicker aluminum, the filter limits the range of intensities that can be detected by the detector. Moreover, it pre-hardens the beam to minimize the hardening that occurs in the patient. The primary drawback of the first-generation scanner was how long it took to finish a single image, i.e., five minutes. The spatial resolution of the image was also subpar.

14.1.2 Second Generation CT

With the introduction of second-generation CT geometry in late 1974, faster scan times were possible. Georgetown University hosted the first waterless full-body CT scanner installation in February 1974, thanks to the efforts of Ledley et al. [9]. A laser indication used to place slices, gantry angulation, table movement via the gantry, and Fourier-based reconstruction were among the advancements presented by the

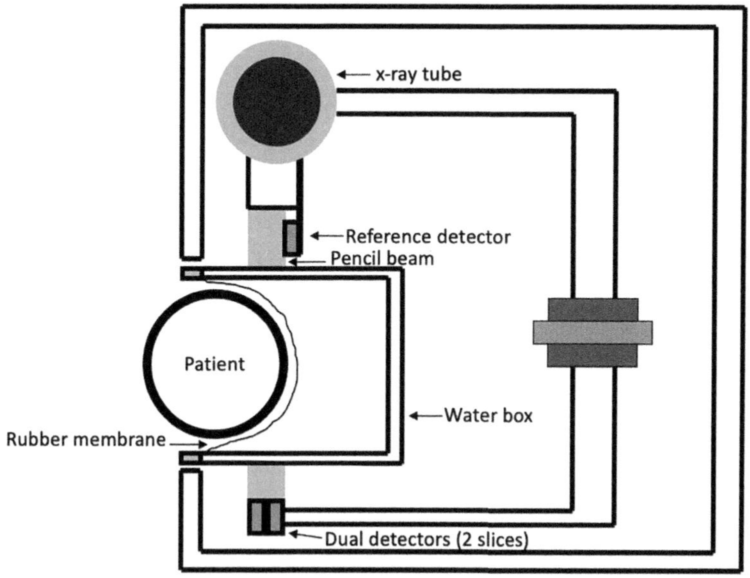

Fig. 14.3 Design of the first-generation water box. Through a rubber membrane, the patient was hollowed into a water box. Beam hardening artifacts were reduced with the use of the water box

device, that are not standard in modern-day CT. Similar to the first-generation architecture, this generation of CT scanners featured several narrow beams, multiple detectors, and rotate-translate motion (Fig. 14.4). When a set of rays measured by each detector is taken into account, the detectors actually obtain independent, comprehensive views from various angles. For instance, when the three thin or narrow beams are angled at 1°, one translation gathers information for views at 30°, 31°, and 32°. The number of translations needed is lowered to one-third, and the scan duration is reduced three-fold, from 5–6 min to just 2 min. With N being the number of detectors, the second-generation CT scan, on average, reduce scan times by a factor of 1/N.

While retaining the same general concepts, second-generation CT was an improvement over first-generation CT. The second-generation CT designs featured more than 20 narrow beams and detectors, lowering scan times to 20 s or less, compared to the first-generation CT. The quality of body scans improved significantly at this point since most patients could have scans completed while holding their breath. While it still uses rotate-translate acquisition, second-generation CT is faster than first-generation design. However, due to convolution of the rotate-translate geometry, further efficiency improvements were limited. Heavy x-ray tubes, the gantries that support them, and the electronics they are connected with must be moved swiftly, precisely, and without producing detectable vibrations. Data would be back projected through the incorrect voxels if there were even slight variations in the scanner hardware position with respect to the reconstruction matrix voxels, severely

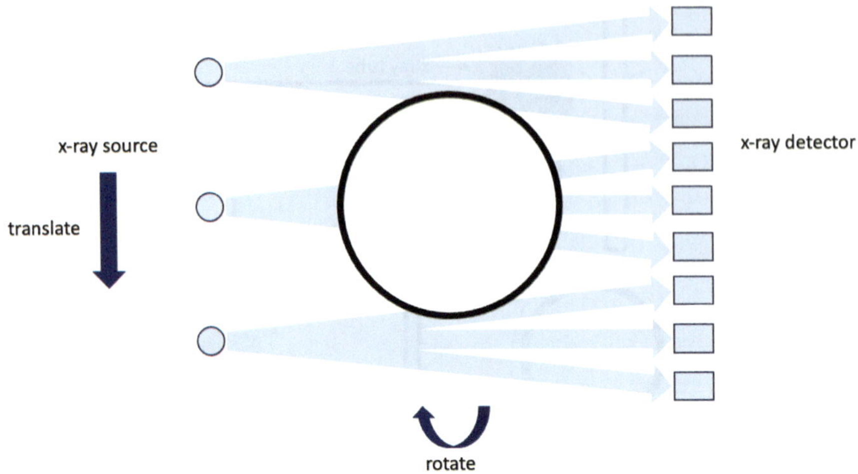

Fig. 14.4 Design of the second-generation CT, displaying the transmission of several thin beams that are detected during each translation by multiple detectors

distorting the images. Given the complexity and mechanical tolerances involved, translation motion had to be eliminated.

14.1.3 Third Generation CT

To achieve faster scan times, translation motion must be replaced with smoother and simpler rotating motion. Instead of a narrow beam, the beam is widened into a fan-shaped beam that covers the whole patient's width and is detected by a variety of detectors. The x-ray tube and detector rotate in unison, all over the patient due to their permanent connection, resulting in the *rotate-rotate* movement of the tube and detectors. As opposed to roughly 250 in the initial models, later revisions incorporated as many as 750 detectors to enable a proper number of measurements to be taken across the scan. Third-generation geometry refers to this system, which is defined by a linked tube and detector arrangement that only experiences rotational motion (Fig. 14.5).

Extremely high detector stability was required for third-generation CT scanners. The third-generation tubes and detectors are fixed in place, which implies that, depending on its location, each detector only detects light that travels a certain distance from the center of rotation. A circular artifact, also known as a *ring artifact*, will emerge from a detector that is out of calibration since it would continuously provide an erroneous reading at each angular position (Fig. 14.6). In theory, ring artifacts are more likely to occur in a scanner using solid-state detectors, as each detector is a separate unit, as opposed to a scanner using gas detectors, where each detector array is a single, xenon-filled chamber split by electrodes [10]. A xenon

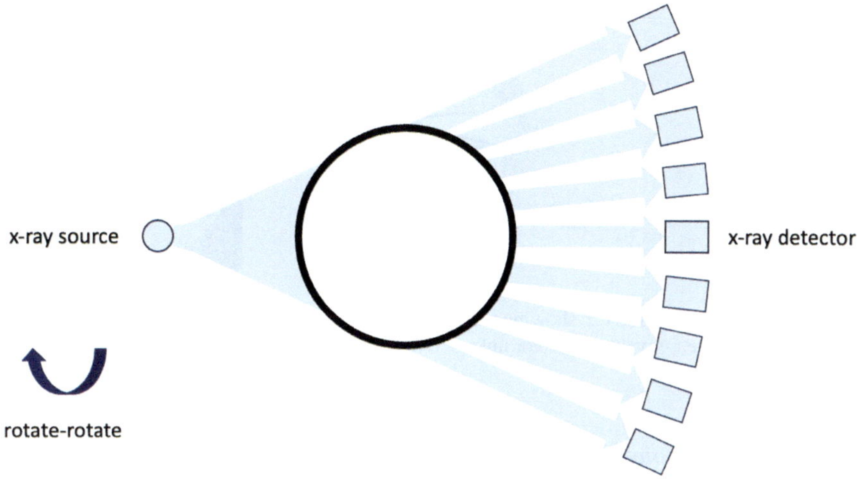

Fig. 14.5 Design of third-generation geometry

array, that is intrinsically stable, was employed to lessen the ring artifacts. A xenon array was made out of a long metal chamber with thin plates, or septa, dividing it into several smaller chambers. A common positive voltage source, which served as collectors for each alternate septum and independently attached to an electronic readout, was connected to each septum. Consequently, every tiny compartment served as an ionization chamber. All elements shared the same factors that affected the detector response, or these factors were fixed across time. X-ray absorption efficiency was increased by using pressurized xenon. Due to their inherent stability, solid-state detectors gradually took the place of xenon arrays. The ring artifacts were never completely eliminated by third-generation CT scans. Regular calibration scans and excellent detector design helped to minimize them. After that, image-processing techniques are used to eliminate residual ring artifacts from the images, which would have otherwise probably resulted in rings appearing in every CT image. However, ring artifacts can occasionally still be caused by detector drift or failure. Another issue with third-generation CT is sampling. The detector design determines the sample size and spacing. As the center of rotation approaches, samples cannot be arranged closer to one another. Multiple smaller detectors would be required to get measurements that are closer in size.

14.1.4 Fourth-Generation CT

This generation CT scanner was developed to principally suppress the ring artifacts. In this scanner, several thousand individual detectors are used (Fig. 14.7). The mechanical motion is the x-ray source that revolves around a stationary detector

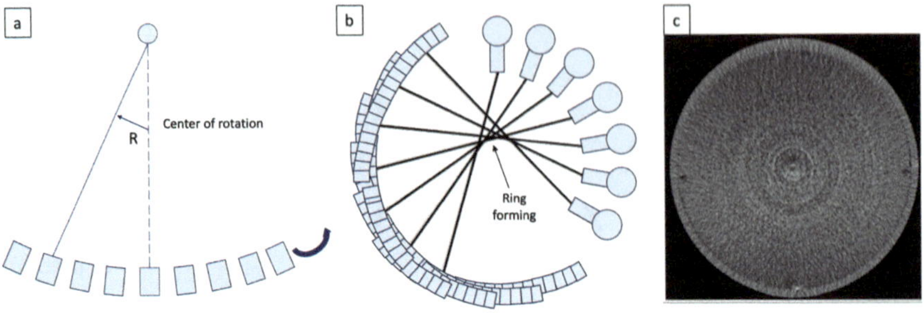

Fig. 14.6 Third-generation CT showing ring artifacts. **a** Rays are measured at a fixed distance by each detector, in the x-ray tube-detector linkage (R). **b** Calibration error, generating the ring artifact in (**c**)

array, or *rotate-stationary*. Images are captured in less than a second when the fan beam moves across each detector. Every detector acquired one-third-generation angular view, or a "fan beam view." When the x-ray tube slides across the detector's field of view, several closely spaced samples were produced because the tube and detectors were not connected, leading to a high sample density [11, 12]. Every fourth-generation detector, as opposed to third-generation detectors, may dynamically calibrate before entering the patient's shadow, hence eliminating ring artifacts. Additionally, it can monitor rays at any angle away from the rotational center. Nevertheless, the size of the matrix was constrained by the number of detectors and, consequently, by the views. Sustaining appropriate tube-skin distances, the detector ring needed to have a large diameter of roughly 170–180 cm, since the tube circled inside it. Subsequently, two rings were utilized: one that was smaller and featured a different design, with the tube going outside the ring, and one that was closer to the patient. The unusual wobbling motion of the ring during tube rotation is called nutation. It occurs when a portion of the ring tilts between the tube and patient, blocking the x-ray beam path. The fourth-generation designs never completely eliminated dispersion and scatter. The level of progress of CT from third and fourth-generations was equal. Fourth-generation CT became outdated after the introduction of multi-slice CT technology. The comparison of first to fourth-generation CT design is mentioned in Fig. 14.8.

14.1.5 Fifth Generation CT

In this generation CT design, the source and detector are both stationary, or *stationary-stationary*. This design consists of a scanning x-ray tube, where the electrons are steered magnetically. The x-ray source is an electron beam that is incident on a tungsten target after being guided and accelerated by microwaves, rather than an x-ray tube. Electron beam CT (EBCT) was meant for faster imaging, i.e., scan time less than 50 ms [13]. 50% of the imaging circle is covered by the target, and the other 50%

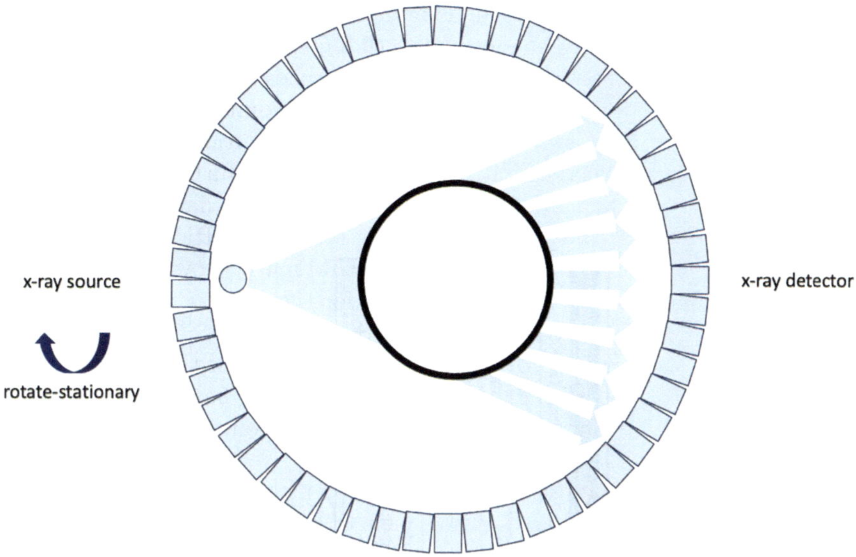

Fig. 14.7 A patient is surrounded by a fixed ring of detectors in a fourth-generation CT scanner system. The detectors are fixed, and only the x-ray source spins with a wide fan beam geometry

is covered by the detector array. A moving source is created by guiding the electron beam along the tungsten target's curvature (Fig. 14.9). EBCT is mainly applied for cardiac imaging. Large bell-shaped x-ray tubes were used in the design to accomplish ultra-fast scan times. After being concentrated into a narrow beam and electrically redirected, the electron steam, that was emitting from the cathode, impinges on a narrow focus region on an annular tungsten target anode, which causes the emission of x-rays. Next, the target is electrically swept by the electron beam in a 360-degree circle, either completely or partially. When the electron beam strikes the annular target, the x-rays are collimated into a fan beam. Now that there are no elements that move, scan times as low as 10–20 ms are possible. While EBCT remained accessible, its use was restricted to cardiac screening since general scanning produced lower-quality images than traditional conventional CT due to low mAs values and higher equipment costs. The future of EBCT is unknown, because multi-slice CT is making advances in cardiac scanning.

14.1.6 Sixth Generation CT

CT technology stabilized after the fourth-generation and continued to do so until 1987. By then, interscan delays accounted for the majority of CT examination times. The cable connects the rotating elements to the gantry and requires to stop and reverse the revolution after every 360° turn. A drum was used to reel in cables,

	1st generation	2nd generation	3rd generation	4th generation
Detectors	one	Multiple (up to 30)	Multiple (up to 1000)	Multiple (>2000)
Type of x-ray beam	Pencil-like	Fan-shaped	Fan-shaped	Fan-shaped
Tube-detector movements	Translate-rotate	Translate-rotate	Rotate-rotate	Rotate-stationary
Duration of scan (average)	25-30 minutes	< 90seconds	5 seconds	Few seconds
Number of slice per rotation	1	1	Multiple	2 or more
Advantages			Better pre-detector collimation; slip ring	Avoid ring artifact; easy detector calibration
Disadvantages			Ring artifact	Higher patient dose

Fig. 14.8 Comparison of first to fourth-generation CT scanner designs

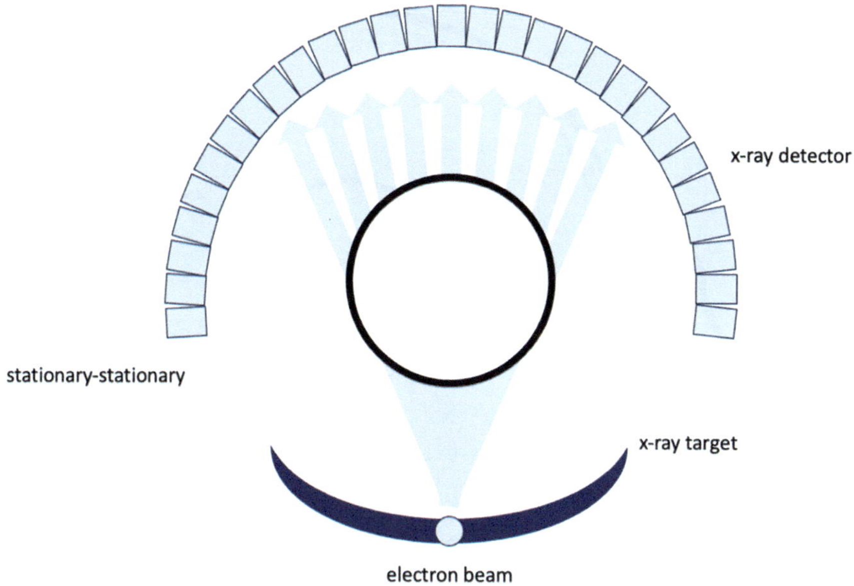

Fig. 14.9 Electron beam CT design

which were subsequently released during rotation and retracted during reverse. It took at least 8–10 s to scan, brake, and reverse; only 1–2 s were needed to collect data. Long process times and limited temporal resolution were the outcomes for dynamic contrast enhancement investigations. The low-voltage slip ring allowed for continuous rotation, which was necessary to eliminate interscan delays. A slip ring eliminates the necessity for permanent connections by allowing electrical power to reach the components that spin [14]. A slip ring is an annulus or drum containing grooves on it, where the electrical conductor brushes slide over (Fig. 14.10). Detectors provide continuous rotation by transmitting the data over many high-capacity wireless protocols. All interscan delays can be completely avoided with slip rings, with the exception of the time needed to transfer the table to the next slice position. Nevertheless, the scan-move-scan method, sometimes referred to as axial step-and-shoot CT, is incredibly inefficient.

There is an alternative method that continuously collects data by tracing a spiral or helical path by the tube and detector trajectory, with respect to the patient, as the table moves through the gantry [15, 16]. A strong concept that permits faster scans of the regions of interest along the Z-axis in one breath hold is *spiral or helical CT* (Fig. 14.11). There are some fundamental distinctions between the concepts of helical CT and axial scanning. Two of the main concepts include the speed of the table and slice interpolation. The speed of the table is determined based on the rotation time and the acquired slice thickness. The slice thickness is divided by the table rotation per rotation to determine the helical pitch. All of the information needed for slice reconstruction for step-and-shoot CT is accumulated at the Z-position of

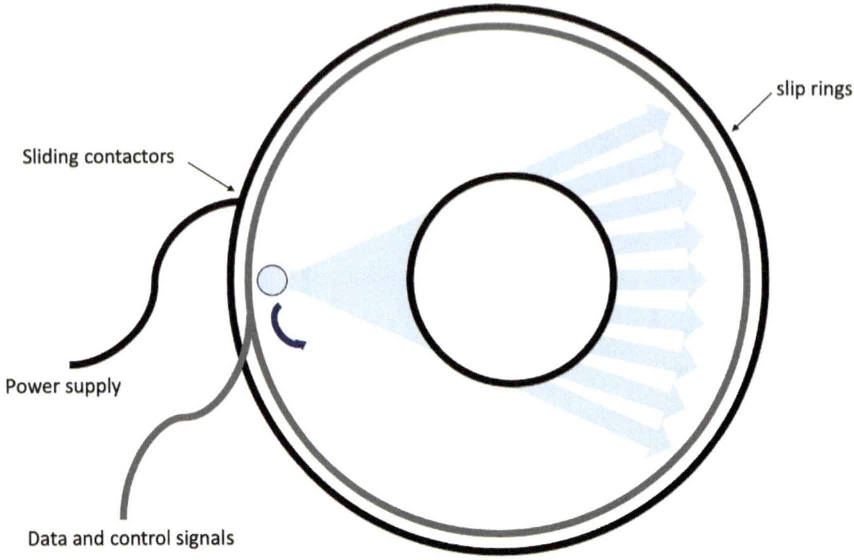

Fig. 14.10 Slip ring configuration. Sliding contactors preserve electrical contact with stationary components while allowing the detectors and x-ray tube to rotate continuously

the slice, prior to moving on to the next location. The reconstructed slice matches the Z-positions, where all the data are obtained. On the other hand, helical CT does not have distinct slice Z-positions; instead, every randomly selected slice plane is analogous to any other. This means that the slices can be reassembled at any Z-position after the helical data has been obtained. To estimate the whole CT data at the desired image location, the measurement details must be interpolated in the Z-direction. Interpolation allows more overlapping images to be created without giving the patient any additional dosage.

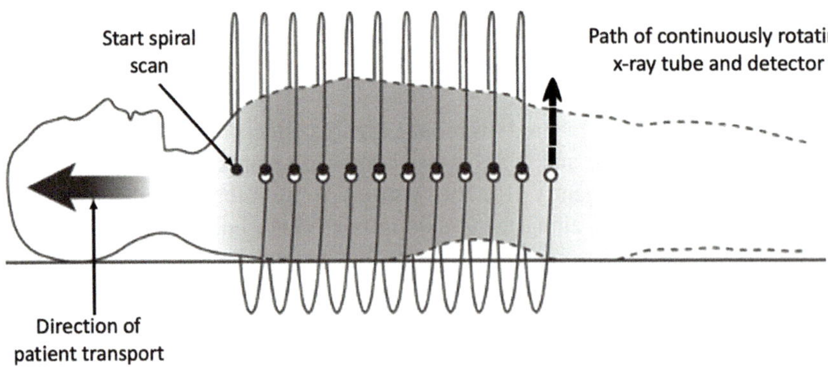

Fig. 14.11 Helical CT diagram

The *pitch* selection is essentially a commutation between precision and patient coverage. More patient coverage can be achieved with larger pitches per breath hold or unit of time; however, because slice data must be interposed between sites that are far apart, there is an increased risk of error. Commonly, pitches between 1 and 1.5 are utilized. Pitch values above 1.5 are rare, whereas pitches over 2 typically produce unsatisfactory outcomes and are not used. Because of double irradiation, single-slice helical CT does not employ pitches less than 1. The primary benefits of sixth generation CT include reduced patient dose, minimal motion artifacts, quicker scan times with larger volumes of data produced, and enhanced temporal and spatial resolution.

14.2 Multi-Slice/Multi-Detector CT

In 1998, the seventh generation of CT, often known as multi-slice or multi-detector CT, was unveiled. Multiple slices can be acquired in a single row using multi-slice CT [17]. Multiple rows of detectors with multiple fan beams scanning simultaneously can speed up the scanning of a body area. Multi-slice CT gathers transmission data for 4 to 64 + slices per revolution of the tube, as opposed to one slice every time the tube revolves around the patient. This is achieved by widening the x-ray beam along the Z-axis and having 4 to 64 + rows of detectors. Based on the number of slices imaged, each rotation grew, the beam was opened and the beam changed from a fan beam to a cone beam. The strength of the x-ray beam is not much decreased and it can interact with a detector array more effectively, as the cone beam does not pass through the narrow collimator (Fig. 14.12). As the cone beam gets bigger, the divergence of the x-rays generates additional distortion toward the top and bottom margins. All things considered, multi-slice CT offers faster scan times, better overall picture quality, and the capacity to carry out additional procedures like CT angiography and cardiac imaging. Additionally, multiple detector arrays or flat panel detectors are modified versions of the linear array of detectors from earlier generations.

What separates single-slice CT imaging from multi-slice CT imaging primarily is how the detector arrays are configured. The one-dimensional single-slice detector arrangement is designed to intercept the x-ray fan beam. They consist of several detector elements (>750) arranged in a single row throughout the irradiated slice. With a length of around 20 mm, the single detector element can intercept the entire x-ray beam width in the direction of the slice thickness, or z. In multi-slice CT, each single detector element is divided into several smaller detector elements in the z-direction, forming a two-dimensional array. There are several parallel rows of detectors rather than just one [18]. A detector design utilizing 16 rows of 1.25 mm detector components, for a total of 20 mm in the z-axis, was seen in one of the earliest modern multi-slice CT scanners. Every one of the 16 detector rows could simultaneously gather data for sixteen slices, each measuring 1.25 mm in thickness (Fig. 14.13). This method would, however, necessitate processing massive volumes of data quickly, since a standard scanner can capture up to 1000 views in a single spin.

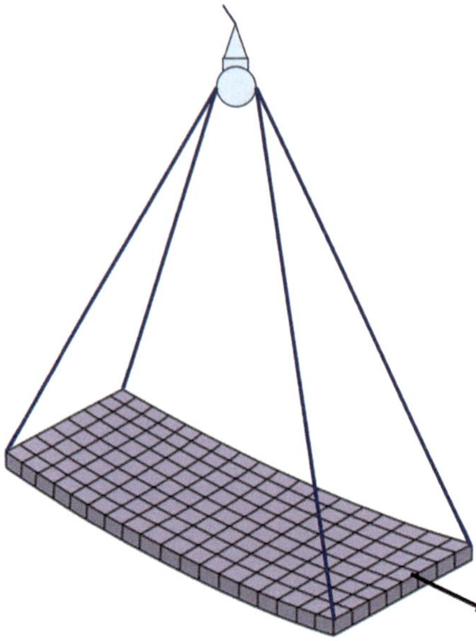

Fig. 14.12 Multi-detector array. Every tube rotation, several rows of detectors of a multi-slice CT detector array gather data for multiple slices

The first iteration of contemporary multi-slice CT limited simultaneous data acquisition due to its original constraints. Data was sent into four parallel data channels via four detector rows, which corresponded to the four slices that were captured concurrently. These 4-slice scanners offered a great deal of configuration flexibility for the detector rows; it is possible to electrically link a group of Z-direction detector units together to make them function as a single, larger detector, giving the four acquired slices of the scanner a great deal of flexibility in terms of slice thickness [19]. As data gathering technology developed, additional data channels became available, which made it possible to acquire more than four slices at once, as seen by scanners with eight and sixteen slices. In 2002, 16 data channels were introduced in multi-slice CT scanners to collect 16 slices simultaneously. The detector arrays installed in the 16-slice scanners were modified to allow for the acquisition of thinner slices in addition to allowing the capture of up to 16 slices concurrently. Multi-slice manufacturers released devices with fewer than or more than 16 channels in 2003 and 2004. Examples of these machines are 6–, 8–, 32–, and 64–slice scanners. In 2005, 64-slice scanners were announced. The primary goals of 64–slice detector array designs were to enlarge the arrays in the Z-direction and to supply all submillimeter detector elements. These days, there are CT scanners with 256–slice and even up to 640–slice options, that provide higher image resolution.

Slice thickness in single-slice CT is determined by the x-ray beam collimators used before and after the patient. In contrast, the slice width in a multi-slice CT is determined by the detector configuration. In computed tomography, the divergent

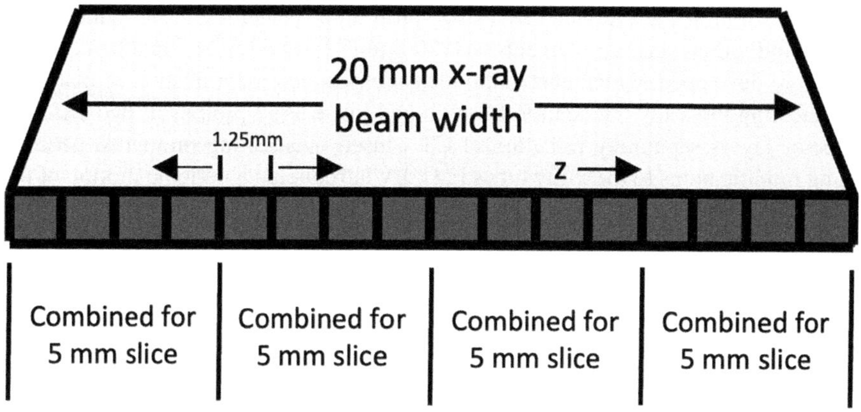

Fig. 14.13 Flexible detector application in 4–slice multi-slice CT scanners

nature of the x-ray beam that exits the x-ray tube is related to cone beam effects. The z-axis of the x-ray beam expands when it exits the patient due to this divergence compared to when it first enters the patient [20]. The main advantages of multi-slice CT include better z-axis resolution, larger coverage, faster scan times, and less motion artifacts.

14.3 Technological Advancements

The advancement of technology, since the inception of CT, has been very noteworthy. Although there are now a lot more adjustable parameters for both acquisition and reconstruction, the result is better image quality with reduced radiation exposure. Because of the interdependence of parameters, optimizing the scanning routine for radiation dose and picture quality is a complex process. Over time, the system's characteristics and related data gathering methods have evolved and grown. Hence, it is critical to understand the technological advancements and innovations that have occurred throughout the evolution of CT, as well as user's influence that goes along with them for protocol optimization.

The user's impact and options regarding protocol optimization are determined by the technological capabilities and system attributes of the CT scanner. The translate-rotate technique was utilized to create the CT images in the first and second generations of CT scanners. To perform a true CT scan examination, multiple breath holds are required due to the wide fan beam and slow spinning detector in the third-generation CT scanning machine. Later, slip ring technology was introduced, supplying electrical power to the rotating gantry component and transmitting measured data to the computer system, allowing the detector and tube to spin constantly [21]. In the first-generation CT scan, there is only one detector with full beam collimation in the Z-direction. Improved longitudinal coverage and better spatial resolution were attained by using spiral multi-detector CT, by reducing the size of individual detector elements to 0.25 mm [22]. In addition, dual layer detectors were developed, which enable the measurement and identification of materials by detecting the x-ray attenuation for high and low energy protons in two different detector layers separately [23]. Spiral CT caused overheating problems, necessitating modifications to the x-ray tubes [24]. By introducing a periodic motion of the focal spot in the Z-direction, which resulted in doubling the measurement location in the longitudinal direction of each rotation, it was possible to increase the spatial resolution and eliminate aliasing artifacts [25].

The automatic exposure control system is comprised of automated tube voltage selection and automated tube current modulation. By regulating the contrast-to-noise ratio, automated tube voltage selection aims to reduce radiation exposure [26]. Using a stationary scanner table, sequential CT scanning rotates the x-ray tube around the subject. Volumetric data could be acquired in spiral CT because the tube could rotate while the scanner table moved across the gantry at a steady pace [27]. Additionally, it presented the idea of user-adjustable pitch. Multi-detector CT is used to intensify the characteristics of the post-processed images by increasing the spatial resolution. Improved temporal resolution with fewer motion artifacts and enhanced clarity of lung and heart imaging are the outcomes of faster gantry rotation times [28]. Cardiac motion restricted the imaging of the heart during the earlier days of CT. Subsequently, significant advancements in technology were made possible by "stop-action" cardiac CT scans and ECG-gated reconstruction, which minimized cardiac motion [29]. The exam times were reduced, contrast volume could be reduced, spatial resolution was improved, and ECG-gated coronary CT angiography became feasible in clinical settings, thanks to multi-detector spiral CT. The 64-row multi-detector CT enabled the possibility of a low heart rate cardiac scan. Adaptive algorithms were developed that could respond to small arrhythmias and variations in heart rate, resulting in a dosage reduction. For ECG-gated data editing possibilities, a retrospective protocol is frequently chosen when the heart rhythm is complex. To guarantee sufficient data for reconstruction, a low pitch for oversampling is necessary in this approach, but it can lead to increased radiation exposure. It is possible for a potential sequence scanning method to contain stack artifacts. Nevertheless, a single heartbeat scan mode, such as a wide area detector scan or a high pitch prospective scan, has no artifacts. A low and steady heart rate is necessary for both single heartbeat techniques [30].

Algebraic reconstruction was used to recreate the images in the initial CT systems. However, due to a lack of processing power, filtered back projection quickly replaced this technique. Using a straight Fourier approach or a convolution method, the filtered back projection images are rebuilt. Nowadays, when it comes to CT image reconstruction, iterative reconstruction techniques are being employed, instead of filtered back projection [31]. The iterative reconstruction technique was primarily applied to an initial image reconstruction from raw data in the image domain, followed by what are known as hybrid reconstructions. Next, a complete model-based iterative reconstruction technique was developed by the reconstruction algorithms [32]. The image matrix size of the first-generation CT scanners was restricted to 80×80 pixels. Currently, 512×512 is the most widely used image matrix size; however, 2048 $\times$

2048 is also an option. Visibility of skin and tissue outside the primary field-of-view is made possible by the reconstructions with extended field-of-view. This is crucial for calculating radiotherapy CT doses and the PET-CT attenuation connection [33, 34]. There are techniques for multi-segment reconstruction that enable scanning of the coronary vessels at increased heart rates feasible. This method may be able to boost temporal resolution even though it results in a very low pitch value and hence raises the radiation dose [35]. Several reconstruction procedures employ noise reduction techniques to raise the spatial resolution and contrast-to-noise ratio. In CT perfusion of the head and body, motion correction algorithms are commonly used to account for minute movements of the head or breathing conditions during the acquisition times. Most algorithms for reducing metal artifacts or beam hardening correction are iterative and require raw data to be used [36].

Both the user and technological advancements have an impact on how well CT protocols are optimized. The goal of the entire scanning protocol process is to enhance and optimize for each patient, taking into consideration the referral question and the particular organ region of interest. Certain technological developments directly impact the radiation dosage administered to the patient [37]. Other advancements, such as automatic voltage or current adaptation in tubes or iterative reconstructions, rely on the awareness and incentive of the user. The best scanning process will result from advances in technology, automated tools, and radiographer attention to the human side of things. Users need to be aware that different CT scanners and institutions may produce varied results when it comes to diagnosis and technological optimization. It is possible to optimize the protocol for each unique patient by tailoring one or more acquisition and reconstruction parameters.

14.4 How Does the Future Look like?

CT is continually developing, bringing with it new advances in technology, innovative approaches, and strategies to healthcare diagnosis. It is important for users to be at the forefront of adopting new technologies and moving toward highly enhanced protocols. Currently, photon counting CT and artificial intelligence (AI) hold great promise for revolutionizing CT. Photon counting CT is expected to provide enhanced low-signal performance, decreased noise and artifacts, high spatial resolution, and intrinsic spectral sensitivity. Moreover, photon counting CT enables multi-energy imaging to be provided with every scan; this is similar to dual-energy CT but only requires one tube voltage. Photon counting CT is able to count every photon that enters the system and measure photon energy. It's possible that less radiation is needed for improved visibility of the iodine contrast [38]. Artificial intelligence has already been used in clinical protocols for image reconstruction and artifact removal. AI can lower radiation exposure without sacrificing image quality [39]. There is yet more to come for an optimal scanning procedure; photon counting CT and AI are certainly not the last of the latest technological advancements in CT.

The application of new machine learning technologies to CT imaging helps in managing the dose and image quality, as well as ensuring operational and clinical factors. Low-dose CT procedure-related image noise has a detrimental knock-on effect on the effectiveness, affordability, and quality of imaging services. The deep learning reconstruction and post-processing methods can reduce the image noise as they can help process CT images faster, in a matter of seconds. These strategies offer benefits in terms of clinical outcomes, operational efficiency, and financial stability. All CT scanners can employ the same image quality, thanks to some of the vendor-neutral technologies. Without having to account for the widespread diversity in image quality that exists now, radiologists may interpret images more rapidly and confidently. As a result, radiologists may be able to interpret exams from specific scanners more quickly, which could lead to an increase in productivity. In order to provide screening services that yield the lowest possible dose, the post-processing methods can be utilized to reduce the patient's concerns about the harmful effects of radiation. Deep learning reconstruction and post-processing methods, powered by AI, have enormous potential to improve the clinical, administrative, and economical facets of CT imaging workflow. It can regularly and drastically lower radiation dose exposure below industry norms, considerably beyond what was previously achievable with iterative reconstruction technologies that are presently in use. The quality, effectiveness, and cost of imaging procedures can be significantly impacted by fully reducing image noise and harmonizing image quality for all CT scanners and CT study types. This is especially true for higher-risk patients and difficult use cases like lung and pediatric screening studies. The application of these novel AI-derived techniques can enhance diagnostic image quality, modality utilization, and radiology workflow efficiency, even in the presence of low-dose regimens. As patients' concerns about potential radiation side effects grow, more are actively looking for screening programs that can provide the lowest possible dose. In terms of image quality, diagnostic accuracy, and patient, technician, and radiologist satisfaction, early adopters of deep learning reconstruction and post-processing technology stand to benefit immediately. Further notable gains in operational and financial performance could be obtained through longer-term, enhanced operational efficiency, and optimized modality use [31].

The AI algorithms in radiology are frequently trained to identify intricate patterns in input data and generate quantitative evaluations of particular imaging parameters. AI is being used in radiology to identify various pathologies, such as, to identify pulmonary nodules in cases of lung cancer and help interpret mammography images for micro-calcifications. Artificial intelligence can be attained in a variety of methods. One approach is through machine learning and its subset, deep learning, which are utilized in CT image reconstruction. A branch of artificial intelligence called machine learning uses vast amounts of structured data to identify patterns. Training machine learning models to eventually be able to make predictions about unlabelled data is the aim. The machine learning model identifies and correlates certain parameters—which are pre-set and engineered by human control—to a specific radiographic outcome. Supervised learning and unsupervised learning are two categories for machine learning. Unsupervised learning refers to models that do not require labels or parameters specified under human supervision during the training process, whereas supervised learning calls for the model to be taught using images or parameters predetermined by humans. Instead, the model picks up the ability to identify particular traits in unlabelled photos on its own. Some of the drawbacks of supervised models with regard to data annotation can be addressed by unsupervised learning models. Data curation is a time-consuming and costly procedure that is frequently carried out by specialists. Conversely, unsupervised learning models can pick up distinctive traits without the assistance of a data curation specialist [40].

Deep learning reconstruction performs better in areas of noise reduction and low-dose image quality preservation. The models of deep learning reconstruction are effective in reducing dosage by enhancing image quality, however, the diagnostic accuracy cannot be directly associated with image quality. As of right now, there is not enough information to establish deep learning reconstruction as a diagnostic accuracy approach that is noticeably better. Numerous factors affect how a reconstruction technique is implemented clinically. Image quality is a crucial consideration as it affects the amount of radiation used during a particular scan. It is necessary to take into account additional variables, too, like reconstruction time, cost, the effect on diagnosis accuracy, and how well it functions in a larger, more complicated clinical setting. There is currently little information available on the complete potential and limitations of deep learning reconstruction, a revolutionary reconstruction technique, as there are only two FDA-approved commercial algorithms available on the market. Before deep learning reconstruction may be used widely in clinical settings, its limitations in sample size, confinement to a single institution, selective data analysis, and insufficient number of observers must be addressed in future research [41]. Since the first CT image reconstruction method was developed, improvements in technology and software have made it possible to minimize radiation doses, scan more quickly, and produce better-quality images.

14.5 Conclusion

Godfrey Hounsfield and Allan Cormack shared the 1979 Nobel Prize in Physiology or Medicine for their discovery of computer-assisted tomography. The announcement reported that, "It is no exaggeration to state that no other method within x-ray diagnostics within such a short period of time has led to such a remarkable advance in research and in a multitude of applications". The field of radiology saw a significant transformation as a result of the work that Hounsfield began. Since the first disruptive breakthrough, scientific innovation and industrial competitiveness have advanced the field, each producing notable incremental advancements that have added together to allow for extraordinarily sophisticated images in the current year. With the advent of CT, diagnostic imaging and radiology as a field saw a significant transition from film-based analog imaging to digital imaging, where algorithms are used to generate pixel values. Patients benefited immensely from this, which served as a blueprint for upcoming possibilities like positron emission tomography and magnetic resonance imaging. Whether be it 16-slice, 64-slice, or beyond, Multi-slice CT will keep becoming more and more important as the primary diagnostic imaging modality.

References

1. J. Ambrose, You never know what is just around the next corner. Riv. Neuroradiol. **9**(4), 399–404 (1996)
2. J. Gawler, G.H. Du Boulay, J.W. Bull, J. Marshall, Computerized Tomography (the EMI scanner): A comparison with pneumoencephalography and ventriculography. J. Neurol., Neurosurg. & Psychiatry. **39**(3), 203–211 (1976)
3. G.D. Rubin, Computed tomography: Revolutionizing the practice of Medicine for 40 Years. Radiology. **273**(2S), (2014)
4. G.R. Jankharia, Commentary 2—radiology in India: The next decade. Indian J. Radiol. Imaging. **18**(03), 189–191 (2008)
5. J.M. Meulepas, C.M. Ronckers, A.M. Smets, R.A. Nievelstein, P. Gradowska, C. Lee et al., Radiation exposure from pediatric CT scans and subsequent cancer risk in the Netherlands. JNCI: J. Natl. Cancer Inst. **111**(3), 256–263 (2018)
6. G.N. Hounsfield, Computerized transverse axial scanning (tomography): Part 1. Description of system. Br. J. Radiology. **46**(552), 1016–1022 (1973)
7. M. Shafiq-ul-Hassan, G.G. Zhang, K. Latifi, G. Ullah, D.C. Hunt, Y. Balagurunathan et al., Intrinsic dependencies of CT radiomic features on voxel size and number of gray levels. Med. Phys. **44**(3), 1050–1062 (2017)
8. R.E. Alexander, R.B. Gunderman, EMI and the first CT Scanner. J. Am. Coll. Radiol. **7**(10), 778–781 (2010)
9. R.S. Ledley, G. Di Chiro, A.J. Luessenhop, H.L. Twigg, Computerized transaxial X-ray tomography of the human body. Science **186**(4160), 207–212 (1974)
10. D. Boyd, J. Coonrod, J. Dehnert, D. Chu, C. Lim, B. Macdonald et al., A high pressure xenon proportional chamber for X-ray laminographic reconstruction using fan beam geometry. IEEE Trans. Nucl. Sci. **21**(1), 184–187 (1974)
11. R.A. Schulz, E.C. Olson, K.S. Han, A comparison of the number of rays versus the number of views in reconstruction tomography. Proc. SPIE **0127**, 313–320 (1977)
12. E.C. McCullough, J.T. Payne, X-ray-transmission computed tomography. Med. Phys. **4**(2), 85–98 (1977)
13. S. Kulkarni, J.A. Rumberger, S. Jha, Electron beam CT: A historical review. Am. J. Roentgenol. **216**(5), 1222–1228 (2021)
14. W.A. Kalender, Technical foundations of spiral computed tomography. Comput. Tomogr. 17–28 (1996)
15. W.A. Kalender, W. Seissler, E. Klotz, P. Vock, Spiral volumetric CT with single breath-hold technique, continuous transport and scanner rotation. Radiology **176**, 181–183 (1990)
16. Y. Toki, Principles of helical scanning, in *Basic principles and clinical applications of helical scan: application of continuous rotation CT*. ed. by K. Kimura, S. Koga (Iryokagakusha, Tokyo, Japan, 1993), pp.110–120
17. H. Hu, Multi-slice helical CT: Scan and reconstruction. Med. Phys. **26**(1), 5–18 (1999)
18. L. Goldman, Principles of CT and evolution of CT technology, in *Categorical course in diagnostic Radiology physics: CT and US cross-sectional imaging*. ed. by L.W. Goldman, J.B. Fowlkes (Radiological Society of North America, Oak Brook, IL, 2000), pp.33–52
19. C. McCollough, F. Zink, Performance evaluation of a multi-slice CT system. Med. Phys. **26**, 2223–2230 (1999)
20. L.W. Goldman, Principles of CT: Multislice CT. J. Nucl. Med. Technol. **36**(2), 57–68 (2008)

21. W.A. Kalender, X-ray computed tomography. Phys. Med. Biology. **51**(13), (2006)
22. H. Takagi, R. Tanaka, K. Nagata, R. Ninomiya, K. Arakita, J.D. Schuijf et al., Diagnostic performance of coronary CT angiography with ultra-high-resolution CT: Comparison with invasive coronary angiography. Eur. J. Radiol. **101**, 30–37 (2018)
23. R. Carmi, Tu–E–210A–03: A double-layer detector, dual–energy CT—principles, advantages and applications. Med. Physics. **36**(6Part24), 2750–2750 (2009)
24. R. Homberg, R. Koeppel, An X-ray tube assembly with rotating anode spiral groove bearing of the 2nd generation. Electromedica. **66**(2), 65–66 (1997)
25. Y. Kyriakou, B.M. Kachelrie, M. Knaup, J.U. Krause, W.A. Kalender, Impact of the Z-flying focal spot on Resolution and Artifact Behavior for a 64-slice spiral CT scanner. Eur. Radiology. **16**(6), 1206–1215 (2006)
26. J.C. Ramirez-Giraldo, M. Fuld, K. Grant, A.N. Primak, T. Flohr, New approaches to reduce radiation while maintaining image quality in multi-detector-computed tomography. Curr. Radiol. Reports. **3**(2), (2015)
27. W.A. Kalender, Principles and applications of Spiral CT. Nucl. Med. Biol. **21**(5), 693–699 (1994)
28. G.D. Rubin, A.N. Leung, V.J. Robertson, P. Stark, Thoracic spiral CT: Influence of subsecond gantry rotation on image quality. Radiology **208**(3), 771–776 (1998)
29. G.S. Harell, D.F. Guthaner, R.S. Breiman, C.C. Morehouse, E.J. Seppi, W.H. Marshall et al., Stop-action cardiac computed tomography. Radiology **123**(2), 515–517 (1977)
30. L.A. Neefjes, A. Rossi, T.S. Genders, K. Nieman, S.L. Papadopoulou, A.S. Dharampal et al., Diagnostic accuracy of 128-slice dual-source CT coronary angiography: A randomized comparison of different acquisition protocols. Eur. Radiol. **23**(3), 614–622 (2012)
31. M.J. Willemink, P.B. Noël, The evolution of image reconstruction for CT—from filtered back projection to artificial intelligence. Eur. Radiol. **29**(5), 2185–2195 (2018)
32. K.W. Shaqdan, A.R. Kambadakone, P. Hahn, D.V. Sahani, Experience with iterative reconstruction techniques for Abdominopelvic computed tomography in morbidly and super obese patients. J. Comput. Assist. Tomogr. **42**(1), 124–132 (2018)
33. A. Hata, M. Yanagawa, O. Honda, N. Kikuchi, T. Miyata, S. Tsukagoshi et al., Effect of matrix size on the image quality of ultra-high-resolution CT of the lung. Acad. Radiol. **25**(7), 869–876 (2018)
34. H. Bruder, C. Suess, K. Stierstorfer, Efficient extended field of view (eFOV) reconstruction techniques for multi-slice helical CT. SPIE Proceedings. (2008)
35. T.H. Flohr, B. Ohnesorge, H. Bruder, K. Stierstorfer, J. Simon, C. Suess et al., Image reconstruction and performance evaluation for ecg-gated spiral scanning with a 16-slice CT system. Med. Phys. **30**(10), 2650–2662 (2003)
36. R.H.H. Wellenberg, E.T. Hakvoort, C.H. Slump, M.F. Boomsma, M. Maas, G.J. Streekstra, Metal artifact reduction techniques in musculoskeletal CT-imaging. Eur. J. Radiol. **107**, 60–69 (2018)
37. A. Christe, CT dose and image quality in the last three scanner generations. World Journal of Radiology. **5**(11), 421 (2013)
38. A. Pourmorteza, R. Symons, D.S. Reich, M. Bagheri, T.E. Cork, S. Kappler et al., Photon-counting CT of the brain: In vivo human results and image-quality assessment. Am. J. Neuroradiol. **38**(12), 2257–2263 (2017)
39. M.M. Lell, M. Kachelrieß, Recent and upcoming technological developments in computed tomography. Invest. Radiol. **55**(1), 8–19 (2019)
40. A. Hosny, C. Parmar, J. Quackenbush, L.H. Schwartz, H.J. Aerts, Artificial intelligence in radiology. Nat. Rev. Cancer **18**(8), 500–510 (2018)
41. M. Akagi, Y. Nakamura, T. Higaki, K. Narita, Y. Honda, J. Zhou et al., Deep learning reconstruction improves image quality of abdominal ultra-high-resolution CT. Eur. Radiol. **29**(11), 6163–6171 (2019)

Index

A. Gogoi and N. Mazumder (eds.), *Biomedical Imaging*, Biological and Medical Physics, Biomedical Engineering, https://doi.org/10.1007/978-981-97-5345-1

GPSR Compliance

The European Union's (EU) General Product Safety Regulation (GPSR) is a set of rules that requires consumer products to be safe and our obligations to ensure this.

If you have any concerns about our products, you can contact us on ProductSafety@springernature.com

In case Publisher is established outside the EU, the EU authorized representative is:

Springer Nature Customer Service Center GmbH
Europaplatz 3
69115 Heidelberg, Germany

Batch number: 10732031

Printed by Printforce, the Netherlands